HISTONE GENES

HISTONE GENES

Structure, Organization, and Regulation

Edited by

GARY S. STEIN
Department of Biochemistry and Molecular Biology
University of Florida College of Medicine

JANET L. STEIN
Department of Immunology and Microbiology
University of Florida College of Medicine

WILLIAM F. MARZLUFF
Department of Chemistry
Florida State University

A Wiley-Interscience Publication

JOHN WILEY & SONS

New York Chichester Brisbane Toronto Singapore

Library of Congress Cataloging in Publication Data:

Main entry under title:
Histone genes.

"A Wiley Interscience publication."
Includes bibliographies and index.
1. Histones. 2. Gene expression. 3. Genetic regulation. I. Stein, Gary S. II. Stein, Janet L. III. Marzluff, William F., 1945– . [DNLM: 1. Histone —Genetics. 2. Gene expression regulation. 3. Cells—Ultrastructure. QU 58 H673]
QH450.H57 1984 574.19′245 83-23421
ISBN 0-471-89615-2

Printed in the United States of America

10 9 8 7 6 5 4 3 2 1

CONTRIBUTORS

Dr. M. Mitchell Smith
Department of Microbiology
University of Virginia School of Medicine
Charlottesville, Virginia

Dr. Michael Grunstein
Dr. Mary Rykowski
Dr. David Kolodrubetz
Dr. Joonho Choe
Dr. John Wallis
Molecular Biology Institute
University of California
Los Angeles, California

Dr. Claus von Holt
Dr. Petronella de Groot
Dr. Sylva Schwager
Dr. Wolf Friederich Brandt
CSIR Chromatin Research Unit
Department of Biochemistry
University of Cape Town
Republic of South Africa

Dr. Lynne M. Angerer
Dr. Kathleen J. Hughes
Dr. Donna V. DeLeon
Dr. Robert C. Angerer
Department of Biology
University of Rochester
Rochester, New York

Dr. Kathryn V. Anderson
Dr. Judith A. Lengyel
Department of Biology and Molecular Biology Institute
University of California at Los Angeles
Los Angeles, California

Dr. Gary A. Bannon
Dr. Martin A. Gorovsky
Department of Biology
University of Rochester
Rochester, New York

Dr. Joseph C. Bagshaw
Dr. Matthew T. Andrews
Dr. Brian A. Perry
Department of Biochemistry
Wayne State University
School of Medicine
Detroit, Michigan

Dr. W. M. A. M. van Dongen
Dr. A. F. M. Moorman
Dr. O. H. J. Destrée
Anatomisch-Embryologisch Laboratorium
Universiteit van Amsterdam
Amsterdam, The Netherlands

Dr. Edwin C. Stephenson
Department of Biology
Indiana University
Bloomington, Indiana

Dr. R. P. Harvey
Dr. J. R. E. Wells
Department of Biochemistry
University of Adelaide
Adelaide, South Australia

Dr. James Douglas Engel
Department of Biochemistry
Molecular Biology and Cell Biology
Northwestern University
Evanston, Illinois

Dr. William F. Marzluff
Dr. Reed A. Graves
Department of Chemistry
Florida State University
Tallahassee, Florida

Dr. Philip Coffino
Department of Medicine
University of California
San Francisco, California

Dr. Eva Stimac
Dr. Vincent E. Groppi
Department of Medicine
Department of Microbiology and Immunology
University of California
San Francisco, California

Dr. David Bieber
Department of Medicine
Department of Biochemistry and Biophysics
University of California
San Francisco, California

Dr. Alfred Zweidler
The Institute for Cancer Research
The Fox Chase Cancer Center
7701 Burholm Avenue
Philadelphia, Pennsylvania

Dr. Ronald W. Lennox
Dr. Leonard H. Cohen
The Institute for Cancer Research
7701 Burholm Avenue
Fox Chase
Philadelphia, Pennsylvania

Dr. Gary S. Stein
Dr. Felipe Sierra
Dr. Mark Plumb
Dr. Farhad Marashi
Lisa Baumbach
Department of Biochemistry and Molecular Biology
University of Florida
College of Medicine
Gainesville, Florida

Dr. Janet L. Stein
Nadine Carozzi
Keith Prokopp
Department of Immunology and Medical Microbiology
University of Florida
College of Medicine
Gainesville, Florida

Dr. Roy S. Wu
Dr. William M. Bonner
Laboratory of Molecular Pharmacology
Division of Cancer Treatment
National Cancer Institute
National Institutes of Health
Bethesda, Maryland

Dr. M. H. P. West
Laboratory of Experimental Pathology
National Institute of Arthritis, Diabetes and Digestive and Kidney Diseases
National Institutes of Health
Bethesda, Maryland

PREFACE

During the last decade, significant advances have been made in our understanding of the eucaryotic chromosome at the molecular level. This increased understanding of the structural and functional properties of the genome has been associated with a major focus of attention on histone genes and the expression of these genetic sequences—largely due to the involvement of histone proteins in DNA packaging, the relationship between expression of histone genes and DNA replication, and the somewhat simplified organization and expression of the histone genes. Thus histone genes have been and continue to be a focal point for cell and molecular biologists because of the regulatory functions of the gene product and the regulated nature of the genetic sequences. It is therefore not surprising that histone genes have recently been the subject of study in a wide variety of organisms.

Historically, histone genes from sea urchins were the first protein coding sequences isolated and cloned. Because of the high degree of sequence conservation in histones from different species, these cloned genes have subsequently been instrumental in the identification and isolation of all the histone genes described in this volume.

Histone mRNAs are unusual among eucaryotic mRNAs. With the exception of those expressed in yeast, *Tetrahymena*, and *Xenopus*, histone mRNAs do not contain the poly A sequence found at the 5′ termini of most eucaryotic mRNAs. Histone genes differ from most other eucaryotic genes. With the exception of one chicken histone H3 gene described by Engel (Chapter 11), histone mRNAs are encoded in the genome by contiguous nucleotide sequences. The attractiveness of histone genes as a model system for studying regulation is further supported by the high gene copy number, abundant mRNA levels, and the sensitive regulation at the levels of both mRNA synthesis and degradation. At least two organisms, sea urchins and chickens, have different gene sets active during different developmental stages.

A further complexity of histone gene structure is the existence of a number of nonallelic protein sequence variants, which are products of independent genes (reviewed in Chapter 3 by von Holt, Chapter 14 by Zweidler, and Chapter 17 by Bonner). Many of these proteins are expressed in the same cell under independent control. The structures of these genes must differ from one another, and we expect progress in the elucidation of the structures and regulation of the different genes in the next few years.

A book containing contributions from laboratories that have done the major work on various aspects of the histone gene problem has not been assembled to date. Chapters written by the actual investigators permit a presentation of specific aspects of the histone gene problem with the type of insight possible only from the people who execute the experiments. We feel that this approach ensures that the book provides an in-depth yet comprehensive treatment of the structure, organization and regulation of histone genes. In preparing this book, we used a phylogenetic presentation of histone gene structure, organization, and regulation. The first section contains chapters on lower eucaryotes and includes coverage of yeast, sea urchins, *Drosophila*, *Artemia*, and *Tetrahymena*. Although an extensive discussion of the structure and organization of sea urchin histone genes has not been included, this topic has been the subject of several comprehensive reviews [C. C. Hentschel and M. L. Birnstiel, *Cell* **25**, 301–313 (1981); L .H. Kedes, *Ann. Rev. Biochem.* **48**, 837–870 (1979)]. The second section includes "intermediate eucaryotes" and covers the work done on *Xenopus* and newt histone genes. The last section focuses on higher eucaryotes—chicken, mouse, and human.

We have included in this book an up-to-date summary of current work involving histone genes and their expression. With the biological systems being studied and the molecular approaches being employed, it is realistic to anticipate that the next several years will bring major breakthroughs in our understanding of the DNA sequences and molecules involved in control of histone gene expression.

GARY S. STEIN
JANET L. STEIN
WILLIAM F. MARZLUFF

Gainesville, Florida
Gainesville, Florida
Tallahassee, Florida
January 1984

CONTENTS

PART 1. LOWER EUCARYOTES

1. The Organization of the Yeast Histone Genes 3
M. Mitchell Smith

2. A Genetic Analysis of Histone Protein Subtypes in Yeast 35
Michael Grunstein, Mary Rykowski, David Kolodrubetz, Joonho Choe, and John Wallis

3. The Structure of Sea Urchin Histones and Considerations on their Function 65
Claus von Holt, Petronella de Groot, Sylva Schwager, and Wolf Friederich Brandt

4. Spatial Aspects of Histone Gene Expression in Sea Urchin Embryos 107
Lynne M. Angerer, Kathleen H. Cox, Donna V. DeLeon, and Robert C. Angerer

5. Histone Gene Expression in *Drosophila* Development: Multiple Levels of Gene Regulation 135
Kathryn V. Anderson and Judith A. Lengyel

6. *Tetrahymena* Histones: Proteins, Messages and Genes 163
Gary A. Bannon and Martin A. Gorovsky

7. Organization and Expression of Histone Genes in the Brine Shrimp, *Artemia* 181
Joseph C. Bagshaw, Matthew T. Andrews, and Brian A. Perry

PART 2. INTERMEDIATE EUCARYOTES

8. DNA Organization and Expression of *Xenopus* Histone Genes 199
W. M. A. M. van Dongen, A. F. M. Moorman, and O. H. J. Destrée

9. Organization and Expression of Newt Histone Genes 225
Edwin C. Stephenson

PART 3. HIGHER EUCARYOTES

10. Chicken Histone Genes and Their Variants 245
R. P. Harvey and J. R. E. Wells

11. Organization and Expression of Chicken Histone Genes 263
James Douglas Engel

12. Organization and Expression of Mouse Histone Genes 281
William F. Marzluff and Reed A. Graves

13. Histone Expression and the Mammalian Cell Cycle 317
Philip Coffino, Eva Stimac, Vincent E. Groppi, and David Bieber

14. Core Histone Variants of the Mouse: Primary Structure and Differential Expression 339
Alfred Zweidler

15. **The H1 Subtypes of Mammals: Metabolic Characteristics and Tissue Distribution** 373
Ronald W. Lennox and Leonard H. Cohen

16. **Organization and Expression of Human Histone Genes** 397
Gary S. Stein, Felipe Sierra, Janet L. Stein, Mark Plumb, Farhad Marashi, Nadine Carozzi, Keith Prokopp, and Lisa Baumbach

17. **Patterns of Histone Gene Expression in Human and Other Mammalian Cells** 457
Roy S. Wu, Michael H. P. West, and William M. Bonner

Index 485

PART 1

LOWER EUCARYOTES

CHAPTER 1

THE ORGANIZATION OF THE YEAST HISTONE GENES

M. MITCHELL SMITH
Department of Microbiology
School of Medicine
University of Virginia
Charlottesville, Virginia

CONTENTS

1. INTRODUCTION
   1.1. Yeast Histones
   1.2. Isolation of the Yeast Histone Genes
   1.3. False Positive Plaque Hybridization

2. ORGANIZATION
   2.1. Local Gene Arrangement
   2.2. Autonomously Replicating Sequences
   2.3. The Genes in the Genome

3. DNA SEQUENCE INFORMATION
   3.1. The Coding Sequences
   3.2. General DNA Sequence Features
   3.3. Architecture of the H3 and H4 Genes
   3.4. Specific DNA Sequence Motifs

4. EVOLUTION OF THE YEAST HISTONE GENES
   4.1. Duplication of the Yeast Genes
   4.2. Histone Protein Evolution

5. CONCLUSION

REFERENCES

1. INTRODUCTION

Saccharomyces cerevisiae is a relative newcomer to the ranks of organisms that have provided experimental systems for studies of the histone genes. The pioneering studies on the histone genes began, of course, with the Echinoderms (1,2). The reiterated organization of the genes expressed in early embryogenesis of the sea urchin (3), physical methods for enriching this DNA (4), and the relative abundance of the early embryonic histone mRNA (5), all contributed to the genes of the sea urchin histone DNA repeat unit being some of the first genomic eucaryotic genes to be isolated by molecular cloning (reviewed in references 1 and 2). By contrast, information on the histone genes of yeast and other single cell eucaryotes, and even the characterization of the histones themselves, has come from much more recent experiments. It is a tribute to the power of recently developed techniques in molecular genetics that these experiments have led to an understanding of the organization of the yeast histone genes in remarkable detail in a remarkably short time. This chapter reviews these investigations on the organization of the yeast histone genes. Functional aspects of the yeast genes are reviewed in Chapter 2.

1.1. Yeast Histones

Although histones have been studied in higher eucaryotes for more than 30 years, purification of the proteins from yeast proved vexingly difficult; a discussion of the problems can be found in Mardian and Isenberg (7). The yeast histones were perhaps first described by Tonino and Rozijn (8) as very basic proteins extracted from yeast chromatin. Various preparations were subsequently described and characterized largely by gel electrophoretic mobility as proteins that migrated similarly to purified calf thymus histones (9–14). All four inner histones, H2A, H2B, H3, and H4, were finally definitively characterized together in single preparations by Mardian and Isenberg(7) and Pastink et al. (15). The amino acid compositions of these purified proteins left little doubt that yeast contained all four of the core proteins. Partial amino acid sequences for the proteins were provided shortly thereafter by Brandt et al. (16).

The yeast histones show strong similarity with those of higher eucaryotes. They display the same pairwise interactions as the histones from calf or chick embryo (7), they undergo modifications *in vivo* such as acetylation (17, 18), and the yeast histones associate with DNA to form nucleosomes and higher-order chromatin structures (14, 19, 20). The yeast histones H2A and H2B even show conserved binding surfaces in interkingdom complexes with calf thymus histones (7).

An exception to this pattern of similarity with higher eukaryotes is the apparent lack of histone H1 from yeast. To date histone H1 remains to be identified in yeast chromatin (21, 7, 16). A number of larger molecular weight basic proteins can be isolated from yeast chromatin; however, amino acid compositions, when available, fail to establish any structural similarity with the H1 family of proteins (21, 16, 15). Clearly, failure to identify histone H1 does not constitute evidence that the protein is absent *in vivo*. Nevertheless, failure to find histone H1 has been sufficiently widespread that it is reasonable to entertain the possibility that it is lacking in yeast chromatin.

1.2. Isolation of the Yeast Histone Genes

A summary of the yeast histone gene organization is presented in Figure 1. At the time work began on the isolation of the yeast histone genes, studies with sea urchin and *Drosophila* had led to expectations that the four (or five) yeast genes would be clustered, and perhaps tandemly reiterated a few times, as in the multicellular eucaryotes. It was something of a surprise to discover that this was not the case. The first yeast histone genes to be isolated were those for H2A and H2B (22). The initial approach taken by Hereford and her colleagues was to search a plasmid library of yeast DNA clones by hybridization with radiolabeled sea urchin histone DNA. This approach led to the isolation of numerous false positive recombinant clones that did not contain histone gene sequences (L. Hereford, personal communication). This type of clone is further discussed later.

Subsequently a second approach was tried. Unlike most histone mRNAs (1, 2), the yeast histone mRNAs appeared to be polyadenylated (later reported in reference 23). A selected set of 100 recombinant clones from the library had already been obtained by virtue of their

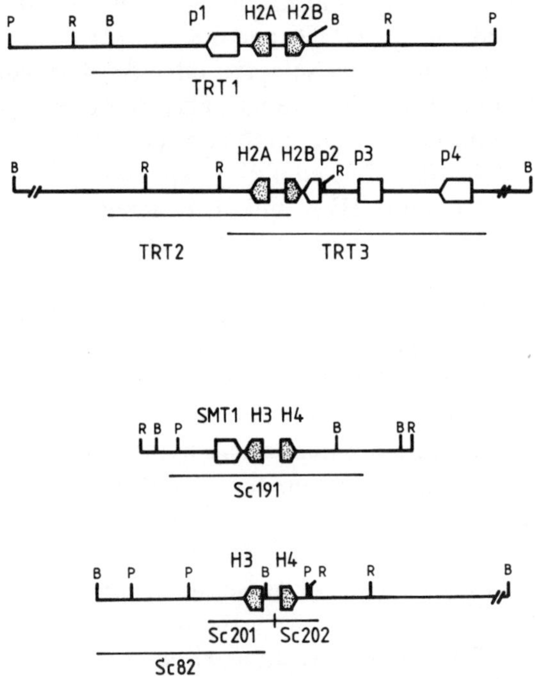

FIGURE 1.1. Structures of the yeast histone gene loci. The histone gene pairs and closely linked genes are indicated on the physical map of the four loci. The boxes are drawn to point in the direction transcription of the genes where this is known. The histone genes are shaded. The genes p1, p2, p3, p4, and *SMT*1 do not code for histone proteins but are closely linked with one of the histone gene pairs. The positions of some restriction enzyme sites are shown for each region. Below each physical map are given the designations and extents of the recombinant clones or fragments representing each locus. The enzyme symbols are as follows: B = BamHI, P = PstI, R = EcoRI. This figure is adapted from Hereford et al. (22) and Smith and Murray (26).

hybridization with small molecular weight poly(A)+ mRNA; this subcollection had been made with the intent of finding yeast ribosomal protein genes (24). Thus these selected clones represented a promising source of yeast histone DNA clones. Purified plasmid DNA from this recombinant collection was used to isolate mRNA by R-loop hybridization (25). The recovered mRNA was subsequently translated *in vitro* and the proteins separated on two-dimensional polyacrylamide gels (7). In this way "the real thing" was finally isolated in three clones designated TRT1, TRT2, and TRT3 (22).

These three plasmids each selected mRNAs that would code for histones H2A and H2B. Two remarkable findings emerged when these three recombinants were analyzed: (1) The three yeast DNA fragments in TRT1, TRT2, and TRT3 represented not one but two nonallelic sets of histone H2A and H2B genes. (2) The genes for histone H3 and H4 were apparently *not* clustered with the H2A and H2B genes.

The genes for histones H3 and H4 were isolated independently by Smith and Murray (26) from bacteriophage lambda genome libraries by hybridization with radiolabeled heterologous histone genes from sea urchin and *Drosophila*. In this case numerous false positive phage recombinants were also obtained. The hybridizing probes in these experiments were complete histone repeat units from either sea urchins or *Drosophila*. The spacer DNA sequences in these repeat units gave extensive cross-hybridization with yeast DNA apparently unrelated to the actual histone genes. For example, when an EcoRI fragment genomic library was screened with the histone genes from the sea urchin *Psammechinus miliaris*, approximately 30 different recombinants were isolated as positive clones. The hybridization of a portion of this collection with *P. miliaris* histone DNA is illustrated in Figure 1.2. If the histone repeat unit from another sea urchin species was used, in this case *Echinus esculentus*, a similar but largely nonoverlapping set of clones was isolated; only two of the clones from the collection made using the *P. miliaris* probe would also hybridize with *E. esculentus* histone DNA. *Drosophila* histone genes gave a similar although smaller collection of false positive clones.

FIGURE 1.2. False positive bacteriophage lambda clones. The blot hybridization of a part of a collection of yeast clones is illustrated. The phage library was made using lambda vector NM641 and a complete EcoRI digestion of yeast DNA. Phage DNA was prepared from plaques selected for their hybridization with *P. miliaris* histone DNA. The DNA was digested with EcoRI and run on agarose gels for blot transfer. The transfers were hybridized with radiolabeled *P. miliaris* histone DNA and autoradiographed. As evident in the autoradiographs, the collection of positive yeast fragments included a wide variety of fragments—none of which contained histone DNA coding sequences.

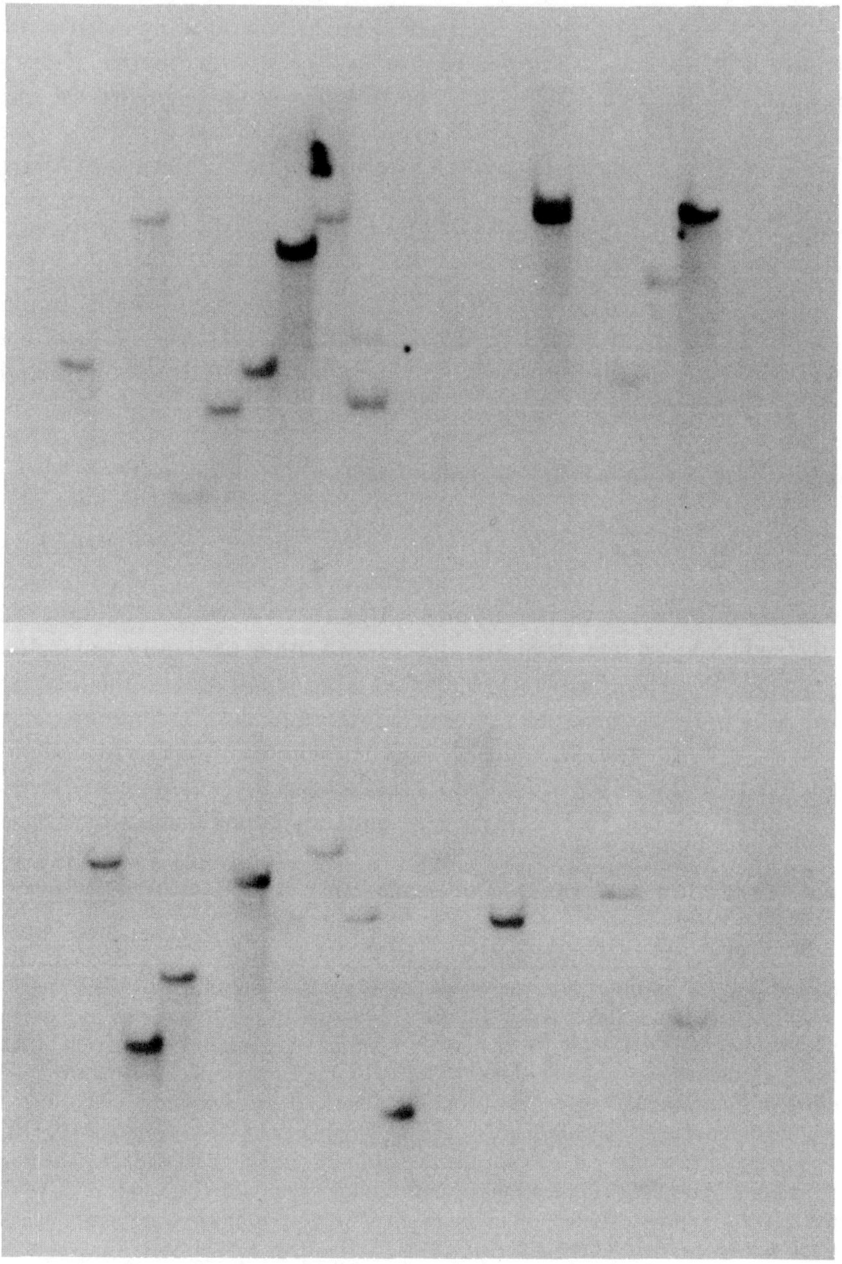

On the assumption that the real yeast histone coding sequences *would* hybridize with the sea urchin or *Drosophila* histone coding sequences, the problem of a large number of false positives was circumvented by a double-screen procedure. The yeast genomic phage libraries were blotted with two nitrocellulose filters. One set of filters was probed with sea urchin histone DNA, and the other set was probed with *Drosophila* histone DNA. Autoradiographs of the two screenings were examined in register to locate a recombinant that would hybridize with both probes. This strategy led to the isolation of a yeast fragment carrying the histone H3 and H4 genes. This clone was subsequently used to isolate the second nonallelic set of H3 and H4 genes depicted in Figure 1.1 (26).

1.3. False Positive Plaque Hybridization

Very little information is available on the false positive hybridizing sequences. Some of the clones are likely the result of hybridization with simple homo- or copolymer sequences known to occur in the spacer DNA sequences of sea urchin and other tandemly reiterated genes (1, 27, 28). The two clones that were positive with the histone DNA of both sea urchin species hybridized to the spacer sequences between the histone H1 and H2A genes where a $5'$-$(C\text{-}T)_{27}-3'$ sequence is located (Figure 1.3). Nevertheless, these two yeast clones contained different EcoRI fragments, and they failed to cross-hybridize with each other. Most of the clones hybridized uniquely with one of the sea urchin or *Drosophila* clones and not the others. Melting

FIGURE 1.3. Cross-hybridization of two false positive clones. Two of the phages from the collection illustrated in Figure 1.2 were also positive for hybridization with the histone DNA from the sea urchin *Echinus esculentus*. The location of the hybridization with sea urchin histone DNA was examined in greater detail. Restriction enzyme digestions of *P. miliaris* histone DNA were blot transferred and hybridized with radiolabeled DNA from the two yeast recombinants λ641Sc31 and λ641Sc144. The blot hybridization patterns for the two probes were identical and are shown in the autoradiographs. The locations of hybridizing sea urchin fragments relative to the *P. miliaris* physical map are given in (*b*). The restriction enzymes used in each track were as follows: (1) HpaII, (2) EcoRI, (3) HaeII, and (4) HindII. Since neither fragment contained H2A coding sequences, both hybridized to regions of the H2A-H1 sea urchin spacer DNA.

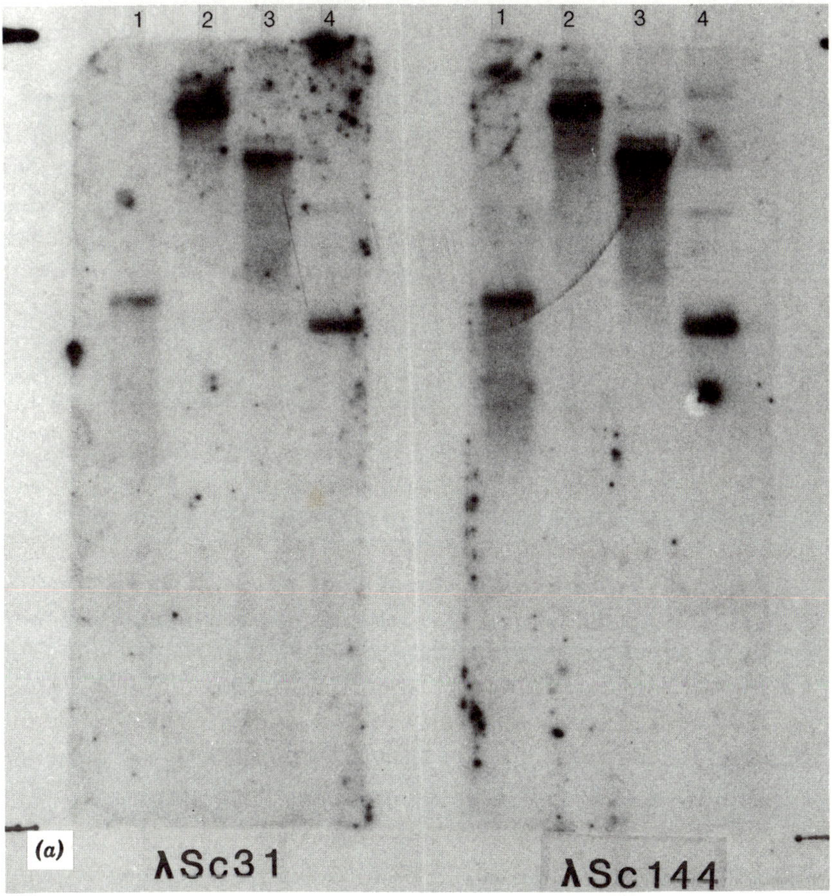
1
2
3
4
1
2
3
4
(a)
λSc31
λSc144

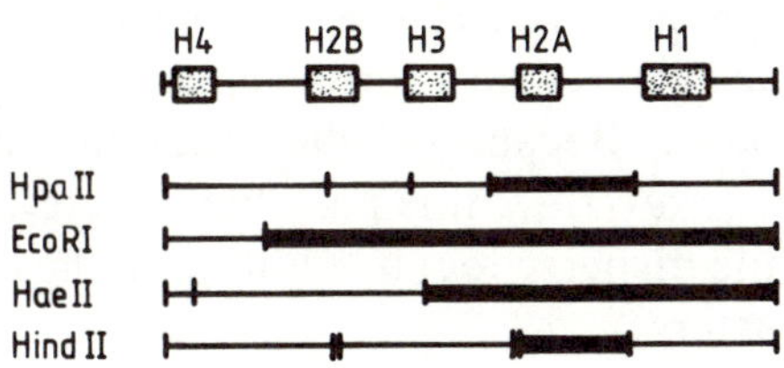
H4
H2B
H3
H2A
H1
Hpa II
EcoRI
Hae II
Hind II

(b)

studies indicate that the hybridization of a number of the clones is of relatively high quality; many of the false positive clones hybridized better with the heterologous histone repeat units than did the actual yeast histone H3 and H4 fragment finally isolated.

False positive hybridization with the spacer sequences of the sea urchin histone repeat unit has been observed in a number of other organisms, including other fungi (B. Klein, personal communication), wheat (A. Gatenby, personal communication), and mouse (K. Peden, personal communication). In the absence of experimental evidence, the relationship of these cross-kingdom hybridizations can only be the subject of speculation. The yeast false positive DNA fragments have no relationship to the yeast histone genes. Thus there is no functional link with the spacer sequences of the sea urchin or *Drosophila* repeats. This seems to rule out any "Garden of Eden" role for these sequences such as is seen with the sex-determination related sequences conserved throughout evolution (29). The spacer sequences between histone genes in tandemly reiterated repeat units are highly divergent even between sea urchin species (1, 2). If these spacer sequences have diverged from a common ancestor, the fact that the yeast fragments hybridized by *E. esculentus* and *P. miliaris* were largely different suggests that the cross-hybridization is simply fortuitous. On the other hand, radical horizontal genetic transfer of sea urchin histone genes has recently been documented (35), and, furthermore, the functional roles of the spacer sequences is just beginning to be understood (30).

2. ORGANIZATION

The yeast genes add to the variety of organizational patterns now recognized for the histone genes. The yeast histone genes are organized as disperse repeated genes with a partial, apparently specific, clustering of coding sequences. Because of the excellent genetic techniques available with *S. cerevisiae*, it is possible to determine the genomic organization of the disperse loci in addition to the local organization of the partial clusters.

2.1. Local Gene Arrangement

There are eight genes per haploid genome encoding the four core histone proteins (Figure 1.1). These eight genes are arranged in four pairs where each pair consists of either one H2A and one H2B gene, or one H3 and one H4 gene. In each pair, the histone genes are arranged so that they are divergently transcribed. The pairs are separated by nontranscribed spacer DNA approximately 600–800 base pairs in length. In the case of the H2A–H2B genes, the two sets are defined by the overlapping clones TRT2 and TRT3 in one case and TRT1 for the other (22). The two H3–H4 gene sets are defined by yeast fragment Sc191 for the copy I set and the two fragments Sc201 and Sc202 for the copy II sets (26). The homologous sets were shown to be nonallelic by restriction site mapping, heteroduplex formation and electron microscopy, and DNA sequence analysis. Each of the gene pairs has been shown to be present at a reiteration frequency of a single copy per haploid genome by reconstruction Southern blot analysis, whole genome restriction site mapping, and genetic mapping.

Heteroduplex analysis first suggested that the sequence homology between the two H2A–H2B gene sets and the two H3–H4 gene sets was confined to the coding sequences of the DNA. This is illustrated in Figure 1.4 for the two H3 genes. The noncoding DNA sequences for the homologous gene pairs are highly divergent. This observation was confirmed at the nucleotide level when the DNA sequences of the genes were determined (31–33). The sequence analysis is considered shortly.

Closely linked to three of the histone gene sets are additional transcribed genes. The genes near the H2A–H2B gene sets have been identified by *in vitro* translation of selected mRNA, by R-loop hybridization analysis using the electron microscope (22), and by northern blot analysis (34). These have been termed protein 1 through protein 4 (22). The closely linked gene in the copy I H3–H4 gene set has been detected by S1 nuclease mapping (26) and by northern blot transcript analysis (unpublished observations). This latter gene has been designated *SMT*1 for S1 mapping transcript. Very little is known about these closely linked genes. They do not code for histone or histone-like proteins. In particular, they do not code for a protein of

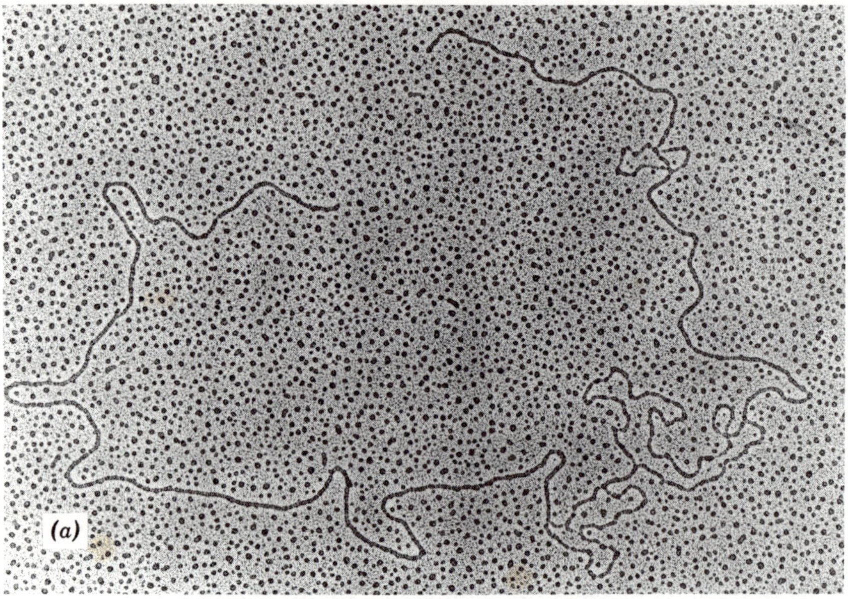

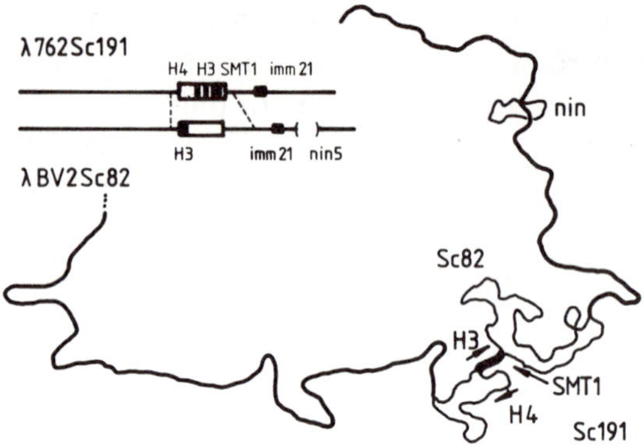

(b)

FIGURE 1.4. (*a*) Heteroduplex of nonallelic H3 genes. Heteroduplex molecules were formed between phage DNA from λ762Sc191, carrying the copy I histone H3 and H4 genes and *SMT*1, and λBV2Sc82, carrying the copy II H3 gene (see Figure 1.1). (*b*) The structures of the two phages carry identical DNA to the left and right of the dashed lines, except for the *nin*5 deletion in λBV2Sc82, but they differ in the middle portion between the dashed lines. A photograph of one heteroduplex seen in the electron microscope is shown, and the structure is interpreted in the diagram. The yeast fragments show homology only in the H3 gene coding DNA regions. This heteroduplex was kindly prepared and photographed by P. Beattie.

the histone H1 family. Thus if yeast does have histone H1, it must be encoded by a gene or genes unlinked to the core histone genes.

One of the nonhistone genes on TRT3 codes for a basic protein of apparent molecular weight of 8000 daltons. The nonhistone gene on TRT1 codes for a protein (protein 1) of 30,000 daltons. Northern blot hybridization measurements have shown that this gene is regulated independently of the histone genes and not under cell division cycle control (34). Expression of the protein 1 gene is also apparently strain-dependent (34). The product of the SMT1 gene has not been studied by *in vitro* translation. Assuming the gene encodes a protein, DNA sequence analysis indicates that there is only one extended open reading frame available for the gene and that it would be transcribed toward the H3 gene. This orientation has been confirmed by single-stranded M13 hybridization probes (unpublished data). The *SMT*1 mRNA appears to be more abundant than the H3 and H4 mRNAs; however, its function and the properties of its encoded protein are unknown. *SMT*1, and probably the genes linked with the H2A-H2B pairs, are single-copy genes.

2.2. Autonomously Replicating Sequences

An intriguing organizational, and undoubtedly functional, feature of the yeast histone gene loci is the recent identification of apparent origins of replication closely linked to each of the four paired genes. Yeast cloning vectors containing a selectable marker, such as *URA*3 or *HIS*3, normally transform host yeast strains for the selectable marker by integration into the host chromosome through homologous recombination (36). Certain DNA fragments, however, have been shown to cause the plasmid to replicate autonomously in the yeast cell as a circular episome (37–39). A consequence of the presence of an autonomously replicating segment, *ars*, on the transforming plasmid is a marked increase in the efficiency of transformation. The transforming plasmid can be detected in the host yeast cell as a circular extrachromosomal DNA molecule, and the high-frequency transformation is presumably owing to the fact that the plasmid need not integrate into the host chromosome to be maintained in the cell. Cells transformed for the selectable marker tend to be unstable when grown in the absence of selective pressure, and the *ars* plasmids, at least as simple bacterial plasmid constructions, are segregated and lost relatively rapidly. The transformants can be stabilized by including cloned, functional, centromeric DNA in the plasmid molecule (40–41).

When plasmid derivatives of the yeast histone genes were used to transform host yeast cells, it was noticed that high frequency transformation was obtained with some of the constructions. Furthermore, the transformants tended to be unstable when grown in the absence of selective pressure. By use of the high frequency transformation assay, *ars* functions have been mapped in at least three of the four histone gene pairs (Hereford, personal communication; Bouton and Smith, unpublished data). In each case the *ars* maps to the 3′ side of one of the histone genes. For the H2A–H2B loci, the *ars* functions map downstream of the H2B genes; for the H3–H4 genes, the *ars* function is located just to the 3′ side of the H4 gene. It seems highly unlikely that this similarity of gene organization is mere coincidence. It is attractive to hypothesize that the cell division cycle control of yeast histone gene transcription is somehow related to the presence of these closely linked origins of replication (66).

2.3. The Genes in the Genome

The four pairs of yeast genes are unlinked at four separate loci in the yeast genome. The first evidence for this conclusion is derived from genomic restriction map analysis. Within the homologous pairs, the H2A–H2B sets and the H3–H4 sets, it was impossible to overlap the genomic restriction maps (22, 26). Thus the two sets of H3–H4 genes had to be at least 15 kb apart and the two sets of H2A–H2B genes 35–60 kb apart. In addition, the genomic maps of the H3–H4 sets could not be overlapped with the H2A–H2B gene clones (Figure 1.2). The H3–H4 genes were isolated from a strain of *S. cerevisiae* different from the H2A–H2B genes, and thus it was possible that the different maps were caused by strain differences. Recently, however, genomic blot hybridization experiments have confirmed, in a single strain, that the four gene pairs do not share common restriction fragments (unpublished data).

The second line of evidence that the gene pairs are unlinked is derived from tetrad genetic mapping (64). Laboratory strains of *S.*

cerevisiae can be found that have restriction site variation within the histone gene regions. Two genomes with such a difference can be distinguished by blot hybridization of restriction digestions of their DNAs. This permits the genetic mapping of the histone genes by following conventional markers using standard nutritional selective plates and following the histone loci by preparing DNA from each spore of a dissected tetrad and hybridizing the digested DNA with the appropriate histone probe. Using this approach, Hereford determined that the two H2A–H2B gene pairs were unlinked (personal communication; reported in reference 22), indicating that they must be at least 235 kb apart, if not on separate chromosomes. Experiments with the two H3–H4 gene sets have given similar results—the two H3–H4 pairs are genetically unlinked (unpublished data). Finally, restriction site polymorphisms in each of the H2A-H2B pairs are unlinked to those in either of the H3-H4 gene pairs. Thus, the four pairs of yeast histone genes map at four genetically unlinked loci.

3. DNA SEQUENCE INFORMATION

S. cerevisiae may be the first organism in which it may be said that the entire genomic complement of histone DNA has been sequenced. This information has provided the predicted amino acid sequences of the histones, a look at the DNA sequence environment of multiple transcription units, possible control signals, and hints on the evolution of the yeast genes.

3.1. The Coding Sequences

The coding DNA sequences of both H2B genes (31) and both H2A genes (32) have been determined, as well as portions of the 5′- and 3′-noncoding flanking regions. For the H3 and H4 genes, the complete DNA sequences of 1800 base pairs spanning the genes in both sets of pairs have been determined and permit a detailed examination and comparison of the two loci (33). Together these DNA sequences provide the basic information necessary for further structural and functional experiments.

The amino acid sequences of the yeast histones may be predicted from the DNA sequences. The predicted sequences for histones H3 and H4 agree very well with the available amino acid composition and sequence information (7, 16, 33, 42); the histone H3 sequences obtained from protein sequencing and predicted from DNA sequencing agree exactly (33, 42). The agreement of the predicted histone H2A and H2B subtypes with amino acid composition and partial sequence information is mixed (7, 16, 31, 32). It is not known whether these differences represent strain differences or errors in sequencing methods. The discussion that follows is based upon the primary amino acid sequences predicted from the DNA sequence data. These sequences are illustrated in Figure 1.5.

The two H2A genes code for two different histone H2A subtypes. Each protein is 131 amino acid residues long; however, they differ from each other by substitutions at two positions: residues 124 and 125 are Ala-Thr in one of the proteins and Thr-Ala in the other subtype. Thus the two histone H2A subtypes have different primary sequences but identical amino acid compositions. Compared with the amino acid sequence of calf thymus histone H2A, there is an 80% homology with the yeast proteins. The distribution of the amino acid substitutions is relatively uniform across the protein except for a highly conserved region from residue 60 to 90.

The two H2B genes also code for two different H2B subtypes. Each H2B protein is 130 amino acids long, and the two subtypes differ by substitutions at four positions, residues 2, 3, 27, and 35. Histone H2B is an evolutionary mosaic (43). In general, the *N*-terminal 20–50 amino acids of histone H2B show rapid evolutionary divergence, and the yeast proteins are no exception; the first 30 residues of the yeast H2B proteins are lysine rich and show limited homology with other histone H2B sequences. However, residues 38–130 are about 78% homologous with the calf thymus H2B amino acid sequence. Again, the distribution of the substitutions with respect to the higher eucaryotic H2B sequences is relatively uniform across the sequence over residues 38–130.

Unlike the H2A and H2B genes, the yeast histone H3 and H4 genes code for identical H3 and H4 proteins. The yeast histone H3 is 135 amino acids long. Compared with calf thymus histone H3, there are 15 amino acid differences in the yeast sequence. Two striking dif-

```
                     1         10        20        30        40        50        60        70        80        90        100       110       120       130
                     .         .         .         .         .         .         .         .         .         .         .         .         .         .
Yeast H2A 1          SG-GK-GGKAGSAAKASQSRSADAGLTFPVGRVHRLLRRGNYAQRIGSGAPVYLTAVLEYLAAEILELAGNAARDNKKTRIIPRHLQLAIRNDDELNKLLGNVTIAQGGVLPNIHQNLLPKKSAKATKASQEL
Yeast H2A 2                                                                                                                                       TA
Calf thymus H2A        R  Q     --R  -KT  SR   Q           K    E V A      A      T                               E       K            QAV     TESHH  KGK-


                     1         10        20        30        40        50        60        70        80        90        100       110       120       130
                     .         .         .         .         .         .         .         .         .         .         .         .         .         .
Yeast H2B 1          SAKAEKKPASKAPAEKKPAAKKTSTST-------------DGKKRSKARKETYSSYIYKVLKQTHPDTGISQKSMSILNSFVNDIFERIATEASKLAAYNKKSTISAREIQTAVRLILPGELAKHAVSEGTRAVTKYSSSTQA
Yeast H2B 2           SA                        V-------------      V
Calf thymus H2B      -----PE  KS    ---  KKGS------KKA--VTKAQKK     KRS   S  V V     V       S A G M             G   R  H   R   TS       L              K     T  K--


                     1         10        20        30        40        50        60        70        80        90        100       110       120       130
                     .         .         .         .         .         .         .         .         .         .         .         .         .         .
Yeast H3             ARTKQTARKSTGGKAPRKQLASKAARKSAPSTGGVKKPHRYKPGTVALREIRRFQKSTELLIRKLPFQRLVREIAQDFKTDLRFQSSAIGALQESVEAYLVSLFEDTNLAAIHAKRVTIQKKDIKLARRLRGERS
Calf thymus H3                            T        A          R           Y                                  VM    AC     G       C         MP    E    I    A


                     1         10        20        30        40        50        60        70        80        90        100
                     .         .         .         .         .         .         .         .         .         .         .
Yeast H4             SGRGKGGKGLGKGGAKRHRKILRDNIQGITKPAIRRLARRGGVKRISGLIYEEVRAVLKSFLESVIRDSVTYTEHAKRKTVTSLDVVYALKRQGRTLYGFGG
Calf thymus H4                           V                                I G   V   N    A             AM
```

FIGURE 1.5. The amino acid sequences of each of the yeast histones is presented in one-letter codes. Below each of the yeast sequences are shown the positions and substitutions in amino acid residues for the calf thymus histones. Dashes indicate residues that had to be deleted to improve alignment. The data are taken from Wallis et al. (31), Choe et al. (32), and Smith and Andresson (33).

ferences are the change of Pro for Lys at position 121 of the yeast sequence and Gln for Lys at position 125. Thus the yeast H3 protein should be more basic than the calf protein, and this is supported by the greater mobility of the yeast protein in acid–urea gels (7, 42). The yeast histone H4 protein is 102 amino acids in length and strongly resembles the H4 proteins of higher eucaryotes. Compared with the prototype calf thymus sequence, there are eight amino acid differences in the yeast protein. The pattern of amino acid differences between the yeast H3 and H4 proteins and those of the higher eucaryotes is very non-uniform. Of the 15 differences in the H3 sequences, 11 occur after residue 80. Similarly for histone H4, all but one of the substitutions occur after position 50.

3.2. General DNA Sequence Features

The relatively small size of the histone gene coding sequences and their organization in yeast offer an opportunity to examine the general DNA sequence environment surrounding more than a single coding region. In addition, the lack of tandem reiteration of the yeast genes means that the sequences will not be biased by any unusual requirements imposed by or resulting from a tandemly repeated multigene family organization.

The DNA sequence confirms the results of heteroduplex analysis; the nonallelic sets of homologous histone loci share little DNA sequence homology other than the coding sequences of the histone genes themselves. This conclusion is illustrated clearly in the pairwise dot matrix analysis of the individual DNA sequences. In this analysis, the sequences are compared as two coordinates of a matrix, and a nucleotide homology is recorded as a dot in the matrix at the coordinates of the match; a matching string of nucleotides in the two sequences is thus visualized as a diagonal line with a negative 45-degree slope (44–46). The dot matrix plots for the two H2A–H2B DNA sequences and for the two H3–H4 sequences are shown in Figures 1.6 and 1.7. The coding sequences stand out dramatically in a sea of random nucleotide matches. It is clear that there are no strong extended sequence homologies of functional or evolutionary significance conserved in the noncoding DNA regions of the two H2A–H2B or the two H3–H4 gene sets.

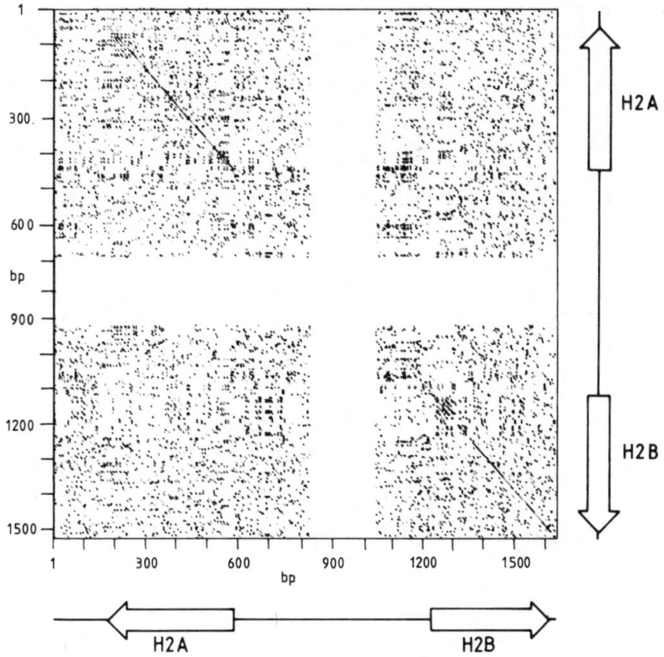

FIGURE 1.6. Dot matrix of the two H2A–H2B DNA sequences. The DNA sequence data from Wallis et al. (31) and Choe et al. (32) for histone H2A and H2B genes were arranged into the structures of the two genetic loci as illustrated in Figure 1.1. The sequence of the middle portions of the intergene regions is not available at present, and these distances have been arbitrarily set as 650 bp between the ATG codons of the gene pairs. Base-pair matches in the sequences of the two loci are recorded as a dot in the matrix if the match is part of four consecutive matches or more.

3.3. Architecture of the H3 and H4 Genes

The H3 and H4 coding sequences account for 1422 out of the 3600 base pairs sequenced for these loci, or approximately 40% of the total. It has been estimated that, on average, about 40% of the yeast genome is transcribed (65). The base compositions of the two sets of DNA sequences are similar and together average 30.7% A, 30.8% T, 17.6% C, and 20.9% G. This is in general agreement with the average base composition of the yeast genome as a whole (47). If

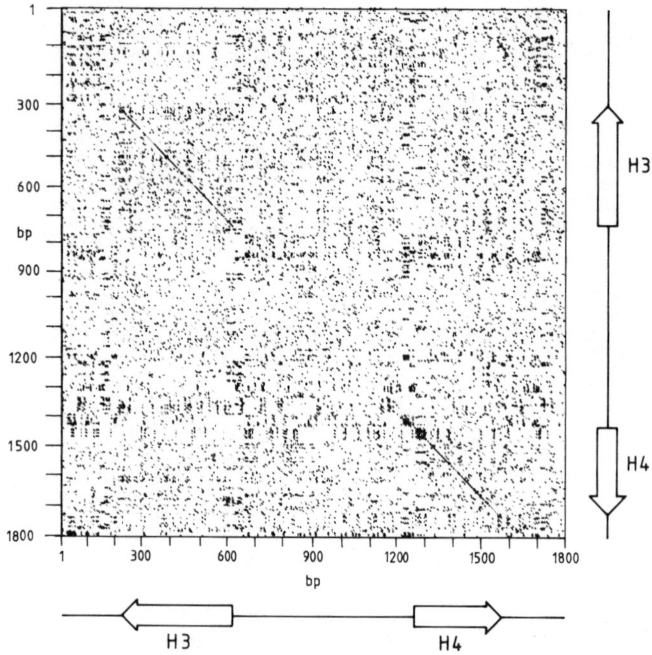

FIGURE 1.7. Dot matrix analysis of the two H3–H4 DNA sequences. The DNA sequences from Smith and Andresson (33) were compared as in Figure 1.6.

there is something special about the yeast histone gene loci, it is not apparent at the level of the overall base composition.

Despite this normal composition, relative to the yeast genome, the distribution of bases is not random. The DNA sequences show an interesting clustering of A + T-rich nucleotides. Each histone coding sequence is surrounded by very A + T-rich regions of 50–200 base pairs in length (33). The intergene DNA is relatively more G + C rich except on the immediate 5′ and 3′ sides of the coding sequences. DNA rich in A + T has been seen previously flanking structural genes; however, the localized and repeating nature of these A + T-rich blocks is much more apparent when the sequence extends across more than a single structural gene (33). The function of these A + T-rich regions is unknown. It is interesting that the 5′ and 3′ termini of the H3 and H4 mRNAs map near the ends of these A + T-rich blocks. It may be possible in the future to engineer mutants in these regions to test if the overall base composition is important. As a last point in this regard, the sequences have relatively more A – A and T – T dinucleotides than is expected by random chance. Thus the A + T-rich regions tend to be composed of short homopolymer stretches of A and T.

3.4. Specific DNA Sequence Motifs

The sequenced regions of the yeast histone genes are expected to contain regulatory sequences required for the proper initiation and termination of gene transcription. At present the termini of the H2B and H2A mRNAs are not known; however, the approximate ends of the steady state H3 and H4 mRNAs have been mapped (26, 33). The positions of these termini and particular sequence features related to them are illustrated in Figure 1.8.

Sequences conforming to the now familiar "TATA" box sequence (reviewed in reference 48) are present in the upstream 5′-flanking DNA of all of the yeast histone genes. This consensus sequence, $5'\text{-TATA}\genfrac{}{}{0pt}{}{\text{A}}{\text{T}}\text{A}\genfrac{}{}{0pt}{}{\text{A}}{\text{T}}\text{-3}'$, normally occurs 20–30 base pairs before the start of mRNA transcription and is thought to play a role in polymerase binding or cap site selection (48). The positions of the H4 TATA boxes are in close agreement with this model. However, the distances for the H3 genes are somewhat more distant than the classic 20–30 base pairs. In the case of the H3 and H4 genes, where 5′-flanking DNA sequence information extends completely across the intergene regions, a second common motif can be found. This consensus sequence, termed the "CAT" box, has the form $5'\text{-GG}\genfrac{}{}{0pt}{}{\text{C}}{\text{T}}\text{CAATCT-3}'$ and is usually observed a further 40–50 base pairs upstream of the TATA box (48). The matches with the H3 and H4 sequences are 6/9 or 7/9 bases.

The DNA sequences near the 5′ ends of the H4 mRNAs are similar, as are those near the 5′ ends of the H3 mRNAs. The H4 mRNAs begin near the sequence 5′-TANGATG-3′ and the 5′ ends of the H3 mRNAs map near the pyrimidine-rich sequence pattern 5′-pyCATTCpu-3′ (33). A common cap site for all the mRNAs, such

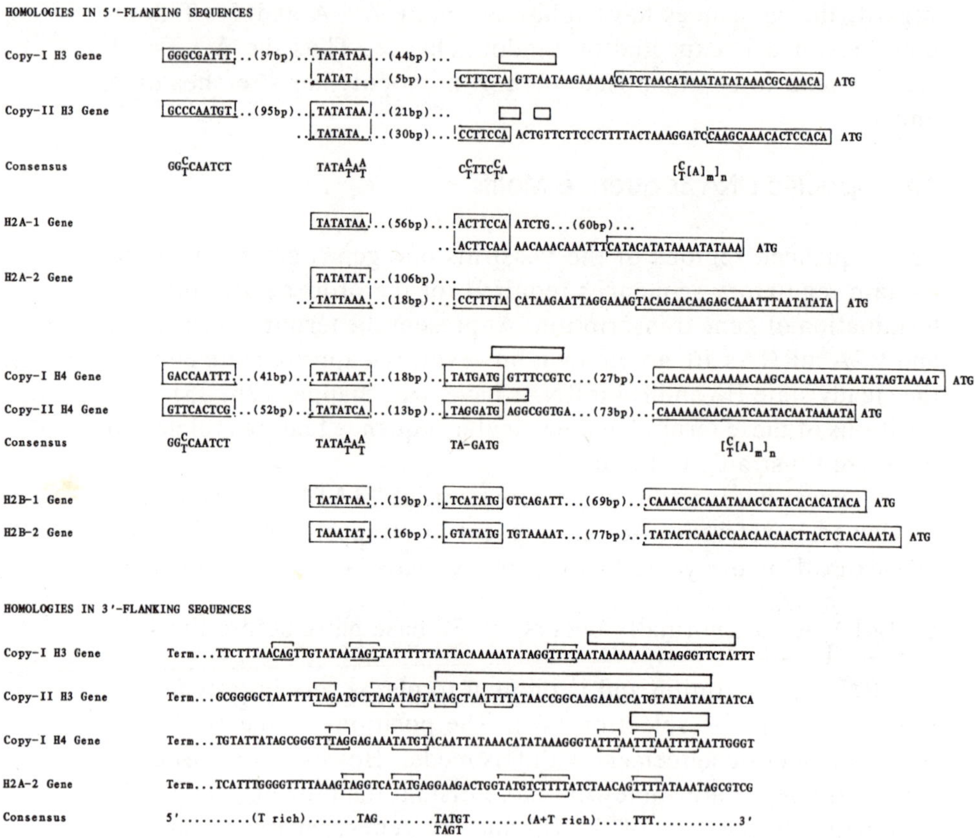

FIGURE 1.8. DNA sequence patterns in the 5′- and 3′-flanking DNA of the histone loci. Particular DNA sequence patterns in the flanking DNA of the histone genes are aligned and marked. The 5′ and 3′ ends of the H3 and H4 mRNAs are indicated by boxes above the sequences. The sequences are folded if there is more than a single occurrence of a prospective pattern. The consensus sequences for each group of genes are written below the alignments. The "TATA" and "CAT" consensus patterns are taken from Breathnach and Chambon (48). The 3′-end processing consensus sequence is taken from Zaret and Sherman (51). The data in this figure are adapted from Smith and Andresson (33).

as is found with the sea urchin early repeat genes (49, 50), is not seen. Similar patterns can be found in appropriate positions in the H2A and H2B genes. The two H2B genes both contain a sequence pattern similar to that at the start of the H4 mRNAs very shortly downstream of the apparent TATA boxes. The H2A genes each contain candidate sites for the pyrimidine-rich pattern found at the starts of the H3 gene mRNAs. These positions are indicated in Figure 1.8. Only the mapping of the termini of the H2A and H2B mRNAs will determine the relevence of these homologies, but in any event the results will be helpful in defining functionally important sequences.

The approximate 3′ ends of the two H3 mRNAs and one of the H4 mRNAs have also been mapped (33). Zaret and Sherman (51) have recently proposed a consensus sequence required for proper 3′-terminal processing of mRNAs based upon an interesting deletion mutant in the *CYC*1 gene and computer-aided DNA sequence comparisons. As previously pointed out by Choe et al. (32), the sequence of the one H2A gene for which 3′-flanking sequence is known contains an appropriate representation of the Zaret–Sherman sequence. Likewise, as indicated in Figure 1.8, a pattern resembling this 3′ motif is present near the termination of each of the H3 genes and the H4 gene. The classical eukaryotic 3′-polyadenylation sequence 5′-AATTAA-3′ is not present in the 3′-flanking DNA of all of the histone mRNAs. Site-directed mutagenesis of these proposed recognition sequences will confirm their importance in the histone gene function.

A final DNA sequence pattern common to the yeast histone genes should be mentioned. Immediately preceding the ATG translational start of all of the genes is an A-rich sequence which conforms to the empirical rule $5'\text{-}(\text{py}(\text{A})_m)_n\text{-}3'$, where py represents a pyrimidine nucleotide, and m and n are integers from about 1 to 10. In Figure 1.8 these regions are blocked in for each gene. As mentioned previously, the general rule is that the DNA sequences of related nonallelic pairs of genes diverge immediately before and after their coding DNA (Figures 1.6 and 1.7). This rule is broken only in the case of the H2B and the H4 genes for this A-rich sequence motif. In each case, the DNA sequence for 20–30 base pairs immediately upstream of the ATG translational start codon are conserved between the nonallelic genes. The sequence motif is present in the two H2A and H3 genes but is not as highly conserved between the related nonallelic genes. The function of this region is unknown.

The yeast genes are missing the conserved 3′-noncoding DNA sequence first identified in the sea urchin embryonic repeat genes (52). This 23 base pair sequence contains within it a 16 base pair sequence which makes up a hyphenated inverted repeat and has now been found in the 3′-flanking DNA of histone genes from a number of different organisms. Birchmeier et al. (30) have recently shown that the inverted repeat is required for correct processing of the 3′ termini of sea urchin H2A mRNAs transcribed *in vivo* when injected into *Xenopus laevis* oocytes. While there is insufficient DNA sequence available at present to rule out this pattern from the H2A or H2B yeast genes, a search of the H3 and H4 genes has failed to identify a related sequence within the transcribed DNA of these genes (33).

4. EVOLUTION OF THE YEAST HISTONE GENES

The organization of the yeast genes provides an opportunity to analyze the molecular evolution of four distinct genetic loci, two sets of related pairs, and eight functionally related genes. The disperse repeated organization poses questions as to the evolution of the genes within *S. cerevisiae* itself and also with respect to the histone genes of multicellular eucaryotes.

4.1. Duplication of the Yeast Genes

Wallis et al. (31) first estimated the time of duplication of the two H2B genes based on the four amino acid changes between the two proteins. These four amino acid substitutions represent approximately a 3.2% difference when corrected for superimposed mutations (31, 53). Assuming that histone H2B proteins evolve at a roughly uniform rate of 60 million years per 1% divergence (54), then the yeast H2B genes would have duplicated about 190 million years ago (Table 1.1). When all of the histone loci and the divergence of the linked pairs is considered, however, a paradox becomes apparent—the genes for each of the four histones appear to have duplicated at very different evolutionary times.

TABLE 1.1. Evolutionary Divergence of Yeast Histone Genes

		Yeast vs. Metazoa		Yeast vs. Yeast	
Yeast Histone Protein	PAM per 100[a] Million Years	Percentage[b] Substitution	Time since Divergence $\times 10^{-9}$	Percentage[b] Substitution	Time since Duplication $\times 10^{-8}$
H2A	0.5–0.83	23.2	2.3–1.4	1.5	1.5–0.9
H2B	0.83	26.4	1.5	3.2	1.9
H3	0.12–0.15	12.0	5.0–4.0		
H4	0.09–0.13	8.3	4.6–3.2		

[a] PAM is the accepted point mutations per 100 amino acid residues given as the number acquired per 100 million years. The value for histone H2B is taken from Wilson et al. (54). For the other histones, the first value is taken from Dayhoff (61) and the second is from Wilson et al. (54).

[b] The percentage of amino acid substitutions has been corrected for superimposed and back mutations by the formula of Kimura (53).

Choe et al. (32) noted that the H2A genes appear to have duplicated much more recently than the H2B genes. First, the H2A proteins are more similar to each other than are the two H2B proteins. The estimated evolutionary time for 1% divergence of histone H2A proteins is approximately the same as that for the H2B genes (54). Thus the two amino acid substitutions between the yeast H2A subtypes would imply that they duplicated only 90 million years ago (Table 1.1). Second, the DNA sequence divergence for the H2B genes is much greater than for the H2A genes (32). In most other systems examined, the rate of silent or neutral base mutations is roughly constant with time, particularly for genes near each other on the chromosome (55–57). The yeast histone genes violate this rule flagrantly. There are 41 silent nucleotide substitutions between the H2B genes (31) and only 22 substitutions between the H2A genes. Approximately half of the substitutions in the H2B are transversions, whereas all of the differences in the H2A genes are transitions (32). The same discrepency is seen between the estimated duplication times for the H3 and H4 genes, with the H4 genes appearing to have duplicated much more recently than the H3 genes. For example, there are only four silent base substitutions between the two H4 genes, while there are 28 silent changes between the two H3 genes (33).

Because of the organization of the genes, simplicity would suggest that the pairs duplicated in a single event, that is, one of the H3–H4 loci duplicated as a unit in a single event. It is more complicated to picture the duplication of a single H3 gene and then, at a much later time in evolution, the duplication of the H4 gene right next to the H3 gene. However, the facts are that genes that logically *should* have duplicated together apparently have *not* duplicated together. What is the solution to this dilemma?

It is possible that, despite the topological horrors, the paired histone genes did not undergo duplication at the same time; perhaps unknown constraints provided that when the second gene of the pair duplicated it had to be near the duplicate of its former associate. A second solution is to accept at face value that the paired histone genes are simply not fixing silent base mutations at identical rates. A number of possible mechanisms could be invoked to explain this discrepancy, such as mRNA structural constraints, for example.

Finally, it is possible to explain the data by postulating that the duplicated histone loci have periodically undergone convergence by recombination or gene conversion of the coding sequences of the nonallelic homologous genes. By this argument, the paired genes duplicated simultaneously in a single event; the H3 genes duplicated as a unit with the H4 genes, for example. The rate of divergence and fixation of silent base mutations has also occurred at roughly identical rates in the pairs of genes. The apparent differences in the rate of divergence of the linked pairs is thus due to the occasional fixation, over evolutionary time, of a recombination or gene conversion between the coding sequences of the related nonallelic genes. The fixation of such a gene conversion, by this hypothesis, would have occurred most recently for the H4 genes.

There are several precedents for recombination between nonallelic genes (58–60). The most relevent for the present discussion is the demonstration by Ernst et al. (60) that a spontaneous revertant of a mutation in the yeast gene coding for iso-1-cytochrome c, *cyc*1-11, occurred by recombination with the nonallelic gene encoding iso-2-cytochrome c. Such events may have larger implications for the maintenance of DNA sequence similarity among other disperse repeated genes. The pattern of base substitutions between related yeast histone genes can be used to support the recombination model. For example, the base substitutions between the H3 genes are rather uniformly distributed across the gene whereas all four of the substitutions in the H4 genes occur in the 3′ half of the sequences; gene conversion at the 5′ ends of the H4 genes may have served to recently eliminate nucleotide divergence within that region. Of course, it must be remembered that other structural and functional constraints could be responsible for this assymetric distribution of base substitutions.

Clearly, if the nonallelic coding sequences undergo recombination or gene conversion, the duplicated yeast histone genes are not evolving independently, and any estimate of their time of duplication based on the coding sequences will be potentially in error.

4.2. Histone Protein Evolution

When the yeast histone amino acid sequences are compared with those of higher eucaryotes, a second paradox is seen. The divergence of the yeast histones H3 and H4 from those of higher eucaryotes is significantly greater than expected based on the accepted rate of protein evolution for these relatively conserved proteins. The arguments have been made by Brandt and von Holt (42) based on the histone H3 primary amino acid sequence. Animals, plants, and fungi are thought to have diverged from a common ancestor about 1.1×10^9 years ago (61). Based upon previous H3 amino acid sequences, the mutation rate for this protein is calculated at 0.135 point mutations per 100 residues per 10^8 years for each line. The yeast and calf histone H3 proteins differ at 15 positions. This represents a 12% divergence when corrected for superimposed mutations (53, 61). At this rate, the fungi would have diverged from the animal and plant evolutionary lines about $4.0–5.0 \times 10^9$ years ago. Similar calculations for histone H4 suggest that the fungi branched from the higher plants and animals $3.2–4.6 \times 10^9$ years ago (Table 1.1). Brandt and von Holt (42) have argued in favor of keeping the current evolutionary clock setting and pushing back the estimated time of divergence of the fungi from the main trunk. Goodman (62) has argued that the earth's age of 4.5 billion years is far too young to account for histone H3 and H4 differences of this magnitude.

The main conclusion to be drawn from such controversy is that there is, at present, far too little available data on which to project evolutionary calculations back such extraordinary lengths of time. The smooth-running molecular clock may not be smooth running. The calculations of Goodman (62) suggest that the rates of protein evolution undergo patterns of acceleration and deceleration that can vary five- to sevenfold, at least in the well-documented case of the globin proteins. It is proposed that these accelerated rates of evolution correlate with the appearance of new protein–protein interactions and cooperative functions and that the substitutions occur at known contact points and functional sites (62). As already mentioned, yeast histones H3 and H4 differ from their higher eukaryotic counterparts in the C-terminal halves of the proteins; these are the regions of strong H3–H4 interactions (reviewed in reference 63), and the properties of the formation of the yeast strong complex *in vitro*, at least, differ from those of calf or pea H3–H4 strong complexes (7). Thus the selection of these altered protein structures may account for an accelerated rate of protein evolution for the yeast histone H3 and H4 genes. On the other hand, interkingdom binding studies with histones H2A and H2B from yeast and calf, referred to earlier, indicate that these proteins have maintained very conserved binding surfaces (7).

Finally, although less likely, the yeast histone genes may not be orthologous with the higher eukaryote histone genes. Since the time of duplication of the yeast genes cannot be determined with accuracy at present, and since it is likely that the duplicated genes have undergone recombination or gene conversion, it is possible that the higher eucaryote histones are derived from a histone gene line that is simply not represented in the *S. cerevisiae* genome.

5. CONCLUSION

Considerable strides have been made in understanding the organization of the yeast histone genes, although several points remain to be determined. With current mapping techniques it should be possible to place each of the histone loci on the yeast genetic map. A question of organization that has not been addressed is the orientation of the histone loci relative to the centromeres of their respective linkage groups. The completed DNA sequences of the intergene and flanking regions of the H2A and H2B genes will aid in the analysis of regulatory signals for the histone genes, as will the mapping of the termini of the H2A and H2B mRNAs. Clearly, there is insufficient data to evaluate properly the evolutionary record of the early histone genes. Thus, while this book testifies to the increasing interest in the histone genes of diverse organisms, it is clearly important to continue to study the organization and DNA sequences of a much wider selection of lower eukaryotes. Only in this way will the histone genes fulfill their potential of providing a window through which to glimpse very early evolutionary events.

ACKNOWLEDGMENTS

I would like to thank my colleagues at the University of Virginia and the University of Edinburgh, particularly Noreen and Kenneth Murray, for help and encouragement, and my "histone gene" colleagues, especially Lynna Hereford, for discussions and personal communications. Stimulating discussions were always found with Max Birnstiel, Larry Kedes, Eric Weinberg, and Mike Grunstein. I am indebted to Jim Murray, University of Virginia, for discussions on the molecular evolution of the genes. The research in the author's laboratory was supported by a University of Virginia Research Development Committee grant and N.I.H. grant 1-R01-GM28920-01.

REFERENCES

1. L. H. Kedes, *Ann. Rev. Biochem.* **48,** 837 (1979).
2. C. C. Hentschell and M. L. Birnstiel, *Cell* **25,** 301 (1981).
3. L. H. Kedes and M. L. Birnstiel, *Nat. New Biol.* **230,** 165 (1976).
4. M. L. Birnstiel, J. Telford, E. S. Weinberg, and D. Stafford, *Proc. Natl. Acad. Sci. USA.* **71,** 2900 (1974).
5. L. H. Kedes and P. Gross, *J. Mol. Biol.* **42,** 559 (1969).
6. M. Grunstein, see Chapter 2.
7. J. K. W. Mardian and I. Isenberg, *Biochemistry* **17,** 3825 (1978).
8. G. J. Tonino and T. H. Rozjin, *Biochim. Biophys. Acta* **124,** 427 (1966).

9. U. Wintersberger, P. Smith, and K. Letnansky, *Eur. J. Biochem.* **33,** 123 (1973).
10. L. Franco, E. W. Johns, and J. M. Navlet, *Eur. J. Biochem.* **45,** 83 (1974).
11. R. Moll and E. Wintersberger, *Proc. Natl. Acad. Sci. USA.* **73,** 1863 (1976).
12. W. F. Brandt and C. von Holt, *FEBS Lett.* **65,** 386 (1976).
13. J. O. Thomas and V. Furber, *FEBS Lett.* **66,** 274 (1976).
14. D. A. Nelson, W. R. Beltz, and R. L. Rill, *Proc. Natl. Acad. Sci. USA.* **74,** 1343 (1977).
15. A. Pastink, T. A. Berkhout, W. H. Mager, and R. J. Planta, *Biochem. J.* **177,** 917 (1979).
16. W. F. Brandt, K. Patterson, and C. von Holt, *Eur. J. Biochem.* **110,** 67 (1980).
17. J. R. Davie, C. A. Saunders, J. M. Walsh, and S. C. Weber, *Nucl. Acids Res.* **9,** 3205 (1981).
18. D. A. Nelson, *J. Biol. Chem.* **257,** 1565 (1982).
19. D. Lohr and K. E. Van Holde, *Science* **188,** 165 (1975).
20. D. Lohr, R. T. Kovacic, and K. E. Van Holde, *Biochemistry* **16,** 463 (1977).
21. A. Sommer, *Mol. Gen. Genet.* **161,** 323 (1978).
22. L. Hereford, K. Fahrner, J. Woolford, Jr., M. Rosbash, and D. B. Kaback, *Cell* **18,** 1261 (1979).
23. K. Fahrner, J. Yarger, and L. Hereford, *Nucl. Acids Res.* **8,** 5725 (1980).
24. J. Woolford, Jr., L. Hereford, and M. Rosbash, *Cell* **18,** 1247 (1979).
25. J. Woolford, Jr. and M. Rosbash, *Nucl. Acids Res.* **6,** 2483 (1979).
26. M. M. Smith and K. Murray, *J. Mol. Biol.* **169,** 641 (1983).
27. W. Schaffner, G. Kunz, H. Daetwyler, J. Telford, H. O. Smith, and M. L. Birnstiel, *Cell* **14,** 655 (1978).
28. I. Sures, J. Lowry, and L. Kedes, *Cell* **15,** 1033 (1978).
29. K. W. Jones and L. Singh, *Hum. Genet.* **58,** 46 (1981).
30. C. Birchmeier, R. Grosschedl, and M. L. Birnstiel, *Cell* **28,** 739 (1982).
31. J. W. Wallis, L. H. Hereford, and M. Grunstein, *Cell* **22,** 799 (1980).
32. J. Choe, D. Kolodrubetz, and M. Grunstein, *Proc. Natl. Acad. Sci. USA* **79,** 1484 (1982).
33. M. M. Smith and O. S. Andresson, *J. Mol. Biol.* **169,** 663 (1983).
34. L. M. Hereford and M. A. Osley, *Cell* **24,** 367 (1981).
35. M. Busslinger, S. Rusconi, and M. L. Birnstiel, *EMBO J.* **1,** 27 (1982).
36. A. Hinnen, J. B. Hicks, and G. R. Fink, *Proc. Natl. Acad. Sci. USA* **75,** 1929 (1978).
37. D. T. Stinchcomb, K. Struhl, and R. W. Davis, *Nature* **282,** 39 (1979).
38. K. Struhl, D. T. Stinchcomb, S. S. Scherer, and R. W. Davis, *Proc. Natl. Acad. Sci. USA* **76,** 1035 (1979).

39. C.-L. Hsiao and J. Carbon, *Proc. Natl. Acad. Sci. USA* **76,** 3829 (1979).
40. L. Clarke and J. Carbon, *Nature* **187,** 504 (1980).
41. C.-L. Hsiao and J. Carbon, *Proc. Natl. Acad. Sci. USA* **78,** 3760 (1981).
42. W. F. Brandt and C. von Holt, *Eur. J. Biochem.* **121,** 501 (1982).
43. I. Isenberg, *Ann. Rev. Biochem.* **48,** 159 (1979).
44. P. A. Hieter, E. E. Max, J. G. Seidman, J. V. Maizel, and P. Leder, *Cell* **22,** 1979 (1980).
45. A. Efstratiadis, J. W. Posakony, T. Mariatis, R. M. Lawn, C. O'Connell, R. A. Spritz, J. K. DeRiel, B. G. Forget, S. M. Weissman, J. L. Slightom, A. E. Blechl, O. Smithies, F. E. Baralle, C. C. Shoulders, and N. J. Proudfoot, *Cell* **21,** 653 (1980).
46. M. Steinmetz, J. G. Frelinger, D. Fisher, T. Hunkapiller, D. Pereira, S. M. Weisman, H. Uehara, S. Nathenson, and L. Hood, *Cell* **24,** 125 (1981).
47. S. Zamenhof and E. Chargaff, *J. Biol. Chem.* **187,** 1 (1950).
48. R. Breathnach and P. Chambon, *Ann. Rev. Biochem.* **50,** 349 (1981).
49. C. Hentschel, J.-C. Irminger, P. Bucher, and M. L. Birnstiel, *Nature* **285,** 147 (1980).
50. I. Sures, S. Levy, and L. H. Kedes, *Proc. Natl. Acad. Sci. USA* **77,** 1265 (1980).
51. K. S. Zaret and F. Sherman, *Cell* **18,** 563 (1982).
52. M. Busslinger, R. Portmann, and M. L. Birnstiel, *Nucl. Acids Res.* **6,** 2997 (1979).
53. M. Kimura, *Nature* **267,** 275 (1977).
54. A. C. Wilson, S. Carlson, and T. White, *Ann. Rev. Biochem.* **46,** 573 (1977).
55. T. Miyata, T. Yasunaga, and T. Nishida, *Proc. Natl. Acad. Sci. USA* **77,** 7328 (1980).
56. T. Miyata and H. Hayashida, *Proc. Natl. Acad. Sci. USA* **78,** 5739 (1981).
57. M. Kimura, *Proc. Natl. Acad. Sci. USA* **78,** 454 (1981).
58. S. Scherer and R. W. Davis, *Science* **209,** 1380 (1980).
59. J. A. Jackson and G. R. Fink, *Nature* **292,** 306 (1981).
60. J. F. Ernst, J. W. Stewart, and F. Sherman, *Proc. Natl. Acad. Sci. USA* **78,** 6334 (1981).
61. M. O. Dayhoff, *Atlas Protein Sequence Struct.* **5** (Suppl. 2), 1 (1976).
62. M. Goodman, *Prog. Biophys. Mol. Biol.* **38,** 105 (1981).
63. J. D. McGhee and G. Felsenfeld, *Ann. Rev. Biochem.* **49,** 1115 (1980).
64. M. V. Olson, K. Laughney, and B. D. Hall, *J. Mol. Biol.* **132,** 387 (1979).
65. L. M. Hereford and M. Rosbash, *Cell* **10,** 453 (1977).
66. M. A. Osley and L. M. Hereford, *Proc. Natl. Acad. Sci. USA*, **79,** 7689 (1982).

CHAPTER 2

A GENETIC ANALYSIS OF HISTONE PROTEIN SUBTYPES IN YEAST

MICHAEL GRUNSTEIN, MARY RYKOWSKI,
DAVID KOLODRUBETZ, JOONHO CHOE,
AND JOHN WALLIS
Molecular Biology Institute and the Department of Biology
University of California, Los Angeles
Los Angeles, California

CONTENTS

1. INTRODUCTION

2. YEAST CONTAINS HISTONE SUBTYPES
   2.1. H2B
   2.2. H2A

3. YEAST HISTONE H2B AND H2A PROTEINS ARE SIMILAR IN STRUCTURE TO THOSE OF OTHER EUKARYOTES

4. INTRODUCTION OF FRAMESHIFT MUTATIONS INTO HISTONE H2B AND H2A GENES

5. THE htb1$^-$ AND htb2$^-$ MUTANTS UNDERGO MEIOSIS

6. THE COMPLETE LACK OF FUNCTIONAL HISTONE H2B IS LETHAL

7. ISOLATION OF HISTONES FROM htb1$^-$ AND htb2$^-$ YEAST

8. HISTONE H2A SUBTYPES ASSOCIATE INTERCHANGEABLY WITH HISTONE H2B SUBTYPES
   8.1. hta1$^-$ Strains Grow More Slowly than hta2$^-$ Strains
      8.1.1. Testing the Interaction of Histone H2A and H2B Subtypes

9. CONCLUSIONS AND FUTURE DIRECTIONS

REFERENCES

1. INTRODUCTION

The five histones H1, H2A, H2B, H3, and H4 in their interaction with DNA provide a basic unit in the formation of the chromosome. However, separation of histones by both size and charge has revealed the presence of histone variants or subtypes in most organisms. Mammals, birds, amphibians, echinoderms, protozoans, and fungi all have histone primary structure subtypes (1–7). These subtypes can differ in very few amino acids. For example, the two yeast H2b subtypes differ at four amino acid residues out of 130 (7). Alternatively, some subtypes, for example, sea urchin sperm versus sea urchin adult histone H2B, differ considerably in both sequence and length. Sperm H2B is approximately 20 amino acids longer than the embryonic protein (8). Some histone subtypes are all synthesized at the same developmental stage, others appear at different stages of embryogenesis (2, 9–12). Different tissues and even different cellular compartments (macronucleus and micronucleus of *Tetrahymena*) (5) can contain subtypes.

Few clues to the function of these variant proteins exist. The findings that in the sea urchin the expression of some histone subtypes follows a developmental pattern and that in *Tetrahymena* the macronuclear and micronuclear H3 proteins differ in primary structure have led to hypotheses that subtypes may have specialized functions affecting gene activity (5, 12, 13). However, it is also possible that the sequence differences between these proteins are functionally neutral and are not the result of evolutionary pressures. In an organism allowing convenient genetic analysis these hypotheses could be tested.

To study histone function and histone subtype function directly, our laboratory has concentrated on the genetic analysis of the yeast, *Saccharomyces cerevisiae*. Yeast contains nucleosomes, the four core histones, and histone subtypes (7, 14). It can exist as one of several cell types (haploid, diploid, spore) and can undergo various transitions (mating, sporulation, and germination). Most important from a technical point of view, it allows a unique opportunity for creating defined changes in a gene *in vitro*, and seeing those changes translated into an altered protein structure *in vivo*. This allows histone genes to be altered and histone function to be probed *in vivo*.

2. YEAST CONTAINS HISTONE SUBTYPES

2.1. H2B

Yeast contains two copies of the genes for each of the four core histones (15). In keeping with the convention in yeast genetics (35), we designate the genes coding for histones H2A, H2B, HTA and HTB, respectively. The organization of the HTA and HTB genes is shown in Figure 2.1. The HTA1–HTB1 genes are genetically unlinked to the HTA2–HTB2 pair (15). In both cases the genes are transcribed divergently [these features are also found for the gene pairs coding for histones H3 and H4 (Mitchell Smith, chapter 1)]. To determine the primary structure of histone H2B and to determine whether the genes were coding for different subtypes, the HTB1 and HTB2 genes were sequenced, and the protein structure was inferred from the DNA sequence (7).

Figure 2.2 is a comparison of the nucleotide sequences and inferred protein sequences of HTB1 and HTB2. These data show that yeast HTB genes code for histone subtypes. The inferred protein sequences

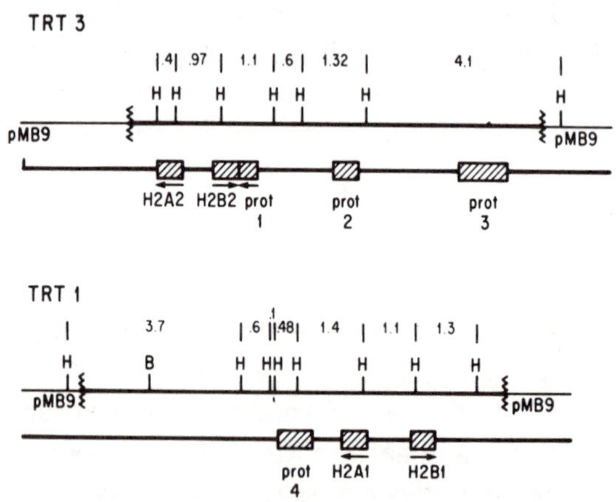

FIGURE 2.1. Restriction maps of plasmid TRT-3 and TRT-1. The upper line in each diagram shows the position of the Hind III (H), Eco R1 (R) and Bam H1 (B) restriction endonuclease cleavage sites, and the lengths of the Hind III fragments in kilobases. The lower line indicates the position of known structural genes and the direction of transcription.

```
TRT1                                                      ATG TCT GCT AAA GCC GAA AAG AAA CCA GCC TCC AAA GCC CCA
TRT3                                                      ATG TCC TCT GCC GCC GAA AAG AAA CCA GCT TCC AAA GCT CCA
yeast                                                   (met) ser ala lys ala glu lys lys pro ala ser lys ala pro
                                                                   ↓   ↓
                                                                  ser ala       5                   10

TRT1   GCT GAA AAG AAA CCA GCC GCT AAA AAG ACT TCC ACT TCC ACT GAT GGT AAG AAG AGA AGC AAG GCT AGA AAG GAA ACA TAC TCT TCT TAC ATT
TRT3   GCT GAA AAG AAG CCA GCT GCC AAG AAA ACA TCA ACC TCC GTC GAT GGT AAG AAG ACA TCT AAG GTT AGA AAG GAG ACC TAT TCC TCT TAT ATC
yeast  ala glu lys lys pro ala ala lys lys thr ser thr ser thr asp gly lys lys arg ser lys ala arg lys glu thr tyr ser ser tyr ile
           15                  20                  25   ↓              30                  ↓                  40
                                                       val                                val

TRT1   TAC AAA GTT TTG AAG CAA ACT CAC CCT GAC ACT GGT ATT TCC CAA AAG TCC ATG TCT ATC TTG AAC TCT TTC GTT AAC GAT ATC TTT GAA AGA
TRT3   TAC AAA GTT TTG AAG CAA ACT CAC CCA GAC ACT GGT ATT TCC CAG AAG TCT ATG TCT ATT TTG AAC TCT TTC GTT AAC GAT ATC TTT GAA AGA
yeast  tyr lys val leu lys gln thr his pro asp thr gly ile ser gln lys ser met ser ile leu asn ser phe val asn asp ile phe glu arg
       45                  50                  55                  60                  65                  70                  75

TRT1   ATC GCT ACT GAA GCT TCT AAA TTG GCT GCG TAT AAC AAG AAG TCT ACT ATC TCT GCT AGA GAA ATT CAA ACC GCT GTT AGA TTG ATC TTA CCA
TRT3   ATT GCT ACT GAA GCT TCT AAA TTG GCC GCT TAT AAC AAG AAA TCC ACT ATT TCT GCT AGA GAA ATC CAA ACA GCC GTT AGA TTG ATC TTA CCT
yeast  ile ala thr glu ala ser lys leu ala ala tyr asn lys lys ser thr ile ser ala arg glu ile gln thr ala val arg leu ile leu pro
                           80                  85                  90                  95                 100                 105

TRT1   GGT GAA TTG GCT AAG CAT GCT GTC TCT GAA GGT ACT AGA GCT GTT ACC AAG TAC TCT TCC TCT ACT CAA GCA TAA TGA AAT CAC TTC
TRT3   GGT GAA TTG GCT AAA CAT GCC GTC TCC GAA GGT ACT AGG GCT GTT ACC AAA TAC TCC TCC TCT ACT CAA GCC TAA GTC ACT CAC TAG
yeast  gly glu leu ala lys his ala val ser glu gly thr arg ala val thr lys tyr ser ser ser thr gln ala term term        term
                   110                 115                 120                 125                 130
```

FIGURE 2.2. Nucleotide sequence of histone HTB: (top line) HTB1: (bottom line) HTB2. Only the mRNA-like strand of each gene is shown. Nucleotide homologies between the DNAs are boxed. A dashed line in the prelude sequence represents a break necessary to align the sequences for greater homology. The inferred amino acid sequence of H2B1 protein is written below the two gene sequences with differences in H2B2 protein indicated at positions 2, 3, 27, and 35.

are 130 amino acids in length and differ by four amino acid residues. The changes between them are ala↔ser, lys↔ala, thr↔val, and ala↔val at amino acid positions 2, 3, 27, and 35, respectively; except for lys↔ala, the other three changes are relatively conservative. A comparison of histone H2B of different organisms reveals that most divergence occurs at the amino terminus (Figure 2.3). The changes between H2B1 and H2B2 are also all at the amino terminus. Both of the HTB genes are expressed as evidenced by the presence of H2B1 and H2B2 mRNAs at the beginning of S phase (16) and by the presence of electrophoretically distinct H2B1 and H2B2 proteins isolated from yeast chromatin (17).

2.2. H2A

A similar sequence analysis (18) of HTA1 and HTA2 is shown in Figure 2.4. The inferred protein sequences show that the two yeast HTA genes also code for subtypes. H2A1 and H2A2 histones are identical in length (131 amino acids) and in amino acid composition but differ by two amino acid residues. Ala↔thr and thr↔ala replacements have occurred at positions 124 and 125, respectively. The proteins are so similar that electrophoretic separation is unlikely, and our data show only one H2A protein band (17). Again, this portion of the H2A molecule, the carboxyl terminus between residues 122 and 131 is evolutionarily the most highly divergent region of H2A (Figure 2.5).

3. YEAST HISTONE H2B AND H2A PROTEINS ARE SIMILAR IN STRUCTURE TO THOSE OF OTHER EUCARYOTES

In Figures 2.3 and 2.5 the yeast histone protein sequences have been aligned with those of other eucaryotes. The length of H2B1 and H2B2 is 130 amino acids compared to an average of 124 amino acids for other H2B proteins shown here. Yeast H2B proteins are not the longest, however. The three H2B protein in sea urchin (*Parenchinus angulosus*) sperm range in length from 143 to 148 amino acids (8). As with other eukaryotic H2B proteins, there are three readily identified regions of yeast H2B. The amino terminus (amino acids 1–29) consists

```
                                                            basic
                ------------ amino terminus -------------- residues -------------------------------- carboxyl region --------------------------------------------
                          10        20                     30        40        50        60        70        80        90         100       110       120    130
YEAST H2B 1     SAKAEKKPASKAPAEKKPAAKKTSTST-------------DGKKRSKARKETYSSYIYKVLKQTHPDTGISQKSMSILNSFVNDIFERIATEASKLAAYNKKSTISAREIQ TAVRLILPGELAKHAVSEGTRAVTKYSSSTQA

YEAST H2B 2      SA                       V-------------       *

TROUT           -----PE  KS  ---- KKGS      KKA--VTKTAGKG    KRS   S AI V      V       S A G M            G S R  H   R   TS          L              K     T K--

HUMAN AND CALF  -----PE  KS  ---- KKGS------KKA--VTKAQKK     KRS   S  V V      V       S A G M            G   R  H   R   TS          L              K     T K--

DROSOPHILA      -----P----------- K---  GKAAKKAGKAQKNITKTD  KKRK   S AI                S A   M            A   R  H   R   TS          L              K     T K--

MOLLUSC         -----P----------- K---V SKGAKKAGKAKAARSG -   KRR   S  I        V     V S A   M            A   R  H   R   TS          L              K     T K--
```

FIGURE 2.3. Comparison of H2B protein sequences. Inferred amino acid sequences of yeast H2B1 and H2B2 are compared to the sequences of H2B proteins isolated from brown trout testes (20), human spleen (21), calf thymus (22), *Drosophila* (23), and the mollusc, *Patella granatina* (24). Only amino acid differences relative to yeast H2B1 are noted. A dashed line in a sequence represents a break necessary to align the sequences for greater homology. The numbering refers to the sequence of the yeast proteins. The following one-letter code for amino acids is used: A—ala, C—cys, D—asp, E—glu, F—phe, G—gly, H—his, I—ile, K—lys, L—leu, M—met, N—asn, P—pro, Q—gln, R—arg, S—ser, T—thr, V—val, Y—tyr.

```
                       -240                -220                -200                -180
H2A 1     TTTCAAAATCTTTTCTCACCCCCTTCCCCGTTCACGAAGCCAGCCAGTGGATCGTAAATACTAGCAATAAGTCTTGACCTAAAAAATATA
H2A 2     ACTACCATTTTCGTTACGTATACTTAGGCCAGAGATTACAACATGACTACTAATATCAAACATAACTCTATATATAAGGGATGAAGATGT

     -160                -140                -120                -100                -80
TAAATAAGACTCCTAATCAGCTTGTAGATTTTCTGGTCTTGTTGAACCATCATCTATTTACTTCCAATCTGTACTTCTCTTCTTGATACTACATCATCATACGGATT
ATGCTTTCTTAGAATTTCAAACATATTCCGTTAAAGTTTTACTTTTCGATTTCAATTTCGACTGCATGATGCTTTTCTTAGGTAGTTTTTTGTTATTAAATAGTATC

 -60                 -40                 -20                             1
TGGTTATTTCTCAGTGAATAAACAACTTCAAAACAAACAAATTTCATACATATAAAATATAAA     ATG TCC GGT GGT AAA GGT GGT AAA GCT GGT
ATAAATTCTTGCCTTTTACATAAGAATTAGGAAAGTACAGAACAAGAGCAAATTTAATATATA     ATG TCC GGT GGT AAA GGT GGT AAA GCT GGT
                                                                (met)ser gly gly lys gly gly lys ala gly

10                                      20                                      30
TCA GCT GCT AAA GCT TCT CAA TCT AGA TCT GCT AAG GCT GGT TTG ACA TTC CCA GTC GGT AGA GTG CAC AGA TTG CTA AGA
TCA GCT GCT AAA GCT TCT CAA TCT AGA TCT GCT AAA GCT GGT TTA ACA TTC CCA GTT GGT AGA GTG CAC AGA TTG CTA AGA
ser ala ala lys ala ser gln ser arg ser ala lys ala gly leu thr phe pro val gly arg val his arg leu leu arg

            40                                      50                                      60
AGA GGT AAC TAC GCC CAA AGA ATT GGT TCT GGT GCT CCA GTC TAC TTG ACT GCT GTC TTG GAA TAT TTG GCC GCT GAA ATT
AGA GGT AAC TAC GCC CAG AGA ATT GGT TCT GGT GCT CCA GTC TAT CTG ACT GCT GTC TTA GAA TAT TTG GCT GCT GAA ATT
arg gly asn tyr ala gln arg ile gly ser gly ala pro val tyr leu thr ala val leu glu tyr leu ala ala glu ile

                        70                                      80                                      90
TTA GAA TTA GCT GGT AAT GCT GCT AGG GAT AAC AAG AAG ACC AGA ATT ATT CCA AGA CAT TTG CAA TTG GCT ATC AGA AAT
TTA GAA TTG GCT GGT AAT GCT GCT AGA GAT AAC AAA AAA ACC AGA ATT ATT CCA AGA CAT TTA CAA TTG GCC ATC AGA AAT
leu glu leu ala gly asn ala ala arg asp asn lys lys thr arg ile ile pro arg his leu gln leu ala ile arg asn

                                    100                                     110
GAT GAC GAA TTG AAC AAG CTA TTG GGT AAC GTT ACC ATT GCC CAA GGT GGT GTT TTG CCA AAC ATC CAT CAA AAC TTG TTG
GAT GAT GAA TTG AAC AAG CTA TTG GGT AAT GTT ACC ATC GCC CAA GGT GGT GTT TTG CCA AAC ATT CAC CAA AAC TTG TTG
asp asp glu leu asn lys leu leu gly asn val thr ile ala gln gly gly val leu pro asn ile his gln asn leu leu

        120                                     130
CCA AAG AAG TCT GCC AAG GCT ACC AAG GCT TCT CAA GAA TTA TAA GAT CGG TTC TGG   TATTTTAAAGAAGGCGGAAGGAATTAAAT
CCA AAG AAG TCT GCC AAG ACT GCC AAA GCT TCT CAA GAA CTG TAA GAA CTG AGT TGA   AAAGAAACAAAGCGGTCAAACATTCTTG
pro lys lys ser ala lys ala thr lys ala ser gln glu leu term            term                            20
                        thr ala

TGCTTCTCCATTGT
TCATTTGGGGTTTTAAAGTAGGTCATATGAGGAAGACTGGTATGTCTTTTATCTAACAGTTTTATAAATAGCGTCGTTTTATTAAAGACAGTGTATAGTATTACTAT
      40                  60                  80                  100                 120

TAATAAAGAAAATAAAAAACGTATCAAGTTTACGG
  140                 160
```

FIGURE 2.4. Nucleotide sequences of histone HTA1 (top lines) and HTA2 (bottom) genes in yeast. Only the nontranscribed DNA strands are shown. Nucleotide differences between the two genes in the protein coding region are boxed. The amino acid sequence of histone H2A1 protein derived from the DNA sequence is shown below the gene sequences. The H2A2 protein differs only at positions 124 and 125.

```
                      1          10        20        30        40        50        60        70        80        90        100       110       120       130
YEAST H2A 1*          SG-GK-GGKAGSAAKASQSRSAKAGLTFPVGRVHRLLRRGNYAQRIGSGAPVYLTAVLEYLAAEILELAGNAARDNKKTRIIPRHLQLAIRNDDELNKLLGNVTIAQGGVLPNIHQNLLPKKSAKATKASQEL
YEAST H2A 2*                                                                                                                                       TA
DROSOPHILA*             R  -   V--KG  -K   NR   Q     I     K    E V A      A  M      V  E                         E      SG            QAV     TE ????????
SEA URCHIN 1            R  - A G--K   -K   SR   Q        F  K    N V A      A                                      E       G            QAV     TGSKSSL----
SEA URCHIN 2*           R  -S   --RT  -KT  SR   Q        F  K    K V G     MA      T                S          V   E       G            QAV     T -SS-----
SEA URCHIN 3*           R  -S   --RT  -KT  SR   Q        F  K    K V G     MA      T                S          V   E       G            QAV     T -SS-----
TROUT                   R  T    --R   -KT  SR   Q           K    E V A      A      T                  -        V   E       G            QAV     TE---  KVAK
CHICKEN*                R  Q    --R   -K   SR   Q           K    E V A      A      T                               E       G            QAV     TDSH-  KAK-
RAT CHLOROLEUKEMIA      R  Q    --R   -K   SR   Q           K    E V A      A      T                               E       K            QAV     TESHH  KGK-
MOUSE                   RE GK   --R   -K   SR   Q           K    E V A     MA      T                               E            E       EAV     TESHH   GK-
FRIEND LEUKEMIA

CALF THYMUS             R  Q    --R   -KT  SR   Q           K    E V A      A      T                               E       K            QAV     TESHH  KGK-
```

FIGURE 2.5. Comparison of amino acid sequences of yeast H2A1 and H2A2 and amino acid sequences of the H2A proteins from *Drosophila* (25), sea urchin 1 gonad (26), sea urchin 2 (27), sea urchin (28), rainbow trout (29), chicken (30, 31), rat leukemia (32), Friend mouse leukemia (33), and calf thymus (34). Sequences marked * were deduced from the DNA sequence of the corresponding H2A gene. The other sequences were derived from amino acid sequence analysis of isolated H2A proteins. Only amino acid substitutions relative to yeast H2A1 are indicated. Letter code for amino acids: A—ala; D—asp; E—glu; F—phe; G—gly; H—his; I—ile; K—lys; L—leu; N—asn; P—pro; Q—gln; R—arg; S—ser; T—thr; V—val; Y—tyr.

largely of basic residues and is the least conserved portion of H2B proteins. Yeast and human H2B are only 28% homologous in this region. This is followed by a block of eight residues, most of which are lys or arg (positions 30–37). From positions 38–130 the yeast protein is hydrophobic in nature (36 of 92 amino acids are hydrophobic. This segment is highly conserved in evolution. Yeast and human H2B are 80% homologous in this region. Most of the carboxyl terminus changes between yeast H2B and other H2B proteins are conservative in nature. There are two nonconservative changes: ser↔ile at position 42, and ala↔his at position 85.

Finally, despite the divergence of the amino termini, they contain a distinct repeating pentapeptide structure beginning with pro and

Trout

PRO GLU PRO ALA LYS SER ALA PRO LYS LYS GLY SER LYS LYS ALA . . .

Drosophila

PRO PRO LYS THR SER GLY LYS ALA ALA LYS LYS ALA GLY LYS ALA . . .

Tetrahymena

PRO LYS LYS ALA PRO ALA ALA ALA GLU LYS LYS VAL LYS LYS ALA PRO ...

Sea urchin

PRO SER GLN LYS SER PRO THR LYS ARG SER PRO THR LYS ARG SER PRO THR LYS ARG SER
PRO GLN LYS GLY GLY ...

Xenopus laevis

PRO GLU PRO ALA LYS SER ALA PRO ALA ALA LYS LYS GLY SER LYS LYS ...

Human, calf, crocodile, chicken

PRO GLU PRO ALA LYS SER ALA PRO ALA PRO LYS LYS GLY SER ...

Mollusc

PRO PRO LYS VAL SER SER LYS GLY ALA LYS LYS ALA GLY LYS ...

Starfish

PRO PRO LYS SER GLY LYS GLY GLN LYS LYS ALA GLY LYS ALA LYS ...

Yeast

SER SER ALA ALA GLU LYS LYS PRO ALA SER LYS ALA PRO ALA GLU LYS LYS PRO ALA ALA LYS LYS

FIGURE 2.6. A repeating pentapeptide structure (underlined) in histone H2B of different organisms.

containing 1–3 charged amino acids. In yeast H2B, this structure occurs at positions 8–12, 13–17, and 18–22. As shown in Figure 2.6, this structure occurs as a repeating unit in H2B proteins from sea urchin, calf, chicken, human, crocodile, *Xenopus*, and trout.

Figure 2.5 compares the primary structures of the H2A proteins from several organisms to those predicted for the yeast H2As. Calf and yeast, the most evolutionarily distant organisms on the list, show 80% homology in the sequences of their H2A proteins. With the exception of a highly conserved stretch of 30 amino acids at positions 60–90, the differences between the yeast and the animal H2As are largely distributed throughout the protein. The most divergent region between the nonyeast H2A proteins is in the 10 amino acids at the carboxy terminus. The two differences between H2A1 and H2A2 occur in the last 10 amino acids.

4. INTRODUCTION OF FRAMESHIFT MUTATIONS INTO HISTONE H2B AND H2A GENES

As described previously, yeast contains histone H2A and H2B proteins with characteristics similar to those of other eukaryotes. Also, like other eukaryotes, yeast contains subtypes in each of these proteins. If one examines the relative amount of H2B1 and H2B2, it is evident that each protein is a major component of chromatin. H2B1 and H2B2 each make up approximately one-half of the H2B protein. With the different processes in which yeast chromosomes are involved, for example, vegetative growth versus sporulation, it was possible to imagine that subtypes could be used differentially. Alternatively, the subtypes could be redundant. If the amino acid changes between H2Bs or between H2As were neutral, then the subtypes might not have separate essential functions.

To test between these extreme possibilities we constructed yeast strains that contain frameshift mutations at each of the HTB loci (HTB1 and HTB2) (17) and at each of the HTA loci (HTA1 and HTA2) (19). The strategy is described here for HTB2.

The plasmid TRT-3, carries HTB2 and HTA2 genes and contains eight Hind III sites. One of these sites was shown by DNA sequencing to occur in the HTB2 gene at the region coding for amino acid 80

(7). TRT-3 DNA was digested with Hind III in the presence of ethidium bromide so that only one random Hind III site was cleaved. When these linear molecules were treated with Klenow DNA polymerase and all four nucleoside triphosphates, four bases were added to each Hind III complementary (sticky) end. When recircularized by ligation the molecules should contain a 4 base pair insertion. The insertion alters the Hind III site so that it is no longer recognized by the restriction enzyme. These molecules were used to transform E. coli, and DNA from the transformants was analyzed for a missing Hind III site using Hind III endonuclease. We identified one plasmid (pYH18) which had lost the 0.97 and 1.1 kb Hind III fragments composing the HTB2 gene (Figure 2.7). Instead this plasmid contained

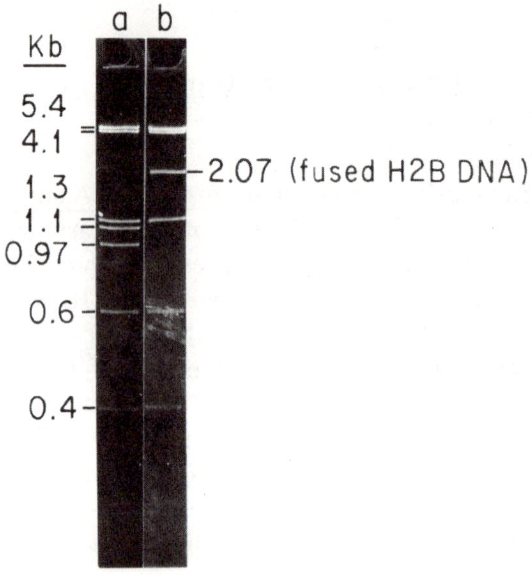

FIGURE 2.7. Comparison of a Hind III restriction digestion of plasmid TRT-3 and the *in vitro* mutagenized plasmid pYH18. The plasmid DNA was digested with Hind III, and the fragments were separated by electrophoresis on an 8% acrylamide gel. The TRT-3 pattern (lane a) shows the presence of the 0.97 and 1.1-kb fragments known to contain the HTB2 gene. These two bands are missing from the pYH18 pattern (lane b), having been replaced by a 2.07-kb band, the result of eliminating the Hind III site within the HTB2 gene.

of the wild $HTB2^+$ gene (legend, Figure 2.10) and the fused 2.07 kb Hind III fragment indicative of the mutant $htb2^-$ gene. DNA from the germinated spores of seven asci (two of which are shown in Figure 2.10) revealed in each case the presence of only the mutant or wild type $HTB2^+$ gene and 2:2 segregation of $htb2^-$:$HTB2^+$. Not only is the $htb2^-$ mutant viable, but also the sizes of the colonies produced from the germinated spores, whether $htb2^-$ or $HTB2^+$, are indistinguishable. The growth rate in liquid culture was also similar for mutant and wild cells. We conclude that the $htb2^-$ mutation has no obvious effect on the viability of a yeast cell undergoing vegetative growth. Similar experiments were done to replace the wild type $HTB1^+$ gene with an $htb1^-$ allele. The insertion in this case was *2bp* producing chain termination at position 105 (Figure 2.11). The results were similar except that the $htb1^-$ spores always formed smaller colonies after germination. In three days, $htb1^-$ spores produced colonies that were approximately the same size as those produced by $HTB1^+$ spores in two days. The small colony size segregated in each case with the $htb1^-$ mutation. We believe that this characteristic, which is closely linked to or a direct result of the $htb1^-$ mutation, affects the cell most obviously after germination (see Section 8.1).

5. THE htb1⁻ AND htb2⁻ MUTANTS UNDERGO MEIOSIS

To investigate the ability of the hisone mutants to mate, undergo mitosis as diploids, and sporulate in the homozygous state, we crossed haploid cells containing the same histone mutation but of opposite

FIGURE 2.9. General strategy for the construction of an $htb2^-$ yeast strain. (*a*) A plasmid containing a mutagenized HTB2 gene and a wild URA3 gene was used to transform a $ura3^-$ yeast strain by selecting for complementation. (*b*) Transformant clones are screened for stability of the Ura^+ phenotype, which indicates that integration has occurred by homologous recombination with the chromosomal HTB2 gene. (*c*) The strain is mated to provide a complementing HTB2 gene. (*d*) The tandem homology permits intrachromosomal recombination, resulting in excision of the plasmid. Depending on which side of the lesion the crossover occurs, either the mutant or the wild type HTB2 gene will be left in the genome. (*e*) Cultures were enriched for ura^- cells, and clones were screened for ura^- phenotype. The ura^- clones were screened to determine which contained the mutation.

mating type to produce homozygous histone mutants. Diploids were produced efficiently under the same conditions as those used for mating wild type strains. The htb1$^-$/htb1$^-$ and htb2$^-$/htb2$^-$ diploids are viable and have wild growth rates. Sporulation efficiency is highly strain dependent, but all heterozygous and homozygous htb1$^-$ and htb2$^-$ diploids sporulated at frequencies (9–43%) within the normal

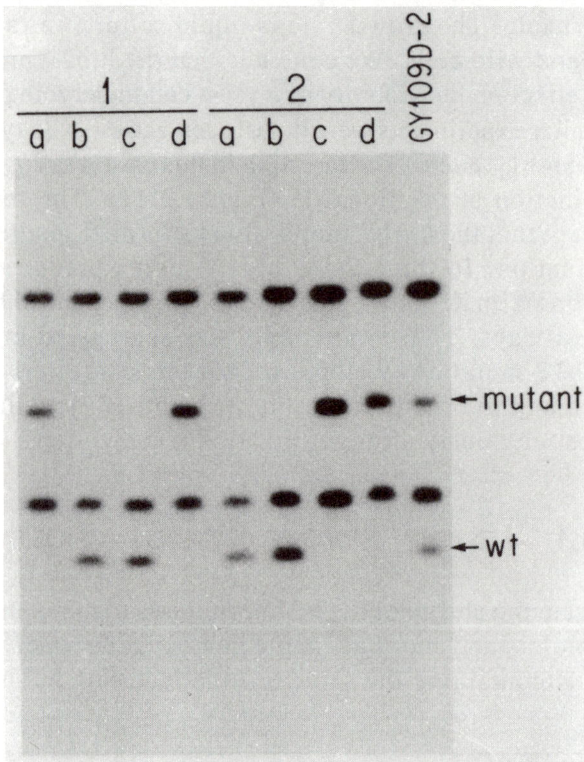

FIGURE 2.10. Segregation analysis of spores derived from htb2$^-$/HTB2$^+$ and h2b2$^-$/HTB2$^+$ yeast strains. Analysis of Hind III-digested DNA from diploid strain TY109D-2 (htb2$^-$/HTB2$^+$) (rightmost lane) and spores of two tetrads derived from it. There is a polymorphism in the wild type H2B2 pattern between TRT-3 (derived from diploid yeast strain D2) and GY109D-2. While the wild type H2B2 gene in TRT-3 results in a 0.97-kb and 1.1-kb fragment when digested with Hind III, the wild type gene in GY109D-2 produces a doublet consisting of two 0.97-kb fragments. When this consideration is taken into account, it is evident that each spore contains either the 0.97-kb wild type doublet (wt) or the 2.1-kb mutant Hind III band. The diploid shows a superposition of the two patterns.

```
h2b1 mutant   ATC GCT ACT GAA GCT CTT CTA AAT TGG ..... GAT TGA
              ile ala thr glu ala leu leu asn trp       glu term
                              80              84        103 104

H2B1 wild     ATC GCT ACT GAA GCT TCT AAA TTG GCT ..... GCA TAA
   type
              ile ala thr glu ala ser lys leu ala ..... ala term
                              80              84        130
```

FIGURE 2.11. Frameshift mutation in histone HTB1 gene.

range. Both homozygotes produced viable haploid progeny, indicating that neither histone is required for the completion of mating or meiosis. From these and the foregoing experiments, we conclude that the complete life cycle of yeast can occur with the presence of either of the wild type H2B protein subtypes.

6. THE COMPLETE LACK OF FUNCTIONAL HISTONE H2B IS LETHAL

The results described show that yeast cells lacking one of the H2B subtypes survive. In each case, these cells still contain the other unmutagenized H2B subtype. We wondered whether yeast could survive with no functional H2B protein. The extent of evolutionary conservation of the histone H2B protein sequence makes this possibility unlikely. Nevertheless, we did an experiment designed to produce cells that lack both histone H2B subtypes.

An htb2$^-$ haploid strain (HTB1$^+$, htb2$^-$) was crossed to an htb1$^-$ haploid strain (htb1$^-$, HTB2$^+$) of opposite mating type to produce a diploid heterozygous at both loci for HTB (HTB1$^+$/htb1$^-$, htb2$^-$/HTB2$^+$). This diploid strain, called 208-1, was sporulated to produce asci for tetrad dissection. Because the HTB1 and HTB2 genes assort independently, one in four spores should have the genotype htb1$^-$ htb2$^-$. Hind III digests of DNA isolated from germinated spore colonies were analyzed on southern blots with both TRT-1 and TRT-3 DNA used as probe. Of 61 colonies, 20 were HTB1$^+$ HTB2$^+$, 21 were htb1$^-$ HTB2$^+$, and 20 were HTB1$^+$ htb2$^-$. None was htb1$^-$ htb2$^-$. We examined the spores under the microscope and found that in 15 of the tetrads dissected, all four spores germinated. Fifteen of these germinated spores gave rise to a single-budded cell while most formed colonies. The genotype of the single-budded cells can be inferred from the genotype of the other spores from the same tetrad. With the assumption of 2:2 segregation of the mutant alleles, the single-budded cell could be inferred to be htb1$^-$ htb2$^-$ in all cases. The results indicate that the double mutant is capable of germination and budding but not of further growth.

Microscopic examination of dissected, germinating spores from this diploid also allowed us to compare the behavior of htb1$^-$ and wild type cells in the same genetic background to define further the phenotype of htb1$^-$ cells. Spores of 208-1 were dissected directly onto germination plates and observed microscopically over a 72-hour period. Most spores budded at approximately the same time (4–7 hours after transfer to germination plates). After this period, the htb1$^-$ colonies increased in size at a slower rate than the htb2$^-$ or wild type colonies.

7. ISOLATION OF HISTONES FROM htb1$^-$ AND htb2$^-$ YEAST

The mutagenesis of HTB1 and HTB2 histone genes described should result in gross alterations in the H2B1 and H2B2 histones. We expected that the results of the mutagenesis would be evident when chromatin proteins of the yeast strains were analyzed by gel electrophoresis.

Histone proteins were extracted from crude chromatin isolated from htb1$^-$, htb2$^-$ and wild type haploid strains, and from commercial baker's yeast. These proteins, along with yeast histone standards, were separated by electrophoresis on an acid–urea polyacrylamide gel. The result is shown in Figure 2.12. The bands have been identified as indicated on the basis of relative mobility.

Wild type yeast histone standards display two protein bands separated by charge in the expected position of the H2B protein (Figure 2.12, lane a). We identified the upper H2B band as H2B2 and the lower band as H2B1 (17). As expected, the chromatin of htb1$^-$ cells lacks the lower band, whereas that of the htb2$^-$ cells lacks the upper band. No lower molecular weight bands that might correspond to the truncated mutant proteins were apparent on longer gels. We concluded

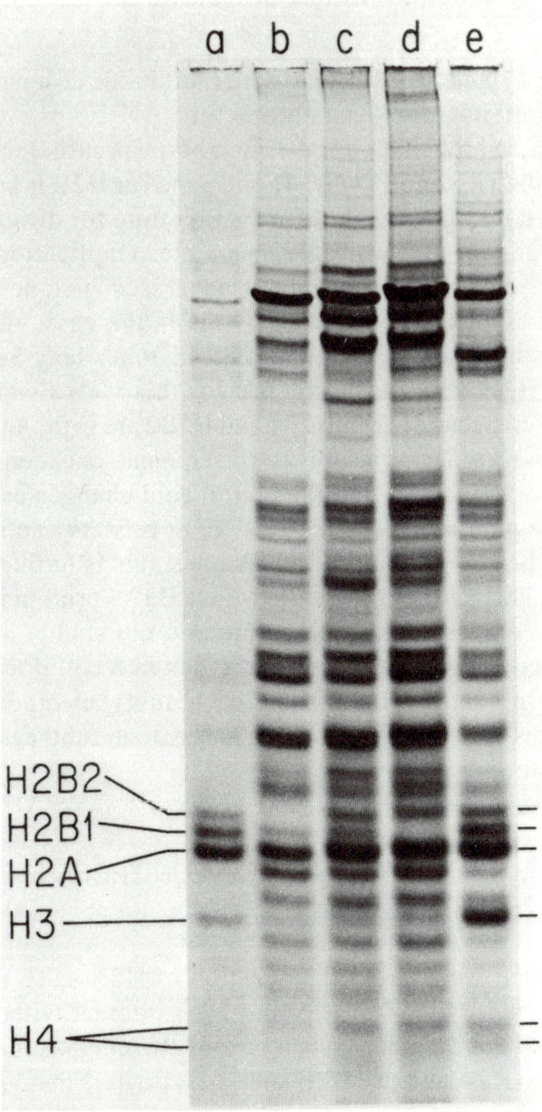

FIGURE 2.12. Polyacrylamide gel electrophoresis of histone proteins from histone mutant and wild type cells. Histone proteins were extracted from crude chromatin and separated by electrophoresis (17). We compare yeast histone standards isolated from baker's yeast, a gift from S. Weber (lane a), to histone from a haploid htb2$^-$ isolate derived from GY109D-2 (lane b), a wild type haploid (lane c), a haploid htb1$^-$ isolate (lane d), and baker's yeast histones isolated in this laboratory (lane e). The histone bands are labeled appropriately (see text). The lower H2B band is missing in the htb1$^-$ strain, and the upper H2B band is missing in the htb2$^-$ strain.

that the htb1$^-$ and htb2$^-$ mutations result in the absence of the H2B1 or H2B2 protein from yeast chromatin.

These data show that either H2B subtype is sufficient for the completion of the yeast cell cycle. If one gene for H2B is sufficient, why does yeast contain two genes for H2B coding for different proteins? At this stage, we can still only speculate. Duplication of genes for H2B could ensure a potential for increased histones synthesis at critical times in the cell cycle. If this is the case, the amino acid differences between H2B1 and H2B2 histones may be neutral with respect to H2B function. Evidence for this view lies in the finding that the differences between H2B1 and H2B2 are in the amino terminus, which shows greatest variability in sequence between species. One could, however, argue that the amino acid changes are not neutral, and that the phenotype of the h2b1$^-$ cells reflects a subtle functional difference between the proteins. We feel this is unlikely owing to a similar phenotype for hta1$^-$ cells whose H2A1 protein is very similar to H2A2 (see Section 8.1). Nevertheless, our studies show only that the presence of histone protein subtypes in a cell does not indicate a different essential function for each subtype; other experiments will be necessary to determine whether histone subtypes display more subtle functional differences.

8. HISTONE H2A SUBTYPES ASSOCIATE INTERCHANGEABLY WITH HISTONE H2B SUBTYPES

In this section we address the question of H2A–H2B subtype interactions *in vivo*. Histones H2A and H2B interact with each other to form part of the octameric structure of the nucleosome core. Since there are two H2A and two H2B subtypes, perhaps there is an exclusive interaction between a specific H2A subtype and a particular H2B subtype. Alternatively, each H2A protein may associate randomly with either H2B subtype to form a functional H2A–H2B dimer. Either possibility is consistent with the experiments showing that cells mutant in HTB1 or HTB2 are viable. For example, as shown in Figure 2.13, in an htb1$^-$ mutant the H2B2 protein could associate with both H2A1 and H2A2. In contrast, it is conceivable that H2B2 only interacts with one of the H2A proteins (H2A2 in this picture).

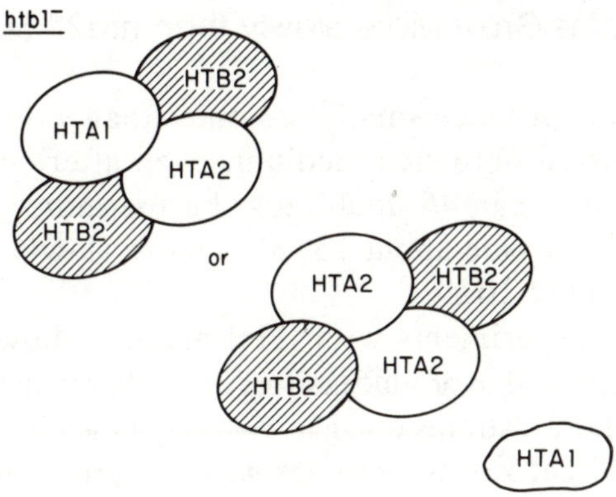

FIGURE 2.13. Potential use of H2B1 protein in nucleosomes of an htb2⁻ yeast strain.

To determine if all combinations of H2A and H2B subtypes allowed cell viability, we also created frameshift mutations in each of the HTA genes (19). The mutant HTA genes each encode polypeptides that are only 18 residues long (Figure 2.14). As with the htb⁻ gene replacements, the mutant hta⁻ genes were used to replace their wild type counterparts (19).

As with the htb mutations, haploids mutant in hta1⁻ (or in hta2⁻) are able to grow and mate. Diploids homozygous for hta1⁻ (or hta2⁻) are capable of growth and sporulation. Therefore neither histone H2A1 nor H2A2 is uniquely required in the yeast life cycle. Also, cells do require at least one functional HTA gene for survival (19). It is interesting to note that cells carrying the hta1⁻ mutation like htb1⁻ cells, do have a phenotype.

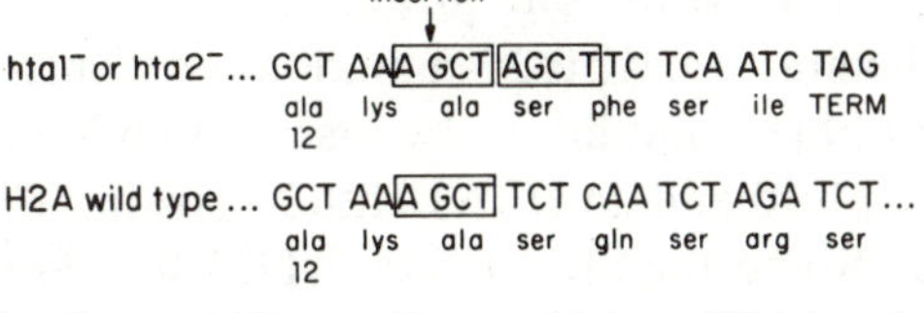

FIGURE 2.14. Frameshift mutations in histone HTA1 and HTAT genes.

8.1. hta1⁻ Strains Grow More Slowly than hta2⁻ Strains

The hta1$^-$ strains produce smaller colonies than wild type or hta2$^-$ strains when plated onto rich medium, even after continuous subculturing for more than 45 doublings. Furthermore, strains homozygous for hta1$^-$ divide about 15–30% more slowly in liquid than do wild type or hta2-strains.

Our previous experiments with HTB mutants showed that htb1$^-$ spores also germinated to produce colonies smaller than those produced by wild type or htb2$^-$ strains on agar. Although the difference between htb1$^-$ and htb2$^-$ cells on agar plates was less pronounced after subculturing for more than 50 doublings, the colony sizes were still different. In liquid culture the differences in growth rate were less evident and more strain dependent. Some htb1$^-$ strains grew as rapidly as htb2$^-$ strains, other continued to grow more slowly.

This phenotype for hta1$^-$ and htb1$^-$ strains may reflect a subtle functional difference between H2A1 and H2A2 proteins and between H2B1 and H2B2 proteins. It is also possible that this phenotypic difference between the two tandem gene pairs results from differences in their mRNA transcription or turnover (16). Our experiments could not distinguish between these possibilities.

8.1.1. Testing the Interaction of Histone H2A and H2B Subtypes Once the HTA mutants were constructed and characterized, we could determine whether or not haploid strains containing a mutation in one HTA gene and one HTB gene are viable. For example, it is possible that the H2A1 protein might form functional nucleosomes only with H2B1. If this were the case, we would expect that a haploid strain containing only wild type H2A1 and wild type H2B2 proteins (hta2$^-$ htb1$^-$) would be nonviable. Likewise, a cell may require some other combination of H2A and H2B subtypes for viability. We therefore asked the question: are there essential interactions between particular H2A and H2B histone protein subtypes?

To combine the hta1$^-$ and htb2$^-$ mutations we constructed a diploid heterozygous for hta1$^-$ and htb2$^-$ by mating two haploid strains, an hta1$^-$ HTB2$^+$ haploid and an HTA1$^+$htb2$^-$ strain. This diploid was then sporulated. Since the HTA1 and HTB2 loci are unlinked, we expect one-fourth of the haploid spores to be hta1$^-$htb2$^-$. Using

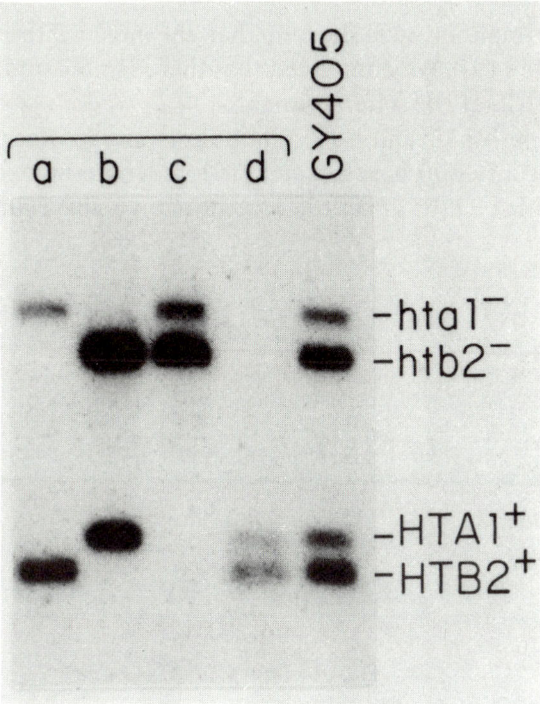

FIGURE 2.15. Southern blot hybridization gels showing viability of hta2$^-$htb1$^-$ haploids. DNA from the appropriate strains was restricted with HIND III restriction endonuclease and analyzed by Southern blot hybridization. The DNA was transferred from the gel onto two nitrocellulose filters. One filter was hybridized to a probe containing the 1.1-kb Hind III fragment from the HTA1-HTB1 region and the other to a probe containing the 0.97 kb Hind III fragment from the HTA2-HTB2 region. The figure shows a photograph of the two resulting autoradiographs superimposed with Southern analysis of haploid spore colonies (lanes a–d) from one tetrad resulting from the dissection of GY405. The locations of the mutant 2.5-kb (hta1$^-$) and 2.1-kb (htb2$^-$) HIND III bands as well as the wild type 1.1-kb (HTA1$^+$) and 0.97-kb (HTB2$^+$) HIND III bands are shown.

southern blot analysis we examined a Hind III digest of DNA from spore clones derived from four-spored tetrads. The presence of both the hta1$^-$ band (2.5 kb) and the htb2$^-$ band (2.07 kb) and the absence of the corresponding wild type bands (Figure 2.15) establishes the viability of the haploid strain hta1$^-$htb2$^-$. Thus H2A2 does not pair only with H2B2, and H2B1 does not interact exclusively with H2A1.

In the same manner, we showed that the hta2$^-$ htb1$^-$ mutant is clearly viable (19). We concluded that there is no exclusive H2A1–H2B2 or H2A2–H2B1 interaction.

Because the hta1$^-$ and htb1$^-$ mutations are so closely linked in the genome (only 800 bases apart) (15), it would be extremely rare to obtain an hta1$^-$ htb1$^-$ haploid recombinant using a similar genetic

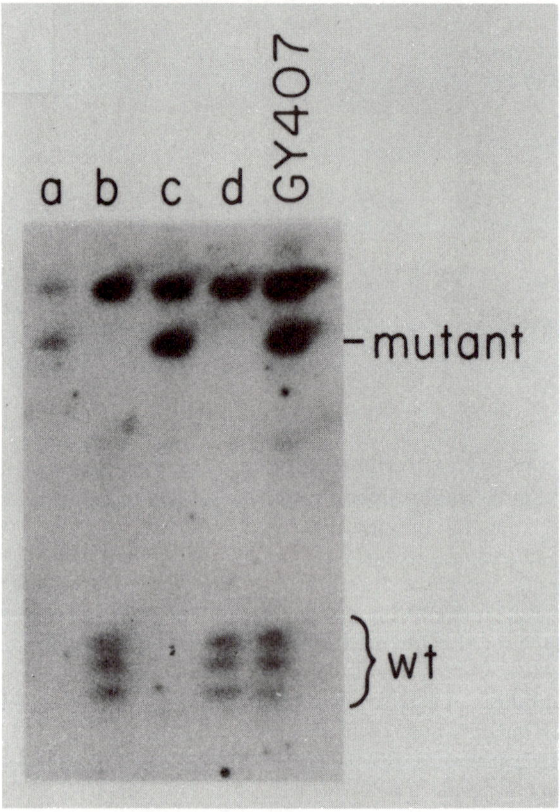

FIGURE 2.16. Viability of hta1$^-$ htb1$^-$ haploids. DNA from the appropriate strains was digested with HIND III restriction endonucease and analyzed by Southern blot hybridization. (a–d) Southern analysis of haploid spores from a tetrad from dissected GY407. The hybridization probe used was the HTA1–HTB1 containing plasmid TRT-1. The bands shown are the mutant 3.8 kb band (hta1$^-$htb1$^-$) and the wild type 1.1-kb band (HTA1$^+$HTB1$^+$). The uppermost band found in all lanes is due to hybridization of sequences that are outside HTA1 and HTB1 but contained in the TRT-1 probe.

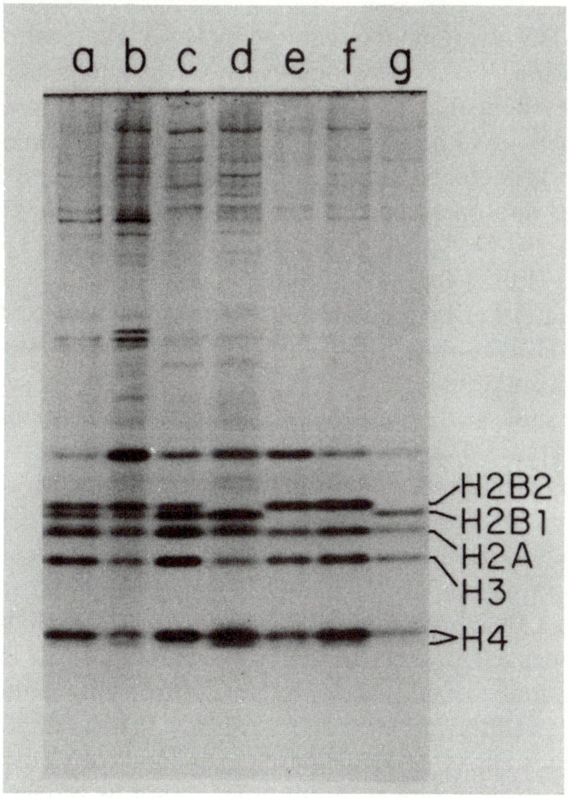

FIGURE 2.17. Electrophoretic comparison of the histones from wild type and mutant cells. Coomassie Blue stained acid–urea gel of the histones isolated from strains with the genotypes indicated. The histone bands are marked appropriately. The first lane (a) displays all four core histones, including both H2B subtypes, from a wild type strain. The next two lanes show that both H2B subtypes are present in cells mutant in HTA1 (b) or HTA2 (c). The last four lanes show that the appropriate H2B subtypes are missing in cells mutant in HTB1 or HTB2: (d) hta1$^-$htb2$^-$; (e) hta2$^-$htb1$^-$; (f) hta1$^-$htb1$^-$; (g) hta2$^-$htb2$^-$.

approach. Therefore this mutant was constructed *in vitro* using plasmid pMR257 which contains both the hta1$^-$ mutation and the htb1$^-$ mutation (19). This plasmid was used to replace the wild type genomic copies of both HTA1 and HTB1 by integration and excision. A Ura$^-$ segregant was then made diploid as previously described, sporulated, and the spore clones analyzed by southern analysis (Figure 2.16).

The 3.8 kb DNA fragment generated by Hind III digestion represents the mutant hta1$^-$ and htb1$^-$ genes, both of which lack a Hind III site. It is evident that there is the expected 2:2 segregation of the mutant (lanes a, c) and wild type (lanes b, d) colonies. From this experiment it is clear that the hta1$^-$htb1$^-$ haploid cell is viable.

A similar approach was used to produce the hta2$^-$htb2$^-$ mutant haploid strain (19). Again, southern analysis of haploid spores showed that an hta2$^-$htb2$^-$ haploid is viable. The genetic experiments demonstrate that either histone H2A subtype can associate productively with either H2B subtype since, in these crosses, the mutant cells are still able to divide mitotically.

These results were confirmed by an analysis of the histones from the HTA mutants (Figure 2.17). If, for example, there is an exclusive interaction between H2A1 and H2B1, then histone proteins isolated from chromatin of an hta2$^-$ strain would contain only H2B1 protein rather than both H2B subtypes. Histone proteins were isolated from an hta1$^-$ strain (lane b) and an hta2$^-$ mutant (lane c) and displayed by acid–urea gel electrophoresis (Figure 2.17). As can be seen, both H2B subtypes are present in the two mutant strains. This confirms the genetic analysis. Figure 2.17 also confirms that mutants in htb1$^-$ (lanes e, f) or htb2$^-$ (lanes d, g) are missing the appropriate H2B protein. Since H2A1 and H2A2 are not separable by electrophoresis, the absence of only one H2A subtype cannot be biochemically corroborated. Thus, although we cannot exclude subtle functional differences between histone subtypes, the genetic and biochemical results lead us to conclude that there is no exclusive interaction between an H2A subtype and H2B subtype.

9. CONCLUSIONS AND FUTURE DIRECTIONS

The experiments reviewed show that H2A and H2B subtypes are dispensable as long as there remains one functional H2A and H2B protein for nucleosome formation. Therefore, since histone subtypes can substitute for each other, we can exclude the possibility that they have essential functions. These experiments focus on yeast histones and do not address the question of functional differences between developmentally regulated histone subtypes of organisms such as the sea urchin. Nevertheless, they illustrate the usefulness of a combined genetic and biochemical analysis in analyzing chromatin structure-function relationships. Although it may be difficult to select for mutations in chromosomal proteins, *in vitro* mutagenesis and replacement of the genes for these proteins allow genetic analyses to be performed.

Similar genetic approaches may be adapted to answer other questions pertaining to chromatin function. With which proteins do histones interact? Why are histones modified during the cell cycle? What is the function of high mobility group (HMG) proteins *in vivo*? Do each of the histones have different functions in the folding of the chromosome? Does the amino terminus of a histone have a different function from its carboxyl terminus? Recent experiments in our laboratory* show that small deletions and insertions in the amino terminus of one of the histones (H2B) can be tolerated in a living yeast cell. This allows a deletion analysis to be done to determine the functions of the various protein domains. The ability to isolate and mutagenize genes for other structural proteins and enzymes of the chromosome should lead to conditional mutations in many chromosomal elements. This should prove important in determining the relationship between chromosome structure and function.

ACKNOWLEDGMENTS

This project was supported by a grant by the National Institutes of Health GM 23674 and by grants from the CICR and UCLA Biomedical Research Support Fund. During the course of this work Michael Grunstein was a recipient of a Faculty Research Award from the American Cancer Society. Mary Rykowski and John Wallis were NIH NRSA predoctoral trainees. David Kolodrubetz was supported in part by a National Research Service Award/National Cancer Institute postdoctoral fellowship and by a National Research Award GM 07872. Joonho Choe was a recipient of a UCLA University Fellowship.

* J. Wallis, unpublished.

REFERENCES

1. J. Ruderman, C. Baglioni, and P. R. Gross, *Nature* **247,** 36 (1974).
2. L. H. Cohen, K. M. Newrock, and A. Zweidler, *Science* **190,** 994 (1975).
3. R. D. Cole, *Molecular Biology of the Mammalian Genetic Apparatus*, Elsevier, Amsterdam, Holland, 1977, pp. 93–104.
4. S. G. Franklin and A. Zweidler, *Nature* **266,** 273 (1977).
5. C. D. Allis, C. V. C. Glover, J. K. Bowen, and M. A. Gorovsky, *Cell* **20,** 609 (1980).
6. J. M. Flynn and H. R. Woodland, *Dev. Biol.* **75,** 222 (1980).
7. J. W. Wallis, L. Hereford, and M. Grunstein, *Cell* **22,** 799 (1980).
8. W. Strickland, M. Strickland, W. Brandt, and C. von Holt, *Eur. J. Biochem.* **77,** 277 (1977).
9. M. Grunstein and P. Schedl, *J. Mol. Biol.* **104,** 323 (1976).
10. M. Grunstein, *Proc. Natl. Acad. Sci. USA* **75,** 4135 (1978).
11. N. S. Kunkel and E. S. Weinberg, *Cell* **14,** 313 (1978).
12. W. Brandt, W. Strickland, M. Strickland, L. Carlisle, D. Woods, and C. von Holt, *Eur. J. Biochem.* **94,** 1 (1979).
13. K. M. NewRock, L. H. Cohen, M. B. Henricks, R. J. Donnelly, and E. S. Weinberg, *Cell* **14,** 327 (1978).
14. I. Isenberg, *Ann. Rev. Biochem.* **48,** 159 (1979).
15. L. Hereford, K. Fahrner, J. Woolford, Jr., M. Rosbash, and D. B. Kaback, *Cell* **18,** 1261 (1979).
16. L. M. Hereford, M. A. Osley, J. R. Ludwig, II, and C. S. McLaughlin, *Cell* **24,** 367 (1981).
17. M. C. Rykowski, J. W. Wallis, J. Choe, and M. Grunstein, *Cell* **25,** 477 (1981).
18. J. Choe, D. Kolodrubetz, and M. Grunstein, *Proc. Natl. Acad. Sci. USA* **79,** 1484 (1982).
19. D. Kolodrubetz, M. C. Rykowski, and M. Grunstein, *Proc. Natl. Acad. Sci. USA* **79,** 7814 (1982).
20. A. Kootstra and G. S. Bailey, *FEBS Lett.* **68,** 76 (1976).
21. Y. Ohe, H. Hatashi, and K. Iwai, *J. Biochem.* **85,** 615 (1979).
22. K. Iwai, H. Hayashi, and K. Ishikawa, *J. Biochem.* **72,** 357 (1972).
23. I. Isenberg, *Ann. Rev. Biochem.* **48,** 159 (1979).
24. P. van Helden, W. Strickland, W. Brandt, and C. von Holt, *Eur. J. Biochem.* **93,** 71 (1979).
25. M. Goldberg, dissertation, Stanford University, Stanford, CA, 1979.
26. D. Wouters, P. Sautiere, and G. Biserte, *Eur. J. Biochem.* **90,** 231 (1978).

27. M. Busslinger, R. Portmann, J. C. Irminger, and M. L. Birnstiel, *Nucleic Acids Res.* **8,** 957 (1980).
28. I. Sures, J. Lowry, and L. H. Kedes, *Cell* **15,** 1033 (1978).
29. G. S. Bailey and G. H. Dixon, *J. Biol. Chem.* **248,** 5463 (1973).
30. B. Laine, D. Kmiecik, P. Sautiere, and G. Biserte, *Biochimie* **60,** 147 (1978).
31. R. D'Andrea, R. Harvey, and J. R. E. Wells, *Nucleic Acids Res.* **9,** 3119 (1981).
32. B. Laine, P. Sautiere, and G. Biserte, *Biochemistry* **15,** 1640 (1976).
33. L. A. Blankstein, B. D. Stollar, S. G. Franklin, A. Zweidler, and S. B. Levy, *Biochemistry* **16,** 4557 (1977).
34. L. C. Yeoman, M. O. J. Olson, N. Sugano, J. J. Jordan, C. W. Taylor, W. C. Starbuck, and H. Busch, *J. Biol. Chem.* **247,** 6018 (1972).
35. J. Strathearn, E. Jones, and J. Broach, Eds., *Molecular Biology of the Yeast Saccharomyces*, Cold Spring Harbor Laboratory, New York, 1981.

CHAPTER 3

THE STRUCTURE OF SEA URCHIN HISTONES AND CONSIDERATIONS ON THEIR FUNCTION

CLAUS VON HOLT, PETRONELLA DE GROOT,
SYLVA SCHWAGER, AND WOLF FRIEDERICH BRANDT
Department of Biochemistry
University of Cape Town
Republic of South Africa

Dedicated to

Jacobus Stephanus de Wet and Arthur Kipps in Gratitude for Many Years of Encouragement

CONTENTS

1. INTRODUCTION
2. THE PURIFICATION OF ISOHISTONES
3. THE ISOHISTONES H2B OF SPERM CELLS
4. THE ISOHISTONES IN THE EARLY SEA URCHIN EMBRYO
5. THE GENERAL STRUCTURE OF H2B HISTONES
6. THE SIGNIFICANCE OF THE PENTAPEPTIDE REPEAT
7. THE SPERM HISTONES H2B, POSSIBLE PRECURSORS OF THE PROTAMINES
8. THE HISTONES, H2A, H3, AND H4
9. THE H1 HISTONES
10. ASSEMBLY OF ISOHISTONES INTO CHROMATIN SUBSTRUCTURES
REFERENCES

1. INTRODUCTION

Any investigation designed to elucidate the structural organization of the eucaryotic genome to establish the framework within which regulation can manifest itself must, as its prime premise, accept the large size of the DNA. The latter must be accommodated within the confines of the nucleus. This requires a considerable degree of compaction, which in the higher eucaryotes in certain stages of the cell cycle may reach the order of 10^4(1). This compaction, however, must be envisaged as an ordered process resulting in morphologically identifiable reproducible structures, the chromosomes. It goes without saying that such ordered macrostructures can be built only if an even higher degree of precision applies to the molecular structures underlying the morphological phenomenon. The unit molecular structure must thus permit the reproducible dismanteling and reconstruction of the chromosome. To allow this the unit structure must be sufficiently small and precise in its dimensions to lend itself to the construction not only of the highly condensed structures employed in the final stages of the mitosis where no transcriptional processes take place but also to the more expanded structures pertaining to the many intermediate stages. The latter are characterized functionally by either active transcription in selected regions or temporary but revocable repression in others, as well as permanent repression of shorter or longer stretches of the genome. It is in this aspect of the structure of chromatin, namely, the provision of a structural framework within which regulatory mechanisms can manifest themselves, where the histones probably play their key role.

The identification by Kornberg (2) of an ubiquitous elementary chromatin subunit, the nucleosome, has put structure-function investigation of chromatin on a rational basis. At the level of the nucleosome the interaction of the major chromosomal proteins, the histones, with themselves and the DNA must be understood in the first instance to get insight into the complex problems related to the structural framework within which regulation of transcription takes place. It is unlikely, in the light of the present evidence, that the histones themselves exert a specific regulatory function. However, different and variable fine structures of the nucleosomes may lend specific properties to the DNA complexed to them on their circumferences. Structurally different isohistones may thus provide a variety of histone octamers to result in specific nucleosome structures, whereas enzymatic modification of the component histones could induce a large number of variable states in such nucleosomes. The combination of these two, together with a nonrandom distribution of the nucleosome on a given stretch of DNA (3), may well provide the structural prerequisites that may render a gene either regulatable or permanently repressed.

A multicellular eucaryotic system, easy to synchronize and lending itself to the mass production of cells in various stages of differentiation, is a suitable experimental system to investigate a number of aspects of the role of histones in chromatin structure and gene expression. Such a system is provided by the developing sea urchin embryo. The histone genes in sea urchins are highly reiterated (4). Prior to the identification of isohistones, this reiteration had been interpreted to assist in the provision of a sufficiently large number of histone mRNAs (5) for the rapid construction of chromatin during the short cell cycle in the early stages of the embryogenesis.

The identification of several isohistones of the H2B type with highly specialized variant structures in the mature haploid sperm cells of several sea urchin species in our laboratory (for review see Reference 6) pointed to the existence of unique nucleosomes tailored for specific purposes during the life cycle of sea urchins. Not only are such specialized isohistones transcribed from the genome during the formation of gametes, but also the entire early embryogenesis of the sea urchin is characterized by the stage-specific appearance of histone-like proteins with typical electrophoretic properties (7). Stage-specific mRNA (8) coding for these proteins has been identified. The ultimate proof that the electrophoretically identified histone-like proteins are indeed structurally different histones came from the partial elucidation of their primary structures (9, 10). It is now firmly established that several isotypes of the histones H1, H2A, and H2B are expressed in a programmed fashion very distinctly during embryogenesis (7, 8, 9, 10) and also as tissue-specific isotypes in mature animals (6).

What appeared as a high degree of reiteration of identical histone gene batteries in the genome of the sea urchin must now be understood as a tandem arrangement of gene batteries, each containing a combination of the isohistones H1, H2A, and H2B together with the standard H3 and H4 types. There is some evidence that even the highly conservative histones H4 and H3 may be present as different structural variants (11). It is conceivable that the histone gene batteries are arranged in the genome in such a way that specific combinations of isohistone genes in a particular tandem are presented for transcription at suitable times during the life cycle of the organism.

2. THE PURIFICATION OF ISOHISTONES

The amount of histone in the sea urchin cell amounts to only approximately 2×10^{-12} g. It follows that for the purpose of preparative isolation of isohistones for primary structure investigations a considerable number of sea urchin embryos must be processed to recover sufficient protein. On the other hand, in the male sea urchin, just prior to spawning, several milligrams of histones can be isolated from the gametes of a single animal. All isolation strategies for sea urchin isohistones must take into account that within a family of isohistones properties of the polypeptide chains, that is, molecular weight, charge, and primary structure are very similar and thus require fairly discriminating purification methods with ensuing low yields of the final product.

We have applied with more or less success a combination of selective extraction (12), chromatography on molecular sieves (13), ion exchange chromatography in high concentrations of urea as introduced for basic ribosomal proteins by Wittman (14), and preparative electrophoresis (11). Any of these methods, at suitable stages of the purification, is preferably to be used in a recycling mode with the result that for most of the isohistones the yields are very low. The establishment by Zweidler (15) of differential nonionic detergent binding to the various histone classes as a basis for PAGE analysis has aided considerably the development of preparative and analytical methodology. The selective extraction as described by Johns (12), though very effective for calf thymus and chicken erythrocyte chromatin, results with sea urchin embryo chromatin as starting material only in an enrichment of the corresponding fractions but not in a clear separation. This probably reflects the variability in the primary

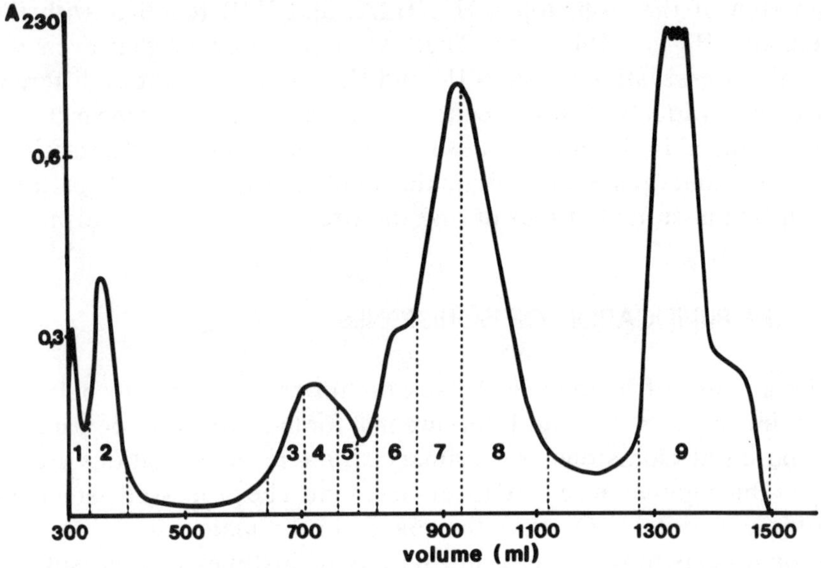

FIGURE 3.1. Chromatography of sea urchin embryo histones on Biogel P60 in 0.02*N* HCl. Fractions cut as indicated, see Figure 3.2.

structure of the many isohistones, with the consequent disappearance of the distinct borderlines with respect to DNA binding and solubility between the standard histones and the isohistones.

Chromatography on certain molecular sieve matrices yields a limited group separation (Figure 3.1). This is, however, clearly not due to a molecular sieve effect only (13). Histone H3 can only be separated from the bulk of the H2B and H2A histones by exclusion chromatography after dimerisation of the histone H3 (12). The histone H4 fraction from the P60 biogel column is usually heavily contaminated with a variety of polypeptide chains which can, however, easily be

FIGURE 3.2. PAGE of sea urchin embryo histones eluted from Biogel P60 (Figure 3.1). PAGE conditions: 15% acrylamide, 0.1% bisacrylamide, 6 mM Triton, 8.5 M urea, 0.1% thiodiglycol, polymerized photochemically with riboflavin as sensitizer, 0.9 M acetic acid as tray buffer, 16 hours constant current 20 mA. (*a*) Numbering of lanes refers to the fraction identification in Figure 3.1. Fraction 1 from Figure 3.1 is not shown. (*b*) Lanes 1, 2, 3, and 5: histones $H2A_{(1)}$, $H2A_{(2)}$, $H2A_{(3)}$, and histones H2B after further recycling and removal of histone H3 through dimerization. Lane 4: histone H3 after dimerization and reduction (13).

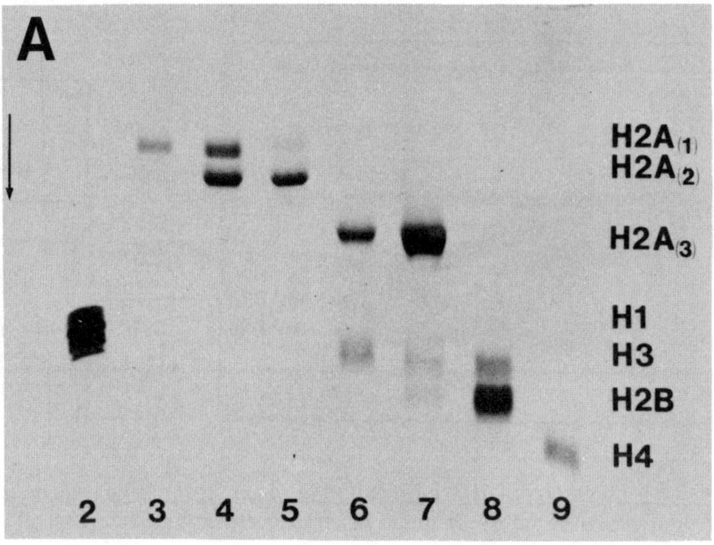
A
H2A(1)
H2A(2)
H2A(3)
H1
H3
H2B
H4
2 3 4 5 6 7 8 9

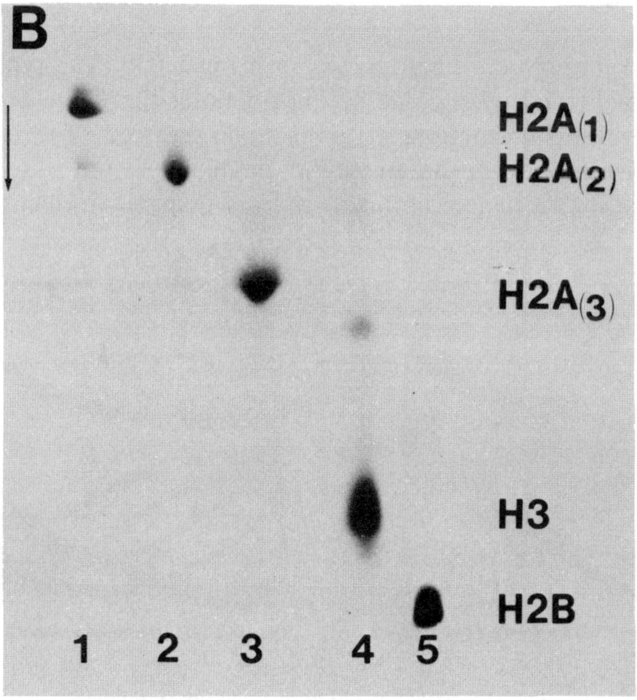
B
H2A(1)
H2A(2)
H2A(3)
H3
H2B
1 2 3 4 5

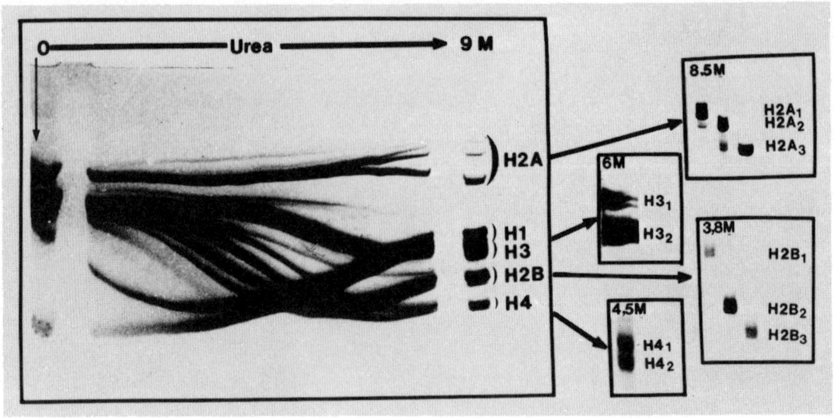

FIGURE 3.3. Preparative PAGE purification of sea urchin embryo isohistones. The main frame shows the embryo histones in a transverse urea gradient in 6 mM Triton (see also Figure 3.2). Histone subfractions were isolated from preparative slab gels at the urea concentrations given in the small frames, which depict the isolated subfractions.

removed through chromatography on Biogel P30. The type of separation achieved by a combination of molecular sieve procedures and recycling of fractions is given in Figure 3.2. The isohistone groups isolated from the molecular sieve can be processed further by way of ion exchange chromatography or preparative electrophoresis (Figure 3.3).

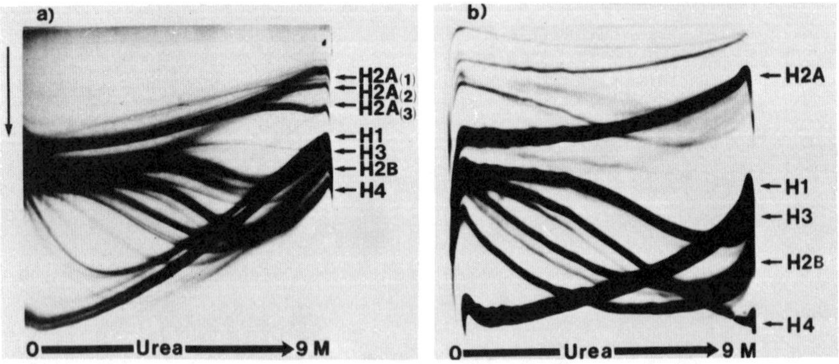

FIGURE 3.4. PAGE of sea urchin histones (*Parechinus angulosus*) in the presence of Triton X-100 in a transverse urea gradient, conditions as in Figures 3.2 and 3.3. (*a*) sea urchin embryo histones; (*b*) sea urchin sperm histones.

TABLE 3.1. Amino Acid Composition of Sea Urchin Embryo Histones

Amino Acid	$H1_{(1)}$ (mol %)	$H1_{(2)}$ (mol %)	$H1_{(3)}$ (mol %)	$H2A_{(1)}$ (mol %)	$H2A_{(2)}$ (mol %)	$H2A_{(3)}$ (mol %)	$H2B_{(1)}$ (mol %)	$H2B_{(2)}$ (mol %)	$H2B_{(3)}$ (mol %)	$H3_{(1)}$ (mol %)	$H3_{(2)}$ (mol %)	$H4_{(1)}$ (mol %)	$H4_{(2)}$ (mol %)
Asp	2.7	2.4	2.2	6.5	6.9	6.5	5.0	5.0	4.8	3.6	4.3	5.4	5.7
Thr	7.9	3.4	4.7	3.7	3.9	3.7	6.5	7.0	6.4	6.8	6.7	5.6	5.1
Ser	3.1	4.1	3.3	5.3	5.4	5.8	7.1	8.7	9.6	5.2	5.6	3.1	4.9
Glu	11.0	5.9	3.9	9.0	8.8	8.5	8.0	8.6	8.0	11.6	11.2	7.1	8.6
Pro	6.6	8.1	7.7	5.7	5.1	5.7	5.4	4.7	3.8	5.8	5.0	1.7	1.8
Gly	2.9	4.6	4.1	11.3	11.4	11.2	8.2	7.5	7.8	6.3	6.6	16.9	16.8
Ala	20.7	25.6	30.1	11.9	12.4	10.9	10.9	10.4	11.2	12.2	12.5	7.0	8.0
Val	4.1	4.1	2.7	6.2	6.5	6.5	6.9	6.7	6.8	5.7	5.6	7.8	7.6
Met	0.8	0.7	0.5	—	0.5	0.8	1.2	1.5	1.5	0.6	0.3	—	—
Ile	2.6	2.0	2.1	4.4	3.9	3.9	5.1	6.1	5.0	5.2	5.1	5.3	5.4
Leu	3.7	2.9	2.8	10.0	10.5	10.2	5.9	5.1	5.1	8.6	8.5	8.1	9.0
Tyr	0.4	1.1	2.2	2.4	2.3	2.4	3.3	3.6	3.8	2.1	2.5	3.5	2.6
Phe	1.3	0.7	0.9	2.0	1.8	1.7	1.9	1.5	1.7	2.3	2.6	2.2	1.3
Lys	32.3	34.1	33.0	11.0	10.4	10.9	13.0	15.0	15.1	10.3	10.7	9.8	9.7
His	0.4	0.6	0.4	1.6	1.5	1.5	1.8	1.4	1.5	1.7	1.3	2.6	2.2
Arg	1.6	1.9	1.8	8.8	8.9	9.5	9.4	6.7	7.6	11.7	11.3	13.5	12.1
Lys/Arg	21.5	17.9	18.3	1.2	1.2	1.2	1.4	2.2	2.0	0.9	1.0	0.7	0.8
$\frac{\text{Lys + His + Arg}}{\text{Glu + Asp}}$	2.5	4.4	5.8	1.2	1.3	1.4	1.9	1.7	1.9	1.6	1.5	2.0	1.7
N terminal	BL	BL	BL	BL	BL	BL	Pro	Ala	Pro	Ala	Ala	BL	BL

TABLE 3.2. Amino Acid Composition of Sea Urchin Sperm Histones

	H1 Mol%	H1 Residues	H2A Mol%	H2A Residues	$H2B_{(1)}$ Mol%	$H2B_{(1)}$ Residues	$H2B_{(2)}$ Mol%	$H2B_{(2)}$ Residues	$H2B_{(3)}$ Mol%	$H2B_{(3)}$ Residues	H3 Mol%	H3 Residues	H4* Residues
Asp	4.0	4	8.7	9	3.3	5	3.4	5	3.4	5	3.8	5	5
Thr	4.6	5	2.3	3	8.0	12	5.6	9	5.1	9	6.2	10	6
Ser	15.6	19	5.4	7	9.5	15	11.3	18	11.9	20	4.3	6	2
Glu	5.2	5	9.5	9	8.0	11	6.0	8	5.7	8	12.0	15	6
Pro	17.6	20	5.1	5	4.3	7	5.5	8	6.4	9	4.7	6	1
Gly	9.8	10	15.3	15	10.5	14	9.4	13	9.0	13	5.9	7	17
Ala	63.1	63	17.8	17	7.2	10	8.7	12	8.3	12	13.3	18	7
Cys	—	—	—	—	—	—	—	—	—	—	0.7	1	1
Val	9.1	9	7.7	8	8.3	12	7.6	11	7.2	11	4.7	6	9
Met	3.9	4	—	—	1.2	2	1.5	2	1.2	2	1.4	2	1
Ile	2.1	2	5.3	6	2.5	4	3.4	5	3.3	5	4.9	7	5–6
Leu	5.1	5	15.5	15	4.2	6	4.4	6	4.2	6	8.6	12	8
Tyr	1.2	2	2.7	3	3.1	5	3.0	4	2.7	4	2.2	3	4
Phe	1.0	1	1.8	2	1.1	2	1.6	2	1.4	2	3.1	4	2
Lys	69.0	69.5	13.3	13	11.3	16	10.1	14	10.9	15	9.7		10
ε-Me-Lys	—	—	—	—	—	—	—	—	—	—	0.6	13	1
His	1.7	2	1.9	2	1.2	2	1.3	2	1.3	2	1.5	2	2
Arg	26.1	26.5	11.7	11	16.3	21	17.7	24	18.9	25	13.0	18	14
Lys/Arg		2.6		1.2		0.2		0.6		0.6		0.7	0.8
(Lys + His + Arg)/(Glu + Asp)		10.8		1.4		2.4		3.1		3.2		1.7	2.5
N Terminal		Pro		Bl		Pro		Pro		Pro		Ala	Bl
C Terminal		Lys		Lys		Arg		Arg		Arg		ND	ND

If applied to mature sperm cells from *Parechinus angulosus*, these methods yield up to three histones of H2B, and one each of H2A, H3, H4, and H1 (Figure 3.4*b*). From the sea urchin embryo, depending on the stage of development, a number of isohistones H1, H2A, and H2B can be isolated (Figure 3.4*a*). Both histone H3 and H4 from sea urchin embryo display electrophoretic heterogeneity (Figure 3.3). It has, however, not yet been firmly established whether in the case of the latter two this is the result of postsynthetic enzymatic modification or of genuine differences in the primary structures. The amino acid compositions typical for the various major isohistones are given in Tables 3.1 and 3.2.

3. THE ISOHISTONES H2B OF SPERM CELLS

The occurrence of isohistones in sperm cells is not confined to one particular species but is widespread. In our laboratory the partial structures of isohistones H2B have been established for *Parechinus angulosus* (16, 17, 18), *Psammechinus miliaris* (19), and *Echinolampas crassa* (6), (Figure 3.5). Similarly the occurrence, by electrophoretic criteria, of isohistones H2B has been reported for a number of other sea urchin species (20, 21). A similar variability appears to apply to the histone H1 in sperm cells. Structurally different isotypes have been identified by us in *Parechinus angulosus* (22, 23), *Echinolampas crassa* (23), and *Spaerechinus angulosus* (25). The sea urchin sperm cells of a single animal (*Parechinus angulosus*) always contain $H2B_1$ and, in addition, either $H2B_2$ or $H2B_3$; in few cases the latter two are replaced by a mixture of approximately equal amounts of $H2B_2$ and $H2B_3$ (26), (Figure 3.6). Whereas the nature of the repeat pentapeptide in the $H2B_{(1)}$ isotype is Pro-Thr-Lys-Arg-Ser-; the $H2B_2$ and the $H2B_3$ isotypes display the residues Pro-Arg-Lys-Gly-Ser in their repeating pentapeptide (Figure 3.5). The repeat number of the pentapeptides in isotypes 2 and 3 is higher than in isotype 1. In addition, isotype 2 and 3 in their 45 *N*-terminal residues resemble each other more than they resemble isotype 1. This may indicate that sperm from a single organism may contain two types of chromatin, that is, two groups of cells, one possessing the isotype 1 and the other either the isotype 2, or 3, or a mixture of 2 and 3. Whether the

```
                                       10                  20                  30                  40                  50                  60
H2B
  (1)Psammechinus sperm     —— P S Q K S P T K R S P T K R S ———————————————— P Q K — G G K ——— G A K R G G K A G K R R R G V A V K R R R R R R E S Y G I
H2B
  (1)Parechinus sperm       —— P S Q K S P T K R S P T K R S P T K R S ——————— P Q K — G G K G G K G A K R G G K A G K R R R G V Q V K R R R R R R E S Y G I
H2B
     Echinolampas sperm     P K S P S K G S P R K G S P R K G S ——————— P T R R G A ——— G G K ——— G A K R A G K G G R R R T X V A - K R R X X R R E S Y G I
H2B
  (2)Psammechinus sperm     P K S P S K S S P R K G S P R K G S P R K G S ——————— P K R G G K ——— G A K R A G K G G R R X X   V - - - - - - - - - - - - -
H2B
  (2)Parechinus sperm       P R S P A K T S P R K G S P R K G S ——————— P S R K A S P K R G G K ——— G A K R A G K G G R R R R — V — V K R R R R R R E S Y G I
H2B
  (3)Parechinus sperm       P R S P A K T S P R K G S P R K G S P R K G S P S R K A S P K R G G K ——— G A K R A G K G G R R R R — V — V K R R R R R R E S Y G I
H2B
  (2)Parechinus (embryo)    ———————————————————————————————————— A P T G Q V A — K K ——— G S K K A V K A P R P S —— G G K K R B X K R K E - - - -
H2B
  (1)Parechinus (embryo)    ——————————————————————————————————— P A K - A Q T A G K K ——— G S K K A - - - - - - - - - - - - - - - - - - - - - - -
H2B
  (3)Parechinus (embryo)    ——————————————————————————————————— P A K T S G K G A K K A —— G —— K A - - - - - - - - - - - - - - - - - - - - - - -
H2B
     Parechinus (gut)       ——————————————————————————————————————————————————————————————————————————————————————————————————————————————————

   70                  80                  90                  100                 110                 120                 130                 140
- - - - - - - - - - - - - - - - - - - - - - N S F V N D V F E X I A - - - - - - - - - - - - - - - - - - - - - - - - - - - - - - - - - - - - - - - - -
Y I Y K V L K Q V H P D T G I S S R A M S V M N S F V N D V F E R I A A E A G R L T T Y N R R S T V S S R E V Q T A V R L L L P G E L A K H A V S E G T K A V T
Y V - - - - - - - - - - - - - - - - M S I M N S F V N D V F E R I A A E A S R L A/S H A/Y N R R S T I S S R X I - - - - - - - - - - - - - - - - - - - - - - - -
- - - - - - - - - - - - - - - - - - - - - - N S F V N D V F E R I A A E A S R L T S A N R R S T I S S R X I - - A V R - - - - - - - - - - - - - - - - - - - -
Y I Y K V L K Q V H P D T G I S S R A M S V M N S F V N D V F E R I A G E A S R L T S A N R R S T V S S R E I Q T A V R L L L P G E L A K H A V S E G T K A V T
Y I Y K V L K Q V H P D T G I S S R A M S V M N S F V N D V F E R I A S E A S R L T S A N R R S T V S S R E I Q T A V R L L L P G E L A K H A V S E G T K A V T
- - - - - - - - - - - - - - - - - - M S I M N S F V N D I/V F E R I A G E A S R - - - - - - - - - - - - - - - - - - - - - - - - - - - - - - - - - - - -
- - - - - - - - - - - - - - - - - - - - - M N S F V N D V F E R - - - - - - - - - - - - - - - - - - - - - - - - - - - - - - - - - - - - - - - - - - - -
- - - - - - - - - - - - - - - - - - - - - M N S F V N D V - - - - - - - - - - - - - - - - - - - - - - - - - - - - - - - - - - - - - - - - - - - - - - -
———————————————————————————— - - M S L M N S F V N D V F E R I A A E A S R L - - - - - - - - - - - - - - - - - - - - - - - - - - - - - - - - - - - - -

    150
- - - - - R
K Y T T S R
- - - - A R
- - - - - -
K Y T T S R
K Y T T S R
- - - - - -
- - - - - -
- - - - - -
- - - - - -
```

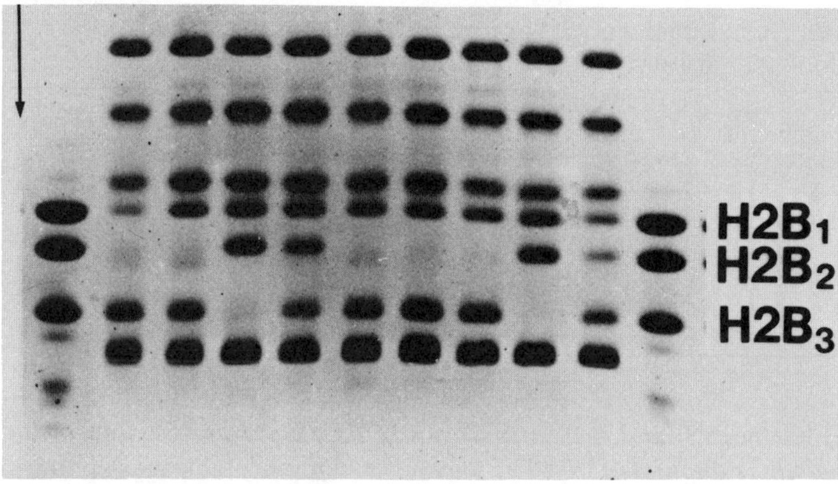

FIGURE 3.6. PAGE of sperm histones of 9 sea urchin individuals (*Parechinus angulosus*) with reference histones $H2B_{(1)}$, $H2B_{(2)}$, and $H2B_{(3)}$ in lanes I and II. PAGE condition is as in Figure 3.2 but at 3.8 M urea.

mode of isohistone occurrence (Table 3.3) in the sperm cells reflects the presence of the X or Y characteristics remains to be seen.

4. THE ISOHISTONES IN THE EARLY SEA URCHIN EMBRYO

The most impressive demonstration of structural specificity can be found in the programmed appearance of isohistone mRNA (8) and its translation products (7, 9, 10) in the synchronized developing sea urchin embryo (Figure 3.7). The programmed appearance of histones during the embryogenesis applies to different species of sea urchins;

FIGURE 3.5. Primary structures of sea urchin histones H2B: *Psammechinus* sperm (19), *Parechinus* sperm (16, 17, 18), *Echinolampas* sperm (6), *Parechinus* embryo (6, 9, 10, 11), *Parechinus* gut (6). Single-letter notation: C—Cys, H—His, I—Ile, M—Met, S—Ser, V—Val, A—Ala, G—Gly, L—Leu, P—Pro, T—Thr, F—Phe, R—Arg, Y—Tyr, W—Trp, D—Asp, N—Asn, B—Asx, E—Glu, Q—Gln, Z—Glx, K—Lys, X—unknown. Gaps to optimize alignment: —— Sequence not yet determined: - - - - -

TABLE 3.3 Distribution of Sperm Cell Isohistones $H2B_{(1)}$, $H2B_{(2)}$, and $H2B_{(3)}$ in Individual Sea Urchins

Number of Animals	Number of Animals Containing Isohistone Type			
	$H2B_{(1)}$	$H2B_{(2)}$	$H2B_{(2)}$ + $H2B_{(3)}$	$H2B_{(3)}$
32	32	28	2	2
35	35	0	5	30
28	28	14	7	7
31	31	29	2	2

it has been found in *Strongylocentrotus purpuratus* (7) and in *Parechinus angulosus* (9). The various developmental stages up to late gastrula are characterized by this isohistone mRNA program (8). That the electrophoretically different proteins translated from these mRNAs isolated from cells at different developmental stages are indeed structural variants and do not represent the result of postsyn-

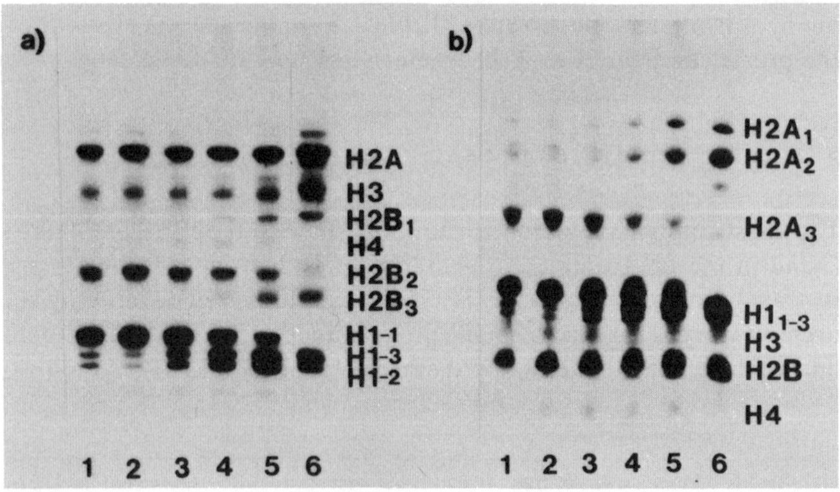

FIGURE 3.7. PAGE-fluorograms of *Parechinus angulosus* histones during embryonic development (26). PAGE conditions are as in Figure 3.2. Embryo cultures have been pulse labeled for 20 minutes and grown to 21 hours when histones were isolated. Lane 1–6: pulses of lysine at 3, 5, 7, 9, 11, and 21 hours after fertilization. (*a*) 3.8 M urea and (*b*) 8.5 M urea gel, respectively.

thetic modification is based on two lines of evidence. First, the mRNAs isolated from *S. pupuratus* are translated by a heterologous translation system (8), in which no postsynthetic modification takes place, into the proteins with the characteristic electrophoretic mobilities identical to those isolated from the growing embryo (6). Furthermore, stage-specific histones have been isolated in sufficient amounts from a different sea urchin species (*Parechinus angulosus*) to allow the determination of the partial structures (9, 10), (see also Figure 3.5, and Figures 3.15, 3.16, 3.17, and 3.18 in Sections 9 and 10.)

There are differences in the two investigated species, *Strongylocentrotus purpuratus* and *Parechinus angulosus*, with respect to the speed by which the developmental program is driven through its various stages (Figure 3.8). In both species the switch to the new set of isohistones has been completed at the late blastula. The switching off of the early genes is, however, a gradual process, at least in

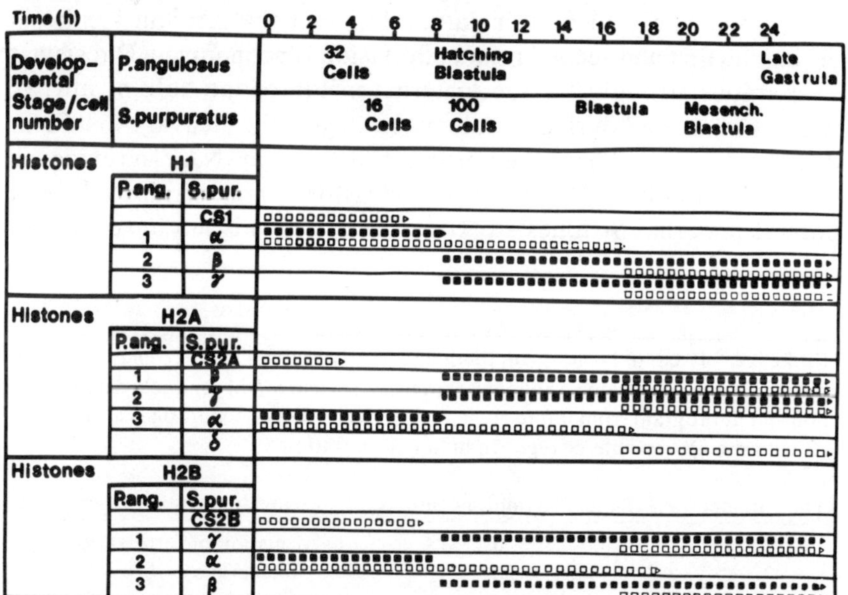

FIGURE 3.8. Developmental program of sea urchin embryo histones. Data are taken from References 6, 9, 11, 27 and 7, 8, 28: CS = cleavage stage histones (28); α, β, γ identification of histones in *S. purpuratus* (7, 8, 28); and 1–3 identification of *P. angulosus* (6, 9, 11, 27), respectively. Histone synthesis for *P. angulosus* is shown by ■ ■ and for *S. purpuratus* by □ □.

Parechinus angulosus. Pulses of sufficiently high specific activity reveal transcription and translation of the early gene set as late as 11 hours (27) (Figure 3.7). In addition to the histone program controlled by the embryo genome with two stages progressing from the one to the other around hatching blastula (Figure 3.7), a third, earlier, set of histones has been described, the cleavage stage histones (28). Whether, in fact, these proteins are under the control of the embryo's genome remains to be seen.

Figure 3.9 shows that in *Parechinus angulosus* prior to the first cleavage radioactive lysine is incorporated into histones (27).

The electrophoretic mobility of these labeled proteins is not significantly different from that of the early set of histones which are synthesized up to the hatching blastula stage. It is well established that the unfertilized egg contains histone mRNA (29, 30), and it appears that this maternal histone mRNA is transferred during the first telophase from its RNP storage form to the polyribosomal fraction (30), presumably to be translated. The fact that protein synthesis during the first and second round of division is actinomycin D resistant (31) argues strongly for the maternal origin of the bulk of mRNA species translated during the cleavage stages; the cleavage histones, therefore, may well have their origin in maternal mRNA. The evidence for the alternative, that the cleavage histones may originate from a pool of maternal histones stored in the egg (32) is not unequivocal.

FIGURE 3.9. PAGE of lysine incorporation into early histones in *Parechinus angulosus* embryo (26). Upper frame *a* and lower frame *b*: 3.8 and 8.5 M urea gels with protein stain and fluorogram from left to right, respectively. PAGE condition is as in Figure 3.2 with the appropriate changes in urea concentration.

Lane Number	Time of Sampling (min.)	Developmental Stage
1	0	Formation of fertilization
2	20	membrane
3	40	1st S-phase
4	60	1 cell
5	80	1st cleavage
6	100	2 cells
7	120	2nd cleavage, 4 cells
		3rd cleavage, 8 cells

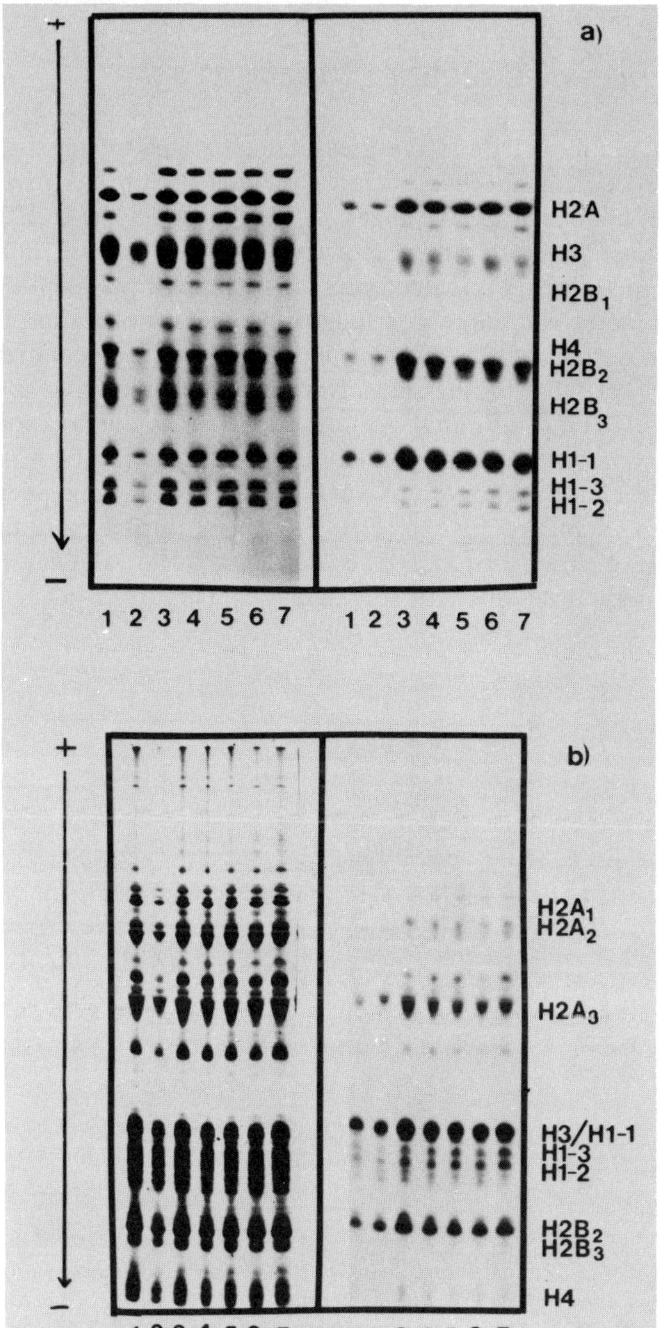
a)
H2A
H3
H2B1
H4
H2B2
H2B3
H1-1
H1-3
H1-2
1 2 3 4 5 6 7
1 2 3 4 5 6 7
b)
H2A1
H2A2
H2A3
H3/H1-1
H1-3
H1-2
H2B2
H2B3
H4
1 2 3 4 5 6 7
1 2 3 4 5 6 7

5. THE GENERAL STRUCTURE OF H2B HISTONES

The elucidation of the structures of isohistones of the H2B class expressed during the life cycle of the sea urchin has revealed an aspect of histone structure not recognized before: The H2B histones can be classified into structural subclasses that exhibit, in addition to point mutations, typical varying length of their polypeptide chains with characteristic sequences suggesting specific domains. Such domains may be present or absent depending on the specialization of the cell type, that is, the transcriptional and replicative activity of the latter. The *N*-terminal regions of two *Parechinus* isohistones H2B compared to that of chicken erythrocytes are given in Figure 3.10. The latter histone represents a type usually considered as a ubiquitous standard histone H2B. From the comparison of these and a number of other histones H2B (33, 34, 35, 36) a general picture of their structural characteristics emerges (Figure 3.11).

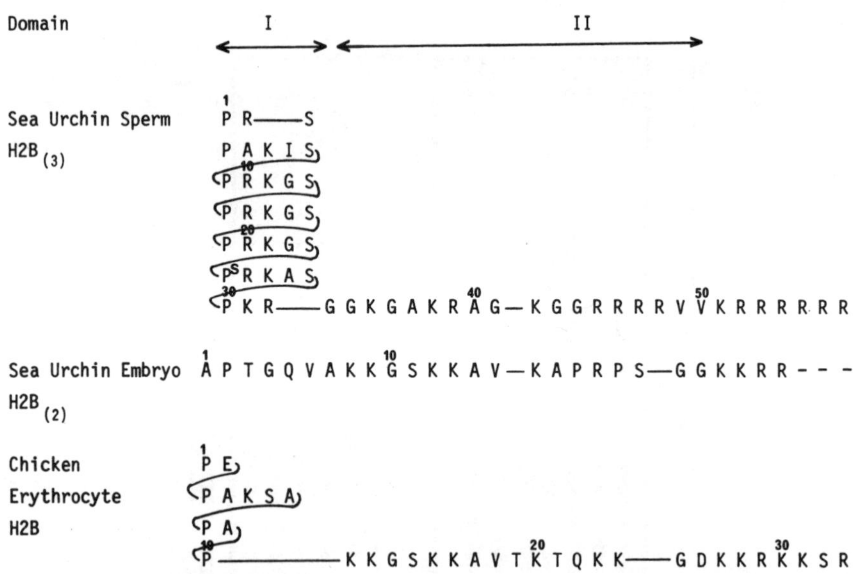

FIGURE 3.10. *N*-terminal area of histones H2B from *Parechinus angulosus* (9, 18) and *Gallus domesticus* (33). The numbering refers to sequence positions of the histone H2B. The areas compared comprise the variable region of domain II and domain I in Figure 3.11.

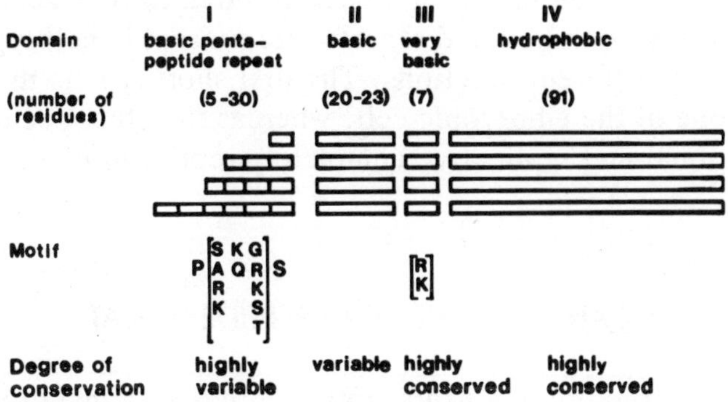

FIGURE 3.11. General structure of histones H2B. The amino acid residues in the motifs are given in order of abundance.

The arrangement and nature of amino acid residues suggests that, obviously, the hydrophobic *C*-terminal area participates in the histone–histone interaction responsible for the octamer formation. This is born out by cross-linking tyrosine residues between histone H2B and H2A with UV at virtually zero length cross-link (37). The *N*-terminal half presents positive charges for histone–DNA interaction. The *N*-terminal domain apparently can undergo a considerable variation (Figures 3.5 and 3.10). Two sequence motifs are varied in this part of the structure. One consists of a pentapeptide where repeat number and actual sequence of the proline–hydroxamino acid delimitated pentapeptide may vary. The second motif appears to consist of a glycine–alanine–basic amino acid rich stretch. The basic *N*-terminal region is terminated toward the center of the chain by a block of seven basic amino acids mostly arginine. This block is common to all histones H2B known at this stage. It is conceivable that this basic block establishes a firm anchorage for the DNA on the surface of the nucleosome and may well have to do with the positioning of a stretch of DNA with respect to the circumference of the nucleosome. The two variable basic regions may interact more reversibly with the DNA. The fact that one phosphorylation site in histone H2B is found in the pentapeptide structure in calf thymus H2B (38) points in this direction.

It is striking that the two extreme structures, namely, the short type with missing or only one *N*-terminal pentapeptide on one side and the long type on the other side with three to four genuine and one or two insertion mutated repeats, are present in cell types with diametrically different functions. The first short type is the typical H2B histone of the embryonic cell, whereas the other occurs in the transcriptional and replicative completely repressed haploid sperm cell.

6. THE SIGNIFICANCE OF THE PENTAPEPTIDE REPEAT

The general structure of a pentapeptide with *N*-terminal proline and a *C*-terminal hydroxy amino acid not only occurs in the sea urchin sperm histones but also in a variety of proteins found in association with both types of nucleic acids (Table 3.4) such as protamines, HMG-proteins, phage-, and virus proteins. The high degree of homology between the various pentapeptides culminating in the identity of a pentapeptide sequence in the DNA-associated sperm histones from sea urchins and the RNA-associated alfalfa virus coat protein is unlikely to be a chance event. The widely spread occurrence of the pentapeptide structure rather suggests that it may well be of general significance for nucleic acid protein interaction.

Considering the nature of the amino acids involved, one of the possible conformation of each pentapeptide in the sperm histone is likely to be an extended chain. Some or all of the central amino acids, depending on their identity, can interact through their basic side chains with the sugar phosphate residues of DNA. Alternatively a conformation similar to bacterial DNA binding proteins may exist (73). The number and nature of the basic amino acids in the central portion of the pentapeptides and the repeat frequency of the latter will determine the strength of the DNA–protein interaction. This, in turn, may determine the degree of accessibility for replicating or transcribing enzymes to the information stored in that particular nucleic acid region. The *C*-terminal hydroxy amino acid in each of the pentapeptides offers a potential site for phosphorylation, as is the case in the somatic histone H2B where serine residue 14 in the pentapeptide extending from position 10–14 can undergo phosphorylation (38). This could very well provide a mechanism to reversibly decrease the polypeptide–nucleic acid interaction. Such a role of the pentapeptide

TABLE 3.4. **Pentapeptide Structures in Nucleic Acid Associated Proteins**[a]

	Eucaryotes	
Histones and Nonhistones		
Trout T	Pro-Lys-Arg-Lys-Ser	(39)
	Pro-Ala-Arg-Arg-Ser	
HMGT	Pro-Lys-Gly-Lys-Thr	(40)
HMG-1	Pro-Lys-Gly-Glu-Thr	(41)
HMG-17	Pro-Gln-Arg-Arg-Ser	(42)
H1 Rabbit	Pro-Ala-Glu-Lys-Ser	(43)
H1 trout	Pro-Lys-Lys-Ala-Thr	(44)
H2B calf	Pro-Lys-Lys-Gly-Ser	(45)
H2B (1) Psammechinus	Pro-Ser-Gln-Lys-Ser	(24)
H2B (1) Parechinus	Pro-Ser-Gln-Lys-Ser	(16)
H2B Echinolampas	Pro-Ser-Lys-Gly-Ser	(6)
H2B (2) Psammechinus	Pro-Ser-Lys-Ser-Ser	(19)
H2B (2) Parechinus	Pro-Ala-Lys-Thr-Ser	(17)
H2B (3) Parechnius		(18)
H2B (1) Psammechinus	Pro-Thr-Lys-Arg-Ser	(19)
H2B (1) Parechinus		(14)
H2B Echinolampas	Pro-Arg-Lys-Gly-Ser	
H2B (2) Psammechinus		(19)
H2B (2) Parechinus		(17)
H2B (3) Parechinus		(18)
Protamines		
Clupein	Pro-Arg-Arg-Val-Ser	(46)
Y[1]II		

Table continues on p. 86.

TABLE 3.4. *(continued)*

	Eucaryotes	
Protamines		
Z	Pro-Arg-Arg-Arg-Thr	
Z^1		
Iridine		(46)
Ia		
Clupein		(46)
Y^1I		
Y II		
	Bacteriophages	
fr	Pro-Lys-Val-Ala-Thr	(43)
F2		
MS2		
R17		
X178	Pro-Leu-Arg-Gly-Thr	(48)
Fd	Pro-Gly-Lys-Tyr-Thr	(47)
M13		
Endolysin	Pro-Lys-Leu-Lys-Ser	(47)
repressor	Pro-Ser-Leu-Arg-Ser	(49)
	Pro-Glu-Leu-Arg-Thr	
Virus		
Alfalfa	Pro-Thr-Lys-Arg-Ser	(50)
Turnip yellow mosaic	Pro-Gln-Asp-Arg-Thr	(47)

structure correlates very well with the occurrence of the isohistone types. The transcriptionally very active embryonic cell contains an H2B in which the pentapeptide is absent. The fully differentiated erythrocyte or calf thymus cell H2B exhibits one full and possibly two degenerate pentapeptides, whereas the fully repressed sea urchin sperm cell protein may contain up to six pentapeptide repeats.

Alternatively, the pentapeptide repeat structure may provide a feature for efficient nucleic acid packing. Because of the recurrence of proline at position one of each in the pentapeptide repeats, it is unlikely that an α-helix similar to the situation envisaged for parts of the salmine molecule (51) exists in the *N*-terminal region induced by the interaction of the phosphate residues in the DNA with the basic residues in the pentapeptide. However, a fairly extended different helical conformation can be envisaged with three pentapeptides per turn (Figure 3.12).

The pitch of this polypeptide helix would be such that the pentapeptides do not easily follow the pitch of either groove of a single

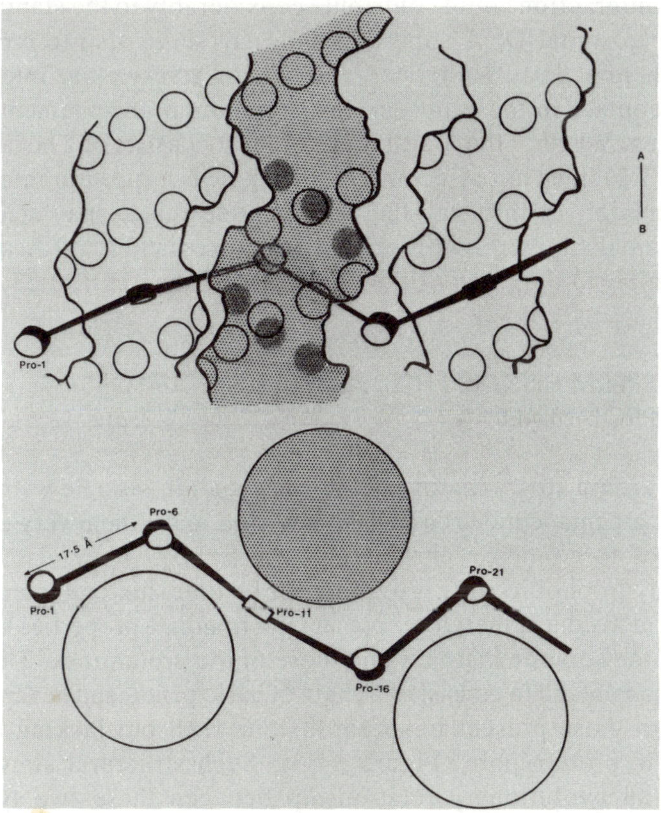

FIGURE 3.12. Model of sperm histone H2B–DNA interaction. The contours of the three DNA dyads and the position of the phosphate groups lining the grooves are shown. The peptide backbone of the extended pentapeptides is indicated by a rod connecting the proline residues. The model is shown in planar view and rotated by about 90° looking down the axis of the DNA dyads from B to A.

DNA dyad. However, the pentapeptide repeat is very well suited to cross-link up to three neighboring DNA columns (Figure 3.12) by aligning itself parallel to one of the grooves of different DNA dyads. The least strained conformation of the pentapeptide region results from an interaction where DNA and pentapeptides are interwoven in such a fashion that one of the DNA dyads is located in a plane different from that of the other two (Figure 3.12). Whereas four repeat units can already lead to a cross-linking of three DNA dyads, five repeats result in an even firmer grip of the third DNA column. Such an interaction could contribute considerably to the stabilization of a nucleosomal DNA supercoil and may serve also to cross-link two such adjacent superhelices, facilitating a very close intersuperhelical contact in the proposed (54) solenoidal arrangement of nucleosomes. Whether the fourth or fifth repeat consists of a hexapeptide (Figure 3.5) does not affect materially the DNA–protein interaction in this model. In this way the pentapeptide repeat may stabilize a nucleosomal supercoil with a pitch of approximately 30 Å, a figure very similar to the pitch of the nucleosomal supercoil of 28 Å (54).

7. THE SPERM HISTONES H2B, POSSIBLE PRECURSORS OF THE PROTAMINES

The sea urchin sperm histones H2B thus contain, as a new structural element, a pentapeptide repeat that may well make them very efficient molecules to condense chromatin more tightly. They also have a typical hydrophobic zone providing for histone–histone interaction. The sperm histones, therefore, present intermediate properties between those of the somatic histones and those of the protamines. The latter may be considered to consist of a chain of basic penta- and hexapeptides similar to those present in sperm histone H2B but lacking the hydrophobic protein core (Figure 3.13). Such structural similarities suggest an evolutionary relationship between these two types of chromosomal proteins.

Considering the codons involved, single-base mutations will lead to arginine virtually in any of the amino acid positions in the histone pentapeptides, arginine being the most abundant basic amino acid in the protamine pentapeptides. However, to base the construction

istone H2B (2) Psammechinus	Pro-Lys-Ser-[Pro]-Ser-Lys-Ser-[Ser]-Pro-Arg-Lys-Gly-[Ser]—-[Pro]-Arg-Lys-Gly-Ser-
lupein YII	[Pro]-Arg-Arg-Arg-[Thr]—-Arg-Arg-Ala-[Ser]-Arg-[Pro]-Val-Arg-Arg-Arg-Arg-
istone H2B (2) Psammechinus	[Pro]-Arg-Lys-Gly-[Ser]-Pro-Lys-Arg-Gly-Gly-Lys-Gly-Ala-Lys-
lupein YII	[Pro]-Arg-Arg-Val-[Ser]-Arg-Arg-Arg-Arg-Ala-Arg-Arg-Arg-Arg-

FIGURE 3.13. *N*-terminal pentapeptide repeat of sea urchin sperm histone and clupein sequence (41).

of lineage between proteins only on single-base mutations of certain codons to and from arginine is a weak argument on its own because of the multiple codons for arginine. The similar spacing of the basic residues between proline and an hydroxy amino acid in the two types of proteins under discussion makes mutation to arginine a stronger argument. In addition, the total length of the protamines with approximately 30 amino acid residues is close to the length of the pentapeptide repeat region realized in the sperm histones H2B (Figure 3.5).

With the advance of evolution beyond the echinoderms, the need for more efficient packing of the increasing genome into the sperm cell must have arisen. To meet this requirement the repeating pentapeptide part of the sea urchin sperm histones, a structure already selected by evolution for packaging DNA, may have evolved into separate genes coding for a new class of proteins, the protamines. Such a pathway of protamine evolution from histones is different from that proposed by Black and Dixon (55) who envisaged an ancestral Ala-Arg-Arg-Arg-Arg-pentapeptide undergoing reiteration and mutation. In view, however, of the late arrival of protamines on the evolutionary scene, and the ample evidence of the presence of the set of histones in early eucaryotic cells (5, 56, 57, 58, 59) the subsequent evolution of the younger protamines from the pentapeptide structure present in the older histones appears to us a plausible pathway. The sperm histones H2B, with their *N*-terminal repeating pentapeptide extension, could thus be considered as the evolutionary link between the histones and the protamines, the two main classes of DNA-associated basic proteins in eucaryotic cells.

The variability (Tables 3.4, Figure 3.5) of the pentapeptide structure in the sperm histones, amounting to six different pentapeptides in three species, and the changing degree of reiteration of this structure (Figure 3.5) indicate that the histones in sea urchin sperm cells are subjected to a considerable selection pressure which has resulted in a comparatively rapid evolution of that cell type specific protein. As a result of this, in the class Echinoidea with approximately 800 species, the total number of variant molecules is likely to be very much higher than the six histone H2B variants hitherto identified in three species from different genera. That such a selection pressure on the structure of the otherwise conservative histones operates successfully on one particular cell type only in a given species is not surprising considering that sperm cells at spawning represent a very large population of nearly identical cells resembling single-cell populations subjected for a short period to a high selection pressure in the form of considerable differences in temperature, hydrostatic pressure, and salinity, depending on the particular habitat, a scenario in which the survival value of any mutation in the population is confirmed immediately in terms of successful fertilization.

8. THE HISTONES H2A, H3, AND H4

Though the three histones H2A from the sea urchin embryo show distinct differences in their detergent binding, resulting in characteristic very pronounced differences in electrophoretic mobilities (Figure 3.4), their amino acid compositions are very similar (Tables 3.1 and 3.2). The sequence of the sperm H2A and the tryptic core of the embryo histone $H2A_{(3)}$ are given in Figure 3.14. The hydrophobic core, usually highly conserved in the histones, has a methionine residue in position 50 of the $H2A_{(3)}$ that has mutated in calf to leucine (45). The histone H2A from sperm in *Parechinus angulosus* (60) and *Psammechinus miliaris* (61) are identical except that residues 9–11 in *Parechinus* are Ala-Lys-Gly, whereas in *Psammechinus* they are Gly-Lys-Ala. When compared to the histone H2A from higher eucaryotes the *Parechinus* histone varies in 14 positions from that of calf (62).

The histone H3 from sperm (63) is for the part sequenced very conservative except for the serine at 96, which in higher species has mutated to cysteine (Figure 3.4). The sperm histone H4, also highly conserved, contains, however, a cysteine in position 73 (64) that in the course of evolution changed to threonine (Figure 3.14).

9. THE H1 ISOHISTONES

The histones H1 of the sea urchin exhibit a considerable heterogeneity. From *Parechinus angulosus* we have identified in our laboratory a

H2A Parechinus (sperm)

```
Ac       10                  20                  30                  40
S G R G K G A K A K G K A K S R S S R A G L Q F P V G R V H R F L R K G N Y A N
                 50                  60                  70                  80
R V G A G A P V Y L A A V L E Y L A A E I L E L A G N A A R D N K K T R I I P R
                 90                 100                 110                 120
H L Q L A I R N D E E L N K L L G G V T I A Q G G V L P N I Q A V L L P K K T G

S K S S K
```

Histone H3 Parechinus (sperm)

```
                 10                  20                  30                  40
A R T K Q T A R K S T G G K A P R K Q L A T K A A R K S A P A T G G V K K P H R
                 50                  60                  70                  80
Y R P G T V A L R E I R R Y Q K S T E L L I R K L P F Q R L V R E I A Q N F K T
                 90                 100                 110                 120
D L R F Q S S A V M A L Q E A S E A Y L V G L F E D T N L C A I H A K R V T I M

P K D I Q L A R R I R G E R A
```

Histone H4 Parechinus (sperm)

```
Ac               10                  20                  30                  40
S G R G K G G K G L G K G G A K R H R K V L R D N I Q G I T K P A I R R L A R R
                 50                  60                  70                  80
G G V K R I S G L I Y E E T R G V L K V F L E N V I R D A V T Y C E H A K R K T
                 90                 100
V T A M D V V Y A L K R Q G R T L Y G F G G
```

FIGURE 3.14. Sequences of sea urchin histones H2A (60), H3 (63), and H4 (69) from sperm cells of *Parechinus angulosus*. For H3 and H4 the sequenced residues are underlined. The remainder of the sequences have been taken from the homologous calf proteins. The sperm H2A from *Psammechinus miliaris* is identical except that residues 9–11 have been reported GKA (61). In *Parechinus* embryo $H2A_{(3)}$ residues 40–81 have been found identical to those in the sperm protein except for residue 51, which is M in embryo (11). Embryo $H3_{(1)}$ and $H3_{(2)}$ have S in position 96 (11).

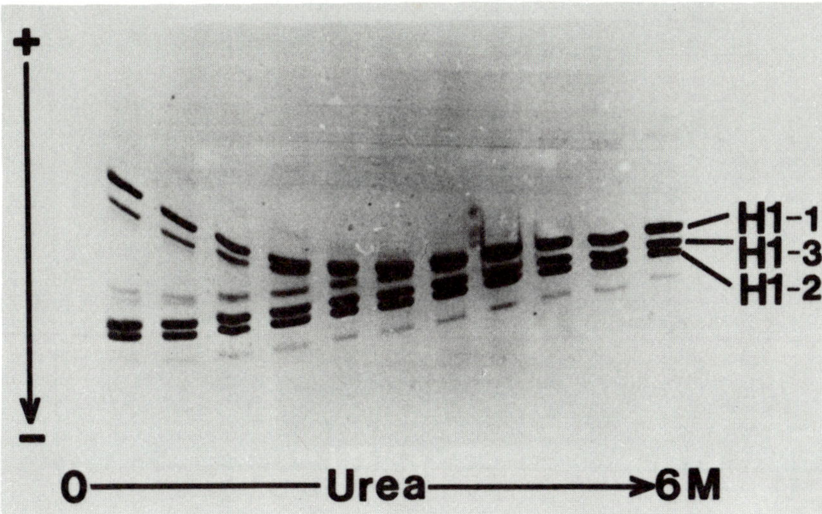

FIGURE 3.15. PAGE of embryo histones H1 (*Parechinus angulosus*) in a transverse urea gradient. The slow moving satellite doublet following $H1_{(3)}$ represents methionine oxidation products of $H1_{(3)}$ resulting from the exposure to HC*l*O during the extraction of the polypeptides from chromatin. Electrophoresis conditions are the same as in Figures 3.2 and 3.3.

minimum of five embryonic histones H1 (65) (Figure 3.15), one sperm H1, and one histone H1 isolated from fully differentiated gut cells (6). The complete or partial structures of some of these are given in Figure 3.16. The complete structure is available only for the sperm isohistone H1, the largest hitherto described histone H1, with 248 residues. In their general construction all the sea urchin histones H1 are designed along the lines recognized by Hartman et al. (66); for the calf thymus H1, namely, there is the attachment of two basic extensions toward the *N*- and *C*-terminal to a potentially globular core. Hartman et al. (66) have described these as nose, tail, and head, respectively. Within that general picture it appears that, similar to the H2B group; the larger nose and tail are characteristic for the replicative and transcriptionally repressed haploid sperm cell, whereas the embryo histones exhibit a shorter nose section (Figure 3.16). The *N*-terminal extension in the sperm histones H1 is the result of the reiteration of a tetrapeptide motif Ser-Pro-Arg-Lys from

```
                                                                          10                  20                   30                  40                  50
H1 Parechinus (Embryo) (1)                                                            Blocked NH2 - - - - - K K V T T K K P A A(H P P)      M V T T A
H1 Parechinus (Embryo) (2)                                                            Blocked NH2 - - - - - - - - - - - - - - - - - - - - - - M V V A A
H1 Parechinus (Embryo) (3)                                                            Blocked NH2 - - - - - - - - - - - - - - - - - - - - - - M V V A A
H1 Strongylocentrotus (Embryo) Gene                                                             M A E K N S S K K V T T K K P A A H P P A A E M V A T A
H1 Parechinus (Sperm)                  P G S P Q K R A A S P R K S P R K—S P K K S P R K A S A S P R R K A K R A R A S T H P P V L E M V Q A A
H1 Echinolampas (Sperm)                A A S P Q K R A A S P R K S P K K—S P R K S P K K K——S P R K R—K—A R S A A H P P V I D M I T A A
H1 Sphaerechinus (Sperm)               P G S P Q K R A A S P R K S P R P G S P K K S P - - - - - - - - - - - - - - - - - - - - - - - - M I R A A

             60                  70                  80                  90                  100                 110                 120                 130
I T E L K E/I R N G S S L N/T A I/V K K Y I E T/V N F D V E M - D R Q L I K I K K A L K S G V E K G K L V Q T K G T G A S G S F K L N V Q A A K A Q A A E K A K K
I T/G A L K E R G G S(S)- - - - - - - - - - - - - - - M - K K Q S V F I K K A L K S G V E K G T L V Q V K G K G A S G S(F)K L G K K P A A G(RT) D A - - - -
I T A L K E R G G S S N Q A L K K Y K A A N - - - - - - - - - - - - - - - - - - - - - - - - - - - - - - - - - - - - - - - - - - - - - - - - - - - - - - -
I T E L K D R N G S S L Q A I K K Y I A T N F D V N M—D R N L L F I K R A L K S G V E K G K L V Q T K G K G A S G S F K V N V Q A A K A Q A S E K A K K
I T A M K E R K G S S A A K I K S Y M A A N Y R V D M—N V L A P H V R R A L R N G V A S G A L K Q V T G T G A S G R F R V G A V A K P K K A K K T S A A
I A A Q K E R R G S S V A K I Q S Y I A A K Y R C D I—N A L N P H I R R A L R N G V K S G A L K Q V S G V G A T G R F R V G A V   K R S   A A S   A N K
I T A M K E R K G S S V A K I K S'Y I A A N Y R V N M T N L Q—P H I R R A L R S G V A S G A L K Q V T G T G A T G R F R V G - - - - - - - - - - - - - -

                 140                 150                 160                 170                 180                 190                 200
E K E K A K L Q A Q R E K A K A K A A A K K E K Q Q K————A T A A K K A K A A P K K K P A K K A V - - - - - - - - - - - - - - - - - - - - E K K E K
- - - - - - - - - - - - - - - - - - - - - - - - - - - - - - - - - - - - - - - - - - - - - - - - - - - - - - - - - - - - - - - - - - - - - - - - - - -
- - - - - - - - - - - - - - - - - - - - - - - - - - - - - - - - - - - - - - - - - - - - - - - - - - - - - - - - - - - - - - - - - - - - - - - - - - -
E K E K A K L L A Q R E K A K E K G C S E E G E T A E————G S R P K K V K A A P K K A L L P V K K————————————————————T T E K K E K
A K A K K A K A A A A K K A R K A K A A A K R K A A L A K K K A A A A K R K A A A K A K K A K K P K K K A A K K A K K P A K K S P K K A K K P A K K S P K
L K A T R E K A R A R A K A K K A K A A A R R K A A A A K R K A A A A K R R A A K K A R K A K A K P - - - - - - - - - - - - - - - - - - - - - - - - -
- - - - - - - - - - - - - - - - - - - - - - - - - - - - - - - - - - - - - - - - - - - - - - - - - - - - - - - - - - - - - - - - - - - - - - - - - - -

  210                 220                 230                 240                 250                 260
K K T P K K A T A A T K K A T P K T V T - - - - - - - - - - - - - - - - - - - - - - - - - - - - - - - - - - - -OH
- - - - - - - - - - - - - - - - - - - - - - - - - - - - - - - - - - - - - - - - - - - - - - - - - - - - - - - -OH
- - - - - - - - - - - - - - - - - - - - - - - - - - - - - - - - - - - - - - - - - - - - - - - - - - - - - - - -OH
K K T P K K A P K K P A A K K S T P K K T P K K A—————A A K K P K T A K P K K P A - K K A A K S K - -OH
K K K A K R————S P K K A K K A A G K R K P————————A A K K A R R S P R K A G K R R S P K K A R K -OH
- - - - - - - - - - - - - - - - - - - - - - - - - - - - - - - - - - - - - - - - - - - - - - - - - - - - - - - -OH
- - - - - - - - - - - - - - - - - - - - - - - - - - - - - - - - - - - - - - - - - - - - - - - - - - - - - - - -OH
```

positions 10–30 with the occasional mutation of the arginine residue to lysine (Figure 3.16). In contrast, the embryo isohistones lack that *N*-terminal tetrapeptide repeat. The *C*-terminal tail region in the sperm histones H1 also exhibits between residues 130 and 248 four different repeats, each reiterated, with some mutations, three to four times. Each of the repeats extends over 7–12 residues (Figure 3.17). The sperm histone is also characterized by a higher arginine abundance in the tail region, whereas the embryonic histones exhibit a more lysine rich *C*-terminal polypeptide chain.

The difference in basicity of the side chains of these two amino acids will result in a stronger interaction of the sperm histone with the DNA as compared with the embryonic histones. This is similar to the chemical characteristics of the respective H2B isohistones from sperm and embryo (Figure 3.5, Tables 3.1 and 3.2). The general construction principle of the sea urchin sperm H1 histones with the longer *N*-terminal extension is also realized in the sperm H1 histones isolated from two other sea urchin species, *Echinolampas crassa* (24) and *Sphaerechinus granulosus* (25), where the *N*-terminal noses are of similar design to *Parechinus* and the peptides from the *C*-terminal region show a similar high abundance of arginine, lysine, and alanine (Figure 3.16).

The unequivocal proof that structural variants of the histones are present as tissue-specific chromatin components and that their synthesis appears to coincide with distinct developmental and terminal differentiation stages has changed the picture of the role of histones in chromatin structure. This picture had been derived mainly from investigations on the histones from cells of fully differentiated higher vertebrates like calf thymus or chicken erythrocytes. From these cells the picture of a rather uniform contribution of the histones to chromatin structure had been conceived. It must now be accepted

FIGURE 3.16. Primary structures of H1 histones from sea urchin: *Parechinus* embryo (65), *Strongylocentrotus* embryo gene (67), *Parechinus* sperm (22, 23), *Echinolampas* sperm (24), and *Sphaerechinus* sperm (25). On sequencing, *parechinus* embryo histones $H1_{(1)}$ and $H1_{(2)}$ proved heterogeneous in a few positions only, whereas $H1_{(3)}$ was heterogeneous for long stretches; only one unequivocal stretch of the major sequence is shown. Residues not yet identified, - - - - -; alignment gaps, ———.

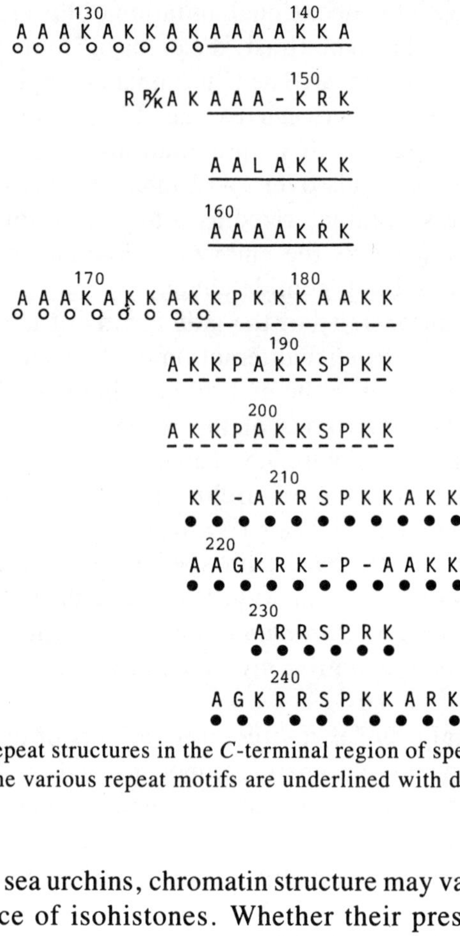

FIGURE 3.17. Repeat structures in the *C*-terminal region of sperm histone H1 from *P. angulosus*. The various repeat motifs are underlined with different symbols.

that, at least in sea urchins, chromatin structure may vary as determined by the presence of isohistones. Whether their presence provides a unique local chromatin structure suitable for specific functional states of chromatin still must be determined.

10. ASSEMBLY OF ISOHISTONES INTO CHROMATIN SUBSTRUCTURES

The results reviewed establish firmly that for the sea urchin a multitude of isohistones rather than a single set provide the structural framework for the compaction of the genome into chromatin. The synthesis of these isohistones is programmed in relation to specific phases of differentiation. It also follows that the fundamental building units, the nucleosomes, must exist in a multitude of structural variants (Figure 3.18). The two extreme types of nucleosomes from sperm and embryo tissue differ on an average by 78 residues in total number. Considering the nature of residues replaced in the different isotypes of H2B, H2A, and H1 a conservative estimate, based on the knowledge of the partial structures only, we put the differences between sperm and embryo types of nucleosome at least 211 residues, or approximately 16% of all nucleosomal amino acid residues. Though the histones H3 and H4, the other partners in the histone octamer, belong to the group of the most conservative proteins, the overall primary structural features of the nucleosomal proteins in the course of evolution have changed considerably (Figure 3.18). So much so that between approximately 100–300 amino acid residues may vary if sea urchin embryo and sea urchin sperm cores or nucleosomes are

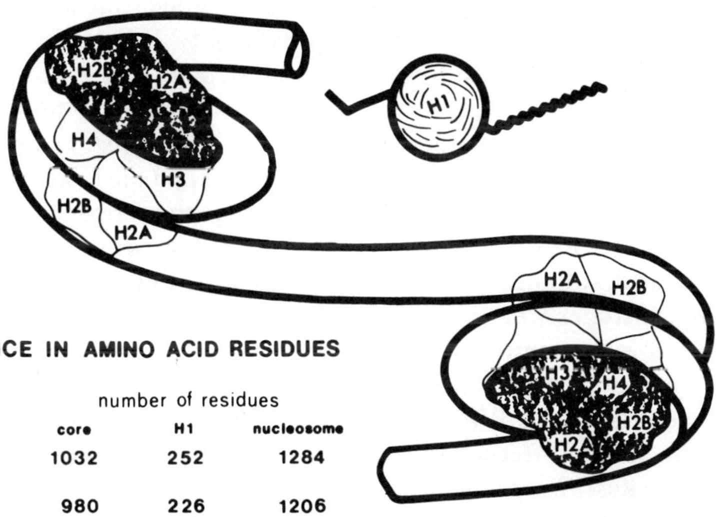

DIFFERENCE IN AMINO ACID RESIDUES

source	number of residues		
	core	H1	nucleosome
sperm	1032	252	1284
embryo	980	226	1206
mammal	988	208	1196
	type of residues		
Δ sperm ↔ embryo	102*	109*	211
Δ embryo ↔ mammal	42*	102*	144
Δ sperm ↔ mammal	124	186	310

*Partial structures only for embryo H1, H2A and H2B

FIGURE 3.18. Variability of nucleosomal histones.

compared to the corresponding structures isolated from a mammalian tissue.

What may be the functional significance of this considerable variation in nucleosomal structure? The timing of synthesis during the sea urchin development and the tissue-specific deposition of uniquely structured histones suggests that the structural features of the isohistones make them particularly suitable to mold the major functional forms of the genome, that is, actively transcribing, temporarily repressed but inducible and permanently repressed chromatin. There is good morphological evidence that functional stages of chromatin are characterized by typical electronmicroscopic structures (68).

It is well established that nucleosomes in transcriptionally active regions are more susceptible to excision by nucleases (69, 70). The histones in these excised nucleosomes are more heavily acetylated (71). The histone H1 and its various isoforms are involved in internucleosomal interactions typical for the condensation of nontranscribed chromatin (68). Assuming that the entire histone content of chromatin in a single cell, amounting to approximately 2×10^{-12}g, is participating in nucleosome formation, we calculate that the total number of nucleosomes in a sea urchin cell would amount to 1.3×10^{7} nucleosomes or histone octamers. If all the isohistones within one histone group were functionally equivalent and the histone gene tandems, reiterated several hundred times, were transcribed as two groups only, early and late genes, one would expect within each group a random distribution of isohistones through the entire chromatin. The early histones would merely be diluted in each new cell generation. If, however, subtle structural differences of the nucleosomes due to the component isohistones were functionally significant, then this should result in a differential distribution of isohistones within chromatin. Such differential distribution should affect the composition of the substructures, that is, core particles, mono-, di-, tri-, oligo-, and polynucleosomes should vary in their component isohistones. The increased nuclease susceptibility of the active regions of the genome as compared to the more condensed repressed regions should thus produce differentially composed substructures. Pulse labeling of sea urhin embryos with subsequent fractionation of chromatin substructures and their component isohistones should give insight into this problem (72).

In the 14-hour embryo, newly synthesized histones appear in varying amounts and proportions in core particles, mono-, di-, tri-, and oligonucleosomes (Figure 3.19). The core particle region contains the largest amount of the newly synthesized histones, and the mononucleosome fraction appears to consist of at least three octamer populations associated with either X, $H1_{(2)}$, or $H1_{(3)}$. Similarly, there are several subpopulations in the dinucleosomal fraction without any of the H1 isotypes or X or, alternatively, with anyone of the extra core histones. Newly synthesized histone H1 variants 3 and 2 are not only assembled into new mono-, and dinucleosomes but also preferentially into higher structures of the tri- and oligonucleosomal type. The core histones in these latter structures are, however, labeled to a much lesser degree, if at all. The component histone octamers of these more complex structures thus must have been assembled earlier and are only now condensed with H1. Core particles and dinucleosomes contain more freshly synthesized histone $H2B_{(2)}$ than the monosomal fraction. The core particle population contains mainly newly synthesized isohistones $H2B_{(1)}$ and $H2B_{(3)}$, $H2A_{(1)}$, and $H2A_{(2)}$ belonging to the late set of histones, though a substantial amount of

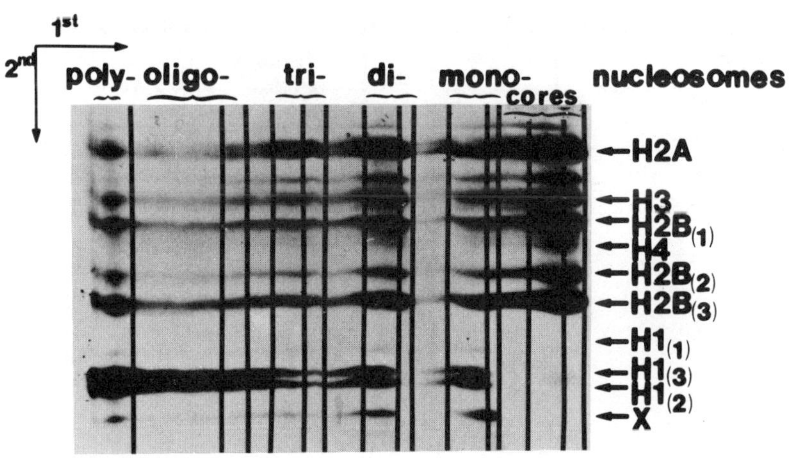

FIGURE 3.19. PAGE of *Staph. aureus* nuclease digest of 9h sea urchin embryo nuclei after a 20-minute pulse with ^3H lysine (71). See also Figure 3.20.

the core particles, the mono and the disomes, have been assembled containing early histone $H2B_{(2)}$ still synthesized at this time.

The acetylation of the isohistones is selective (Figure 3.20). It is a recently assembled core particle population containing the $H2B_{(3)}$, $H2B_{(1)}$, $H2A_{(1)}$, and $H2A_{(2)}$ subtypes which becomes heavily acetylated, whereas recently assembled core particles containing the isotype $H2B_{(2)}$ undergo very little labeling, particularly in the mononucleosomal region. $H2A_{(3)}$ does not become acetylated at all. The higher-order supranucleosomal structures of the tri- and oligosome type that contain selectively large amounts of freshly synthesized histone $H1_{(3)}$ and $H1_{(2)}$ undergo, however, less acetylation. Histones synthesized from the early gene set in a 20-minute pulse at 9 hours are still present at 21 hours. In particular, the H2A is retained in the tri-, oligo-, and polynucleosome fraction, though the other core histones have already been diluted out (Figure 3.21).

From these results it appears that in the nucleosomally organized chromatin of the hatching blastula, consisting predominantly of the early set of isohistones, stable supranucleosomal structures containing any of the H1 isotypes are not generated during the pulse. On the

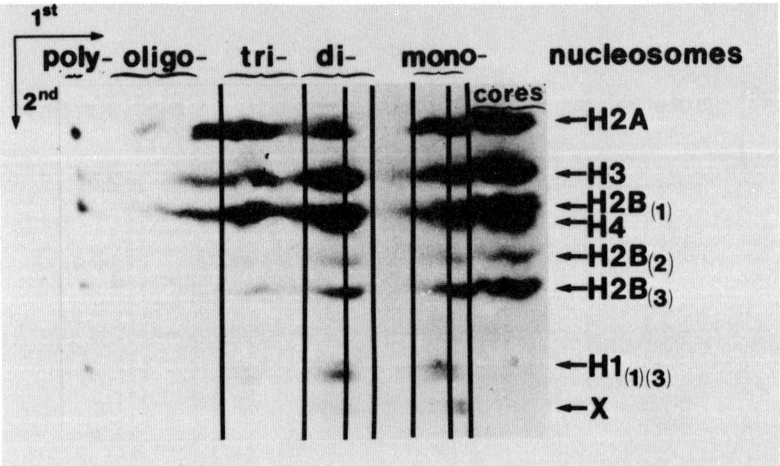

FIGURE 3.20. PAGE of *Staph. aureus* nuclease digest of 14h sea urchin embryo nuclei after a 20-minute pulse with ^3H lysine. 1st dimension: Polyacrylamide gradient gels to separate chromatin substructures. 2nd dimension to separate component histones, conditions as in Fig. 2 at 3.8 M urea (71).

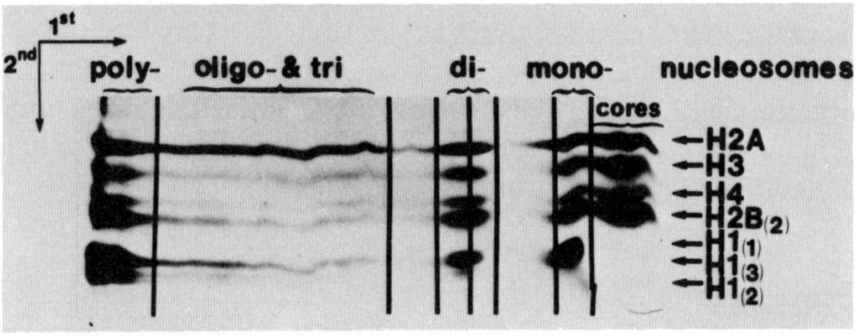

FIGURE 3.21. PAGE of *Staph. aureus* nuclease digest of 14h sea urchin embryo nuclei after a 20-minute pulse with ^{3}H acetate. Electrophoresis conditions are the same as in Figure 3.20 (71).

other hand, the more mature chromatin of the 14-hour embryo has a different appearance altogether. The oligonucleosome fraction consists largely of histone octamers that have been synthesized prior to the radioactive pulse. However, these supranucleosomal chromatin stretches are stabilized to a more condensed form with newly synthesized H1 isotypes 2 and 3. The thus-generated structures, then, no longer participate in transcription, as indicated by the low degree of acetylation. The core particles, the mono-, the di-, and the trinucleosomes, contain subpopulations with respect to octamer composition from recently synthesized isohistones as well as to their acetylation. This indicates that such subpopulations may be involved to varying degrees in transcription.

The selective deposition of H1 isotypes on previously assesmbled histone cores to result in oligonucleosomes, the differential incorporation of core histones isotypes into chromatin substructures, as well as the differential acetylation of isohistones in the various substructures is an expression for the selective utilization of the isohistones in the construction of actively transcribing or repressed genome regions.

ACKNOWLEDGMENTS

The work reported in this chapter has been supported by grants from the C.S.I.R., Republic of South Africa and the University of Cape

Town Research Committee to Claus von Holt. We thank Faezah Davids and Madhu Chauhan for expert assistance; Robin Warren, Patricia Wormald, and Jennifer Rees for help in the collection of sea urchins and in the laboratory; Louise Baker and Maureen Behrens for typing the manuscript and preparing figures.

REFERENCES

1. E. J. Du Praw, *Cold Spring Harbor Symp. Quant. Biol.* **38,** 87–98 (1973).
2. R. D. Kornberg, *Ann. Rev. Biochem.* **46,** 931–954 (1977).
3. S. Wittig and B. Wittig, *Nature* **297,** 31–38 (1982).
4. L. H. Kedes and M. L. Birnstiel, *Nature New Biol.* **230,** 165–169 (1971).
5. S. C. R. Elgin and H. Weintraub, *Ann. Rev. Biochem.* **44,** 725–774 (1975).
6. C. von Holt, W. N. Stickland, W. F. Brandt, and M. S. Strickland, *FEBS Lett.* **100,** 201–218 (1979).
7. L. H. Cohen, K. M. Newrock, and A. Zweidler, *Science* (Washington, D.C.) **190**, 994–997 (1975).
8. L. H. Cohen, K. M. Newrock, M. B. Hendricks, R. J. Donnelly, and E. S. Weinberg, *Cell* **14,** 327–336 (1978).
9. W. F. Brandt, W. N. Strickland, M. Strickland, L. Carlisle, D. Woods, and C. von Holt, *Eur. J. Biochem.* **94,** 1–10 (1979).
10. W. F. Brandt and C. von Holt, *Biochim. Biophys. Acta* **537,** 177–181 (1978).
11. S. Schwager, W. F. Brandt, and C. von Holt, *Biochim. Biophys. Acta*, in press (1983).
12. E. W. Johns, *Methods Cell Biol.* **16,** 183–203 (1977).
13. C. von Holt and W. F. Brandt, *Methods Cell Biol.* **16,** 205–225 (1977).
14. I. Hindennach, G. Stöffler, and H. G. Wittmann, *Eur. J. Biochem.* **23,** 7–11 (1971).
15. A. Zweidler, *Methods Cell Biol.* **17,** 223–233 (1977).
16. M. Strickland, W. N. Strickland, W. F. Brandt, and C. von Holt, *Eur. J. Biochem.* **77,** 263–275 (1977).
17. W. N. Strickland, M. Strickland, W. F. Brandt, and C. von Holt, *Eur. J. Biochem.* **77,** 277–286 (1977).
18. M. Strickland, W. N. Strickland, W. F. Brandt, C. von Holt, B. Wittmann-Liebold, and A. Lehmann, *Eur. J. Biochem.* **89,** 443–452 (1978).
19. M. Strickland, W. N. Strickland, W. F. Brandt, and C. von Holt, *Biochim. Biophys. Acta* **536,** 289–297 (1978).
20. I. A. Zalenskaya and A. O. Zalensky, *Comp. Biochem. Physiol.* **65B,** 369–373 (1980).

21. B. de Petrocellis, L. de Petrocellis, M. Lancieri, and G. Geraci, *Cell. Diff.* **9,** 195–202 (1980).
22. W. N. Strickland, M. Strickland, P. de Groot, C. von Holt, and B. Wittmann-Liebold, *Eur. J. Biochem.* **104,** 559–566 (1980).
23. W. N. Strickland, M. Strickland, W. F. Brandt, C. von Holt, A. Lehmann, and B. Wittman-Liebold, *Eur. J. Biochem.* **104,** 567–578 (1980).
24. W. N. Strickland, M. Strickland, and C. von Holt, *Biochim. Biophys. Acta* **700,** 127–129 (1982).
25. W. N. Strickland, M. Strickland, C. von Holt, and V. Giancotti, *Biochim. Biophys. Acta* **703,** 95–100 (1982).
26. M. Strickland, W. N. Strickland, and C. von Holt, *FEBS Lett.* **135,** 86–88 (1981).
27. P. de Groot and C. von Holt, unpublished, 1982.
28. K. M. Newrock, C. R. Alfageme, R. V. Nardi, and L. H. Cohen, *Cold Spring Harbor Symp. Quant. Biol.* **42,** 421–431, (1977).
29. K. W. Gross, M. Jacobs-Lorena, C. Baglioni, and P. R. Gross, *Proc. Natl. Acad. Sci. (USA)* **70,** 2614–2618 (1973).
30. D. E. Woods and W. Fitschen, *Cell Diff.* **7,** 103–114 (1978).
31. L. H. Kedes, B. Hogan, G. Cognetti, S. Selvig, P. Yanover, P. R. Gross, Cold Spring Harbor, *Symp. Quant. Biol.* **34,** 717–723 (1969).
32. D. Poccia, J. Salik, and G. Krystal, *Dev. Biol.* **82,** 287–296 (1981).
33. P. D. van Helden, W. N. Strickland, M. Strickland, and C. von Holt, *Biochim. Biophys. Acta* **703,** 17–20 (1982).
34. P. D. van Helden, W. N. Strickland, W. F. Brandt, and C. von Holt, *Eur. J. Biochem.* **93,** 71–78 (1979).
35. M. S. Strickland, W. N. Strickland, and C. von Holt, *Eur. J. Biochem.* **106,** 541–548 (1980).
36. S. C. R. Elgin, J. Schilling, and L. E. Hood, *Biochemistry* **18,** 5679–5685 (1979).
37. R. J. De Lange, L. C. Williams, and H. G. Martinson, *Biochemistry* **18,** 1942–1946 (1979).
38. S. V. Shlyapnikov, A. A. Arutyunyan, S. N. Kurochkin, L. V. Memelova, M. V. Nesterova, L. P. Sashchenko, and E. S. Severin, *FEBS Lett.* **53,** 316–319 (1975).
39. G. H. Huntley, and G. H. Dixon, *J. Biol. Chem.* **247,** 4916–4919 (1972).
40. D. C. Watson, E. H. Peters, and G. H. Dixon, *Eur. J. Biochem.* **74,** 53–60 (1977).
41. J. M. Walker, G. H. Goodwin, E. W. Johns, P. Wietzes, and W. Gaastra, *Int. J. Peptide Protein Res.* **9,** 220–223 (1977).
42. J. M. Walker, J. R. B. Hastings, E. W. Johns, and W. Gasstra, *Biochem. Biophys. Res. Comm.* **73,** 72–78 (1976).

43. S. C. Rall and R. D. Cole, *J. Biol. Chem.* **246,** 7175–7190 (1971).
44. G. H. Dixon, E. P. M. Candido, B. M. Honda, A. J. Louie, A. R. MacLeod, and M. T. Sung, *The Structure and Function of Chromatin*, Ciba Foundation Symposium **28,** Associated Scientific Publ., Amsterdam, 1975, pp. 229–250.
45. K. Iwai, H. Hayashi, and K. Ishikawa, *J. Biochem. (Tokyo)* **72,** 357–367 (1972).
46. T. Ando, M. Yamasaki, and K. Suzuki, *Protamines, Molecular Biology and Biophysics*, Vol. 12, Springer-Verlag, Berlin, Heidelberg; 1973, pp. 58–80.
47. M. O. Dayhoff, *Atlas of Protein Sequence and Structure*, Vol. 5 (1972), D379; Supplement 1 (1973)S 30; Supplement 2 (1976) 230, 232; *National Biomedical Research Foundation*, Washington, D.C.
48. D. K. Freymeyer, P. R. Shank, M. H. Edgell, C. A. Hutchinson, and T. C Vanaman, *Biochemistry* **16,** 4550–4556 (1977).
49. R. T. Sauer, and R. Anderegg, *Biochemistry* **17,** 1092–1100 (1978).
50. G. M. A. van Beynum, J. M. de Graaf, A. Castel, B. Kraal, and L. Bosch, *Eur. J. Biochem.* **72,** 63–78 (1977).
51. R. W. Warrant, and S. H. Kim, *Nature (London)* **271,** 130–135 (1978).
52. P. Y. Chou and G. D. Fasman, *Biochemistry* **13,** 222–245 (1974).
53. C. Ramakrishnan and G. N. Ramachandran *Biophys. J.* **5,** 909–933 (1965).
54. J. T. Finch and A. Klug, *Proc. Natl. Acad. Sci. USA* **73,** 1897–1901 (1976).
55. J. A. Black and G. H. Dixon, *Nature (London)* **216,** 152–154 (1967).
56. W. F. Brandt, K. Patterson, and C. von Holt, *Eur. J. Biochem.* **110,** 67–76 (1980).
57. W. F. Brandt and C. von Holt, *Eur. J. Biochem.* **121,** 501–510 (1982).
58. C. V. C. Glover and M. A. Gorovski, *Proc. Nat. Acad. Sci. USA* **76,** 585–589 (1979).
59. H. Hayashi, M. Nomoto, and K. Iwai, *Proc. Jap. Acad.* **56,** Ser. B 579–584 (1980).
60. W. N. Strickland, M. S. Strickland, P. C. de Groot, and C. von Holt, *Eur. J. Biochem.* **109,** 151–158 (1980).
61. D. Wouters, P. Sautière, and G. Biserte, *Eur. J. Biochem.* **90,** 231–239 (1978).
62. L. C. Yeoman, M. O. J. Olson, N. Sugano, J. J. Jordan, C. W. Taylor, W. C. Starbuck, and H. Busch, *J. Biol. Chem.* **247,** 6018–6023 (1972).
63. W. F. Brandt, W. N. Strickland, M. Morgan, and C. von Holt, *FEBS Lett.* **40,** 167–172 (1974).
64. M. Strickland, W. N. Strickland, W. F. Brandt, and C. von Holt, *FEBS Lett.* **40,** 346–348 (1974).
65. P. de Groot and C. von Holt, *Biochim. Biophys. Acta*, in press (1983).
66. P. G. Hartman, G. E. Chapman, T. Moss, and E. M. Bradbury, *Eur. J. Biochem.* **77,** 45–51 (1977).

67. S. Levy, J. Sures, and L. Kedes, *J. Biol. Chem.* **257,** 9438–9443 (1982).

68. U. Scheer, *Biology* in J. B. Le Poole et al., Eds., *Electron Microscopy 1982*, Vol. 3, pp. 419–426. Tenth Int. Cong. Electron Microscopy, Deutische Gesellschaft fur Electronmikroskopie eV.

69. T. Igo-Kemenes, W. Hörz, and H. G. Zachau, *Ann. Rev. Biochem.* **51,** 89–121 (1982).

70. M. Bellard, M. T. Kuo, G. Dretzen, and P. Chambon, *Nucl. Acid Res.* **8,** 2737–2750 (1980).

71. J. D. McGhee and G. Felsenfeld, *Ann. Rev. Biochem.* **49,** 1115–1156 (1980).

72. C. von Holt, J. Retief, D. Landsman, S. Schwager, and P. de Groot, unpublished (1982).

73. Y. Takeda, D. H. Ohlendorf, W. F. Anderson and B. W. Matthews, *Science* **221,** 1020–1026 (1983).

CHAPTER **4**

SPATIAL ASPECTS OF HISTONE GENE EXPRESSION IN SEA URCHIN EMBRYOS

LYNNE M. ANGERER, KATHLEEN H. COX,
DONNA V. DELEON, AND ROBERT C. ANGERER
Department of Biology
University of Rochester
Rochester, New York

CONTENTS

1. INTRODUCTION

2. IN SITU HYBRIDIZATION
   - 2.1. Factors Affecting Hybridization Efficiency
   - 2.2. Specificity

3. HISTONE TRANSCRIPTS IN SEA URCHIN EGGS

4. IN SITU HYBRIDIZATION AS A TOOL FOR STUDYING THE EARLY-LATE HISTONE GENE TRANSCRIPTIONAL SWITCH IN DEVELOPING SEA URCHIN EMBRYOS

5. CONCLUSIONS AND FUTURE DIRECTIONS

REFERENCES

1. INTRODUCTION

In the somatic cells of many eucaryotes, histone gene expression appears to be tightly coupled to DNA replication. Evidence for this comes largely from the observations that synthesis of histone proteins occurs only during S phase of the cell cycle (1–5) and is sensitive to inhibitors of DNA replication (1, 6–10). In cleaving sea urchin embryos, as in mammalian tissue culture cells and yeast, the appearance of newly transcribed histone mRNA in polysomes occurs only during DNA replication and can be blocked with hydroxyurea or cytosine arabinoside, inhibitors of DNA replication (9, 11). The tight cell cycle regulation of histone synthesis can probably be attributed to controls operating at both transcriptional and post-transcriptional levels. Recently, this has been elegantly demonstrated in yeast in which changing levels of histone mRNA during the cell cycle result from periodic transcription coupled with rapid turnover of cytoplasmic RNAs (12). Furthermore, when extra histone gene copies are introduced into yeast, the resulting increased mRNA synthesis is compensated by an increased rate of mRNA turnover (13).

In eggs and very early embryos of echinoderms and amphibians, the production of histone proteins is clearly not coupled to DNA replication or under direct transcriptional control. During oogenesis and maturation, large reservoirs of histone mRNAs (sea urchins) or proteins (amphibians) are accumulated for subsequent use during early embryogenesis (for reviews, see References 1 and 14), a time of extremely rapid cell division when embryonic transcription alone would be insufficient to satisfy the enormous demand for these proteins. In sea urchin eggs, the existence of maternal histone mRNA has been clearly demonstrated both by hybridization (15) and *in vitro* translation assays (16) and inferred from the fact that histone synthesis is insensitive to actinomycin D during early cleavage (7, 9, 17). As development proceeds into the fifth and sixth cleavages, these maternal stores of histone mRNA begin to be replenished and augmented by new transcription which becomes regulated in a cell-cycle specific fashion (9, 11, 18, 19). The synthesis of histone proteins then, is first translationally controlled by recruitment of preexisting maternal RNAs into polysomes after fertilization and subsequently transcriptionally controlled by increasing rates of mRNA synthesis per embryo.

Early studies suggested that at least 95% of maternal histone mRNAs in sea urchin eggs reside in mRNP particles found in the postmitochondrial supernatants of egg extracts (15, 16, 19–21). Speculations on the mechanism of suppression of their translation before fertilization have ranged from deficiencies in the egg's translational machinery to inaccessibility of mRNA due to its association with some unknown masking factor (for reviews, see References 21, 22). Recent studies from our laboratory using *in situ* hybridization of recombinant probes to cellular RNAs and biochemical measurements by Raff and co-workers (24) on the distribution of RNAs in stratified and surgically dissected eggs (reviewed here) suggest that mRNA availability is indeed a major factor regulating the large increase in translation of histone mRNA shortly after fertilization and that the masking factor may simply be the confinement of specific histone mRNAs to the egg pronucleus. These studies are discussed in Section 3.

In later stages of sea urchin embryogenesis, the transcription of the set of genes coding for early variant histones diminishes while that of a second set coding for the late variants increases (1, 25–27). The early gene set consists of 400-fold repeated tandem arrays where each array contains one gene for each of the five histone classes separated by spacer DNA sequences (for reviews see References 1, 28). In contrast, the late histone genes are not clustered and are much less repetitive (1, 29). This transcriptional switch occurs primarily during the blastula stage, a time when cell division rates markedly decline and morphogenesis and cell differentiation commence.

The dramatic developmental regulation of these genes, which code for a set of highly abundant mRNAs in sea urchin embryos, has made this a particularly attractive system for studying control of gene activity. Although we know a great deal about the structure of the genes, particularly that of the early set (for reviews see References 1 and 28), and their expression at different developmental stages (25–27), such studies on whole embryos cannot approach one fundamental question: Is the rather gradual switch in expression from early to late histone genes a progressive regulation within each cell of the embryo, or does it result from more discrete switches whose timing is variable in different cell lineages? To approach this question we are utilizing *in situ* hybridization techniques to monitor the concentration per cell of histone mRNAs during development. In Section 2 we show that the accuracy, specificity, and sensitivity of the *in situ* hybridization method we have developed are sufficient to enable us to discriminate between early and late histone gene transcripts and to measure their relative concentrations in different regions of developing embryos.

2. *IN SITU* HYBRIDIZATION

2.1. Factors Affecting Hybridization Efficiency

Using [^{3}H] poly U to detect poly A tracts in sea urchin eggs and cleaving embryos, we initially established that several aspects of tissue preparation were crucial for improving sensitivity (30). First, in agreement with the studies of Godard and Jones (31), we found that fixation with glutaraldehyde instead of a more commonly used fixative, ethanol:acetic acid, afforded not only superior morphological preservation but also significantly higher retention of cellular RNA. Second, mild digestion of glutaraldehyde-fixed tissue with proteinase K resulted in dramatic increases in signal (eight- to tenfold), presumably by rendering target sequences more accessible to hybridization probes. Proteinase K treatment does not affect retention of target RNAs in the tissue (30). We have observed increases in signals for histone and other mRNAs using these modifications.

The second major factor affecting signals involves the choice of probe, that is, double- (symmetric) versus single-stranded (asymmetric), and RNA versus DNA. Initially we chose to use nick-translated symmetric DNA probes for two reasons: (1) We suspected that higher signals might result by way of formation of probe hyperpolymers on RNA targets, and (2) it was the most convenient way to prepare large amounts of probe from existing cloned sequences. While we succeeded in obtaining specific hybridization and clear signals using exposure times of less than one week (see Reference 32, Figures 1 and 4), the calculated efficiency was about tenfold lower than that obtained for poly U, a single-stranded RNA (30). To prepare an analogous probe representing complements to histone mRNAs, we employed the following strategy. cRNA was synthesized *in vitro* from recombinants in which the sequence of interest [in this case, the 6.3 kilobase early histone repeat unit contained in pCO2, isolated and kindly supplied by E. Weinberg and co-workers (33)] was inserted into a transcription vector. This vector, R7Δ7, constructed in T. Maniatis' laboratory by E. Butler and P. Little, contains a promoter from the *Salmonella* phage, SP6, and a downstream multiple cloning site. One major advantage of this system is that transcription by SP6 RNA polymerase is completely asymmetric and highly efficient. Furthermore, by isolating recombinants that contain the insert in both orientations relative to the promoter, it is possible to obtain both coding strand (mRNA-complementary) and mRNA strand (control) probes. To test directly the efficiency of symmetric versus asymmetric probes, we prepared a symmetric RNA probe by annealing coding and mRNA strand transcripts in solution and trimming the duplexes with RNase. The results of hybridization of double- and single-stranded probes to sections of 12-hour embryos under identical conditions are shown in Figure 4.1. At apparent saturation, the single-stranded probe hybridizes to 8–10 times more target sequences. We believe the relatively low signal obtained with symmetric probes can largely be attributed to competing probe self-reassociation in solution. First collision duplexes formed in solution are probably too large to penentrate the tissue efficiently. This idea is supported by our observations (30) and those of Brahic and Haase (34) that shorter probes (50–200 nucleotides) yield significantly higher signals than longer ones (500 nucleotides).

Although the magnitude of the signal obtainable is clearly an important factor in the potential sensitivity of this approach, the actual limits are currently set by the noise. In this regard, we believe that RNA probes would offer a distinct advantage over single-stranded DNA probes since, in our hands, posthybridization digestion with RNase reduces nonspecific RNA binding close to background levels without major reduction in signal. Our attempts to remove nonspecifically bound DNA probe by S1 nuclease digestion have been unsuccessful since conditions that gave improved signal/noise ratios resulted in substantial reduction of signal. Extensive washing under stringent conditions is not nearly as effective as enzymatic treatment. It is important to note that background levels are linearly dependent on probe input. Thus an additional advantage of this transcription

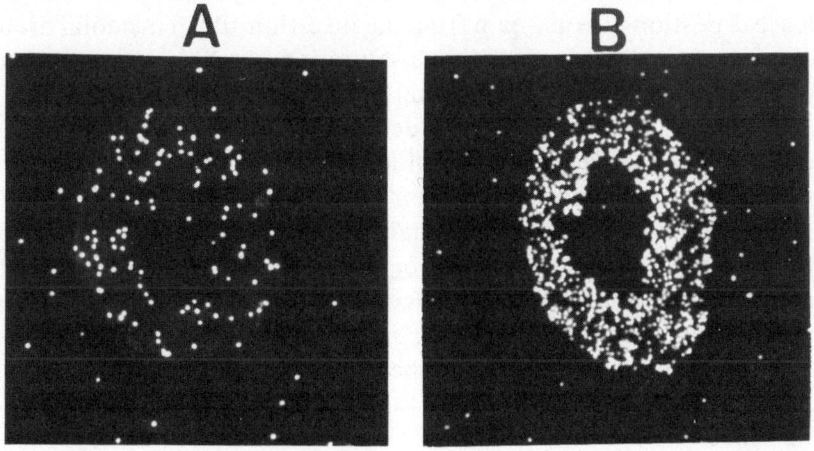

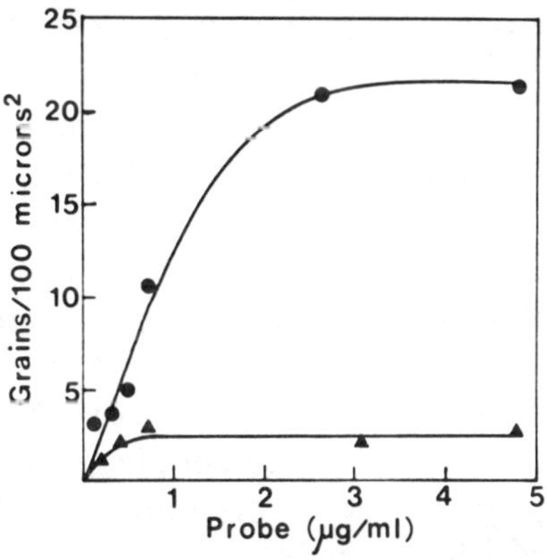

(*b*)

FIGURE 4.1. (*a*) *In situ* hybridization to sections of 12-hour embryos using either (A) symmetric (denatured double-stranded) or (B) asymmetric (single-stranded) RNA probes representing the entire early histone repeat unit (from pCO2). Both probes were of equal specific activity (1.8×10^7 dpm/μg) and fragment length (150 nucleotides). Slides were exposed for 9 hours; dark-field illumination. (*b*) Comparison of saturation of histone mRNA targets in 12-hour embryo sections with either (●) asymmetric or (▲) symmetric RNA probes. The probe properties and exposure times were as described (*c*).

system is that run-off transcripts largely devoid of vector sequences can be synthesized from templates truncated by restriction endonuclease digestion downstream from the insertion site. Thus total probe concentrations necessary to reach saturation of target sequences can be minimized by excluding virtually all vector sequences.

At present, the method allows detection, for example, of mRNAs, 1500 nucleotides long, using a full length ^{3}H-labeled probe of specific activity 2×10^8 dpm/μg, if their concentration is only 0.05% of total mRNA (34). Sufficient signal is obtained in about two weeks, and the signal/noise is at least 3 for randomly distributed mRNA and higher if they are localized. Since histone mRNAs comprise from 2–20% of cellular mRNAs during embryogenesis (36), strong signals and high signal/noise ratios can be obtained.

2.2. Specificity

The high degree of specificity achieved using our hybridization conditions is demonstrated by the following four criteria: (1) Only homologous and not heterologous probes bind (see Section 3 and references 30, 32, 36). (2) The target sequences are RNAs since hybridization signals are sensitive to prehybridization treatments with RNase but not DNase (30, 32). (3) The thermal denaturation curve of hybrids formed *in situ* is sharp, and the *Tm* is only 3–5°C below the solution *Tm* for these RNA–RNA duplexes (44). Furthermore, *in situ Tm*'s reflect the extent of known sequence mismatch between probe and target sequence (see Section 4). (4) *In situ* signals obtained at different developmental stages are proportional to target sequence concentrations at those stages as measured by standard hybridization techniques (30; Figure 4.2). For example, when RNA transcripts representing the early histone repeat unit are hybridized *in situ* to sections of 2-cell and 12-hour embryos, an average 8.8-fold difference in grain density was observed in six separate comparisons (Figure 4.2). Since the volume of the embryo does not change between these two stages, the relative grain density is a direct measure of the difference in target sequence concentration per embryo. Mauron et al. (36) have measured a similar tenfold difference by solution titration of purified RNAs. Our ability to determine accurately concentrations of specific sequences at different developmental stages implies that

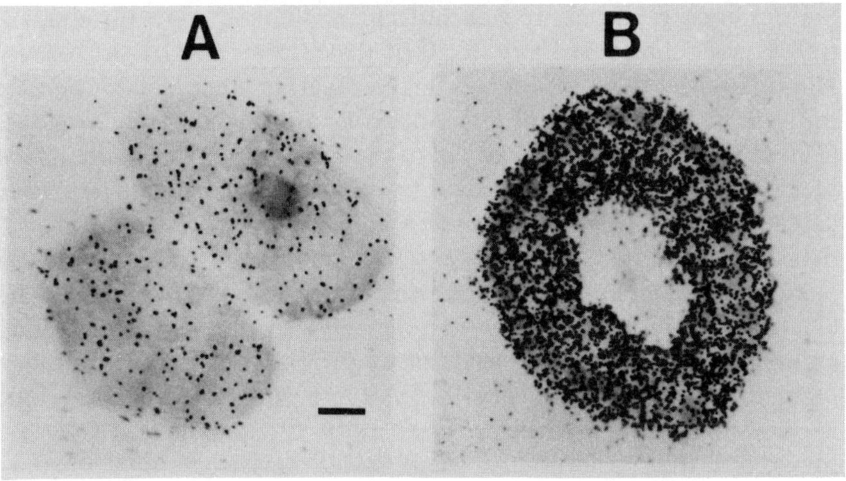

FIGURE 4.2. *In situ* hybridization of pCO2R + transcripts to sections of 2-cell and 12-hour embryos. The specific activity of this probe was 9.4×10^6 dpm/μg, and the exposure time was 4 days. The bar equals 10 μm.

we can also measure relative concentrations in individual cells of the embryo, assuming that the target sequences are equally accessible to probe in all cells.

3. HISTONE TRANSCRIPTS IN SEA URCHIN EGGS

Early *in situ* hybridizations with nick-translated pCO2, which contains the entire early histone repeat unit from S. purpuratus, revealed signals over egg pronuclei much higher than over surrounding cytoplasm (32). This signal was sensitive to prehybridization digestion with RNase A, and several control probes (including nick-translated pBR322) did not bind to pronuclei.

Rough estimates based on multiplying grain densities over nuclei and cytoplasm by their relative volumes indicated that 10–15% of the total early histone mRNA in the egg was pronuclear. This calculation assumed similar hybridization and autoradiographic efficiencies for nuclear and cytoplasmic sequences. Since the probe contained not only histone genes but also spacer DNA sequences and the noncoding strand, it was not clear whether target pronuclear transcripts represented any or all of the histone mRNA sequences. Although the search for high molecular weight precursors of histone mRNAs in cleaving sea urchin embryos has yielded conflicting data (37–40), recent reports have led to the general conclusion that such molecules, if they exist, turn over too rapidly to accumulate to readily detectable levels (37). Since most studies have examined newly synthesized transcripts, there are little data available that characterize bulk maternal histone RNAs. Lifton and Kedes (41) compared the electrophoretic mobilities of histone mRNAs in eggs and cleaving embryos. Their results showed that egg RNAs only in the size range of mature mRNAs were able to compete with labeled embryonic histone mRNA for hybridization to DNA filters. These studies did not exclude the possibility that small, stable transcripts complementary to spacer DNA might exist, which could account for much of the pronuclear transcripts. More recently, Mauron et al. (37) used the more sensitive assay of titrating egg RNA with separated strands of several spacer or gene probes. Only low levels of weakly homologous hybridization could be detected between total egg RNA and parts of two spacers surrounding the H2B gene. Although these results suggested that at least some noncoding sequences in the early histone repeat were not represented in total egg RNA, we felt it was essential to establish more precisely the sequence content of pronuclear target RNAs. For these studies, we have used probes representing either strand of the entire repeat unit as well as nine subcloned regions of pCO2 that include each of the five genes and three different spacer DNA sequences.

The demonstration that the vast majority of pronuclear histone transcripts are derived from coding strand is shown in Figure 4.3. cRNAs synthesized from R7Δ7 recombinants carrying the early histone repeat unit in either orientation were separately hybridized to sections of eggs. The signal obtained with the pCO2R+ probe was at least 300 times greater than that obtained with the pCO2R− probe. The failure of the mRNA strand probe to bind to pronuclei proves that the target sequences are not DNA sequences. The marked asymmetry of these pronuclear target sequences implies that the signal is not due to cross-hybridization with random sequences related to the spacers which might be represented in other nuclear

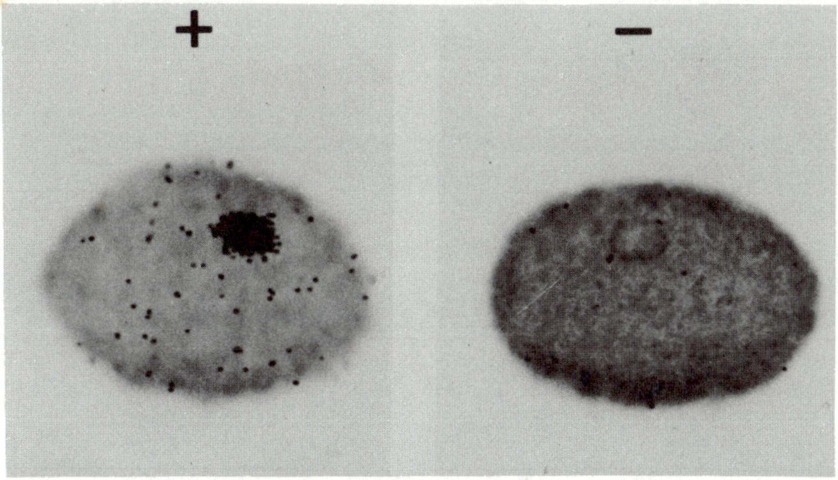

FIGURE 4.3. *In situ* hybridization of either pCO2R+ or pCO2R− transcripts to sections of unfertilized egg. The probe-specific activity was 9.4 × 10^6 dpm/μg, and the exposure time was 2 days. Courtesy of *Developmental Biology* **100**, 197–207 (1983). Copyright held by Academic Press.

RNAs. This is implied by the fact that in all maternal and nuclear RNAs where individual repeated sequences have been found on multiple transcripts, their representation is symmetric (42, 43). Thus asymmetric transcription of spacer sequences included on unrelated nuclear RNAs would require that all such sequences be situated in the same orientation relative to the direction of transcription.

The nine subclones derived from the early histone repeat unit used in this analysis are defined in the diagram at the bottom of Figure 4.4. For some hybridizations we have used nick-translated symmetric DNA probes (denoted by a D); for others, in which greater sensitivity was desired, we have synthesized asymmetric RNA probes from R7Δ7 recombinants (denoted by an R).

The *in situ* hybridizations shown in Figure 4.4 indicate that all five of the probes that contain sequences complementary to mature mRNAs, and only those probes, yield pronuclear signals; three spacer-specific probes (Figure 4.4: 3, 5, 8) gave hybridization levels indistinguishable from background levels although these hybridizations were carried out with RNA probes of maximum specific activity. Furthermore, the thermal stability of hybrids formed *in situ* between

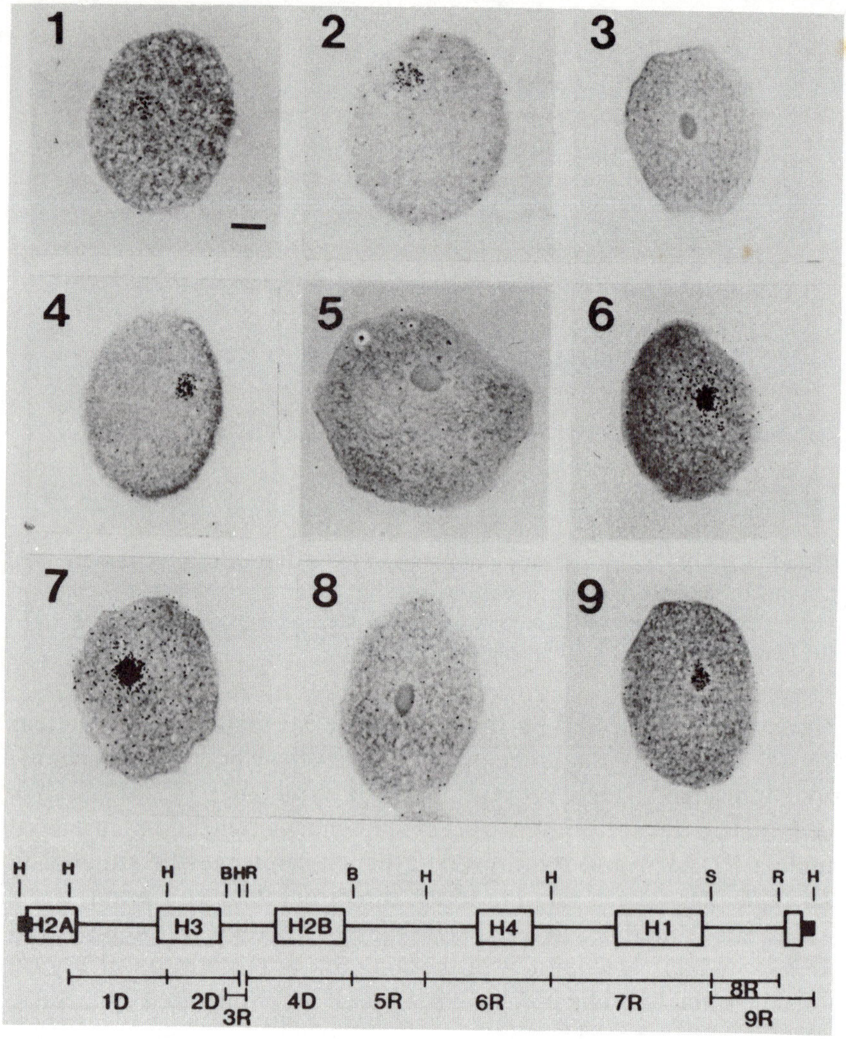

FIGURE 4.4. *In situ* hybridization of probes representing nine subregions of the early histone repeat unit. The sequence content of probes 1–9 is diagrammed in the restriction map at bottom. All DNA probe hybridizations (denoted by a Da) were done at apparent saturation as described previously (32). The specific activities of nick-translated probes 1, 2, and 4 were 9.3×10^7 dpm/μg, 7×10^7 dpm/μg, and 3.5×10^7 dpm/μg, respectively, and the exposure time was 5 days. All RNA probe hybridizations (denoted by an R) were also done at apparent saturation as described in Hughes et al. (44). The specific activity of all RNA probes was 1.4×10^8 dpm/μg, and the exposure time was 4 days. Courtesy of *Developmental Biology* **100**,197–206 (1983). Copyright held by Academic Press.

pronuclear RNAs and probe RNA is identical to that of hybrids between the same probe and cytoplasmic mRNAs of either 2-cell or 12-hour embryos (49). This result indicates that, as in these last two cases, the target sequences are authentic transcripts of the early histone repeat unit and not weakly homologous RNAs that might be found in eggs and cleaving embryos (37, 38). Finally, the strongest argument that the bulk of pronuclear target sequences corresponds to mRNA length molecules derives from the combination of *in situ* and northern blot analyses. When northern blots of total egg RNA are hybridized with RNA representing the entire early repeat unit and then briefly digested with RNase under conditions (1 μg/ml, 15 minutes, 25°C) even milder than those used for *in situ* experiments (20 μg/ml, 30 minutes, 37°C), only mature mRNAs are observed (49). Since a high proportion of total histone transcripts is pronuclear, then it follows that the bulk of pronuclear histone RNAs are primarily the length of mature mRNAs.

Since no estimates are currently available for hybridization and autoradiographic efficiencies to nuclear targets, we have made indirect estimates of the fraction of histone mRNA present in pronuclei from grain densities over egg and 2-cell cytoplasm. For this determination, we assume equal efficiency of hybridization in both cases [about 90%, (44)]. The pronuclear content of histone RNA transcripts decreases dramatically around first cleavage (see later discussion illustrated by Figure 4.6). Thus the increase in cytoplasmic grains between egg and 2-cell can be attributed to sequences in the pronucleus since there is no change in total histone content or cytoplasmic volume between these two stages (18, 21). The data from several such comparisons are given in Table 4.1. When comparisons are made between the cytoplasmic grain densities of egg sections containing a pronucleus and 2-cell sections, all measurements but one indicate that ≥50% of histone mRNAs are localized in the pronucleus. The accuracy of these calculations depends critically upon the assignment of pronuclear borders, which is difficult under the conditions of extremely high pronuclear labeling necessary to measure a significant number of cytoplasmic grains. Most cytoplasmic grains are found near the nucleus and can be accounted for by a small fraction of decays from a nuclear-bound probe (45). Egg–two-cell comparisons made using nonnucleated egg sections lead to much higher estimates of pronuclear histone

TABLE 4.1. Indirect Estimates of Pronuclear Histone mRNA Content

		Nucleated Sections				Nonnucleated Sections			
		Grains/100 μ^2			Fraction[a]	Grains/100 μ^2			Fraction[a]
Probe	Expt.	2-cell	Egg	2-cell/egg	Sequestered	2-cell	Egg	2-cell/egg	Sequestered
H1	1	11	4	2.6	0.6	11	0.5	22	0.95
H1	2	12	8	1.5	0.3				
H4	1	19	4	4.7	0.8	19	−0.1	—[b]	1.0
H4	2	16	5	3.3	0.7				
H2A	1	3.6	1.7	2.1	0.5	N.D.	N.D.	—	—
Whole Repeat	2	68	7.5	8.3	0.9	68	1.1	62	0.98

[a] R-1/R where R is the ratio of 2-cell to egg cytoplasmic grain density.

[b] This ratio is large but cannot be precisely determined, since the signal over egg cytoplasm was not distinguishable from background. From DeLeon et al., 1983 (49).

mRNA content (95–100%), since essentially no signals are observed over the cytoplasm of such egg sections. We conclude that the simplest explanation of all these data is that most, if not virtually all, of the early histone mRNAs found in eggs are confined to their pronuclei.

Recently, Showman et al. (24) have used quite a different approach to measure the subcellular distribution of histone mRNA in sea urchin eggs. They isolated total RNA from nucleate or anucleate halves or quarters of centrifugally stratified eggs and quantitated the amount of several different histone mRNAs by northern blot analysis using cloned probes. These experiments demonstrated that more than 98% of histone mRNA fractionated with nucleated egg fragments while bulk mRNA and mRNA coding for actin and tubulin were distributed in proportion to the cytoplasmic volume of each fragment (Table 4.2). Recently, similar results have been obtained using surgically dissected egg fragments. Showman et al. have concluded that most histone mRNA is either within or associated with pronuclei.

That a substantial fraction of early variant histone mRNA could be pronuclear can also be inferred from experiments carried out in eggs of *Arbacia punctulata* by Arceci (45). He observed that in artificially activated enucleate egg fragments the amount of histone

TABLE 4.2. Distribution of RNAs in Centrifugally Split Egg Fragments[a]

	Whole Egg	Anucleate Half	Nucleate Half	Anucleate Quarter	Nucleate Quarter
Relative volume	1.0	0.39	0.61	0.21	0.40
Total RNA	1.0	0.33	0.67	0.38	0.29
Poly(A)$^+$	1.0	0.31	0.69	0.39	0.30
Actin mRNA	1.0	0.33	0.67	(0.29)	(0.38)
α-Tubulin mRNA	1.0	0.30	0.70	0.46	0.24
Histone mRNAs	1.0	0.02	0.98	0.01	0.98
Mitochondrial RNA	1.0	0.77	0.23	0.23	0.005

[a] Egg and fragment volumes calculated using optical measurements of diameter of cells. Total RNA amounts were determined by absorbance at 260 nm. Poly (A)$^+$ RNA values were obtained by hybridization with ^{3}H-polyuridylate. Histone mRNA probes used: H1, H2B, H3. Specific RNA measurements were done using ^{32}P nick-translated sequence specific probes to measure the amount of each RNA present. All values are expressed as fraction of total per egg and are the average of two or more separate experiments. The values in parentheses were determined from a single experiment. From Showman et al., 1982 (24).

synthesis was significantly depressed as compared to that in nucleated controls.

Several laboratories, including ours, have attempted to demonstrate the subcellular distribution of histone mRNA by cell fractionation procedures (15, 16, 20, 24). So far, an enrichment for histone RNAs in isolated pronuclei has not been obtained. In light of all the results cited, we believe that leakage of these RNAs from the extremely fragile pronucleus occurs during standard nuclear isolation procedures. Association of histone mRNAs only with the nuclear membrane can be ruled out since this would lead to peripheral grain distributions around nuclei cut at the upper surface of sections. This distribution has never been observed in many thousands of sections.

The accumulation of histone mRNAs in the pronucleus apparently does not begin until a time during or after maturation. Figure 4.5 shows *in situ* hybridizations of RNA probes representing the entire early histone repeat unit to sections of a full-grown oocyte (A) and a mature egg (B). Our preliminary estimates indicate that, after correction for the difference in volume between the germinal vesicle and the egg pronucleus, the difference in histone mRNA content is about tenfold.

The high pronuclear histone mRNA content persists only until first cleavage, as shown by the time course in Figure 4.6. At that time

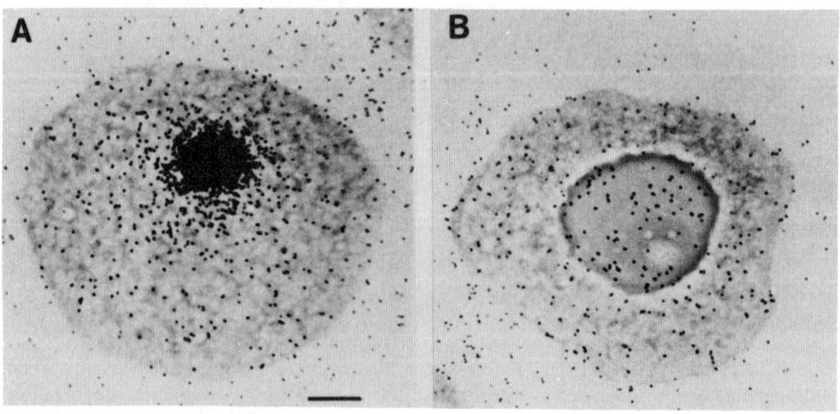

FIGURE 4.5. *In situ* hybridization of pCO2R+ transcripts to sections of (A) unfertilized egg and (B) vitellogenic oocyte. The probe-specific activity was 1.7×10^8 dpm/μg, and the exposure time was 5 days. The bar corresponds to 10 μm.

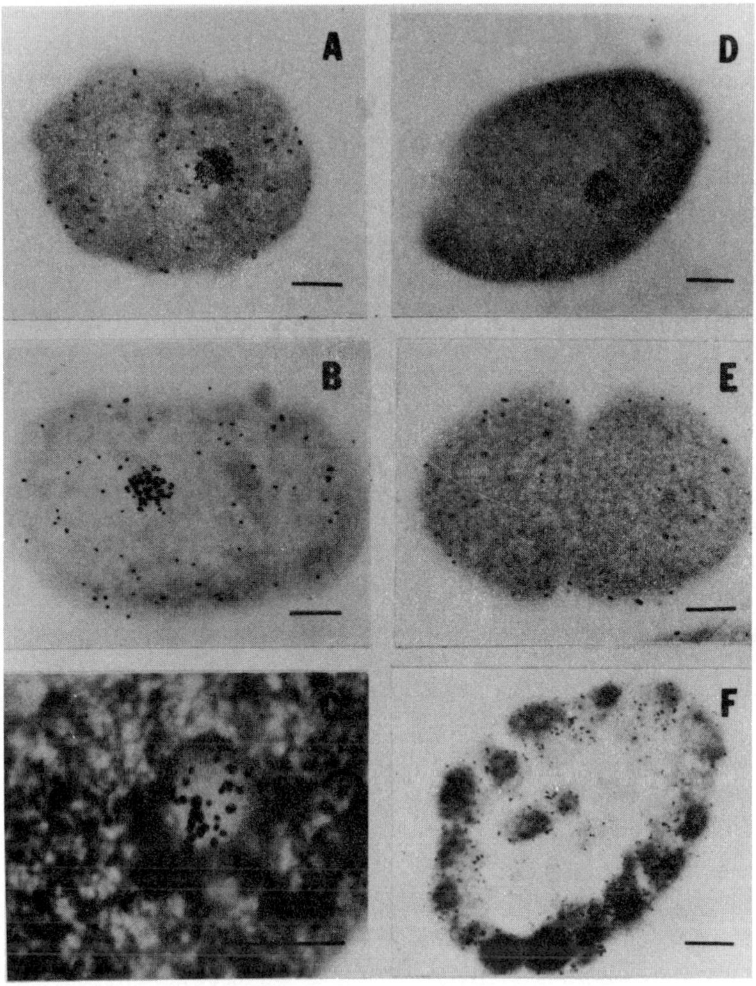

FIGURE 4.6. Disappearance of nuclear histone repeat transcripts at first cleavage. Nick-translated pCO2 DNA ($5.5–7.5 \times 10^7$ dpm/μg) was hybridized to sections of (A) unfertilized egg, (B) and (C) zygote, 15 minutes after fertilization near the time of pronuclear fusion; the arrow points to the male pronucleus, (D) zygote, 60 minutes after fertilization, (E) 2-cell embryo, approximately 30 minutes after first cleavage, (F) 9-hour embryo. Bars = 10 μm. (Courtesy of *Cell* **24**, 385–91 (1981); copyright held by M.I.T.)

these mRNAs are either released to the cytoplasm or rapidly degraded. Because of the reciprocal and simultaneous increase in two-cell cytoplasmic grain density (49), we believe that the first alternative is much more likely. Figure 4.6, F also demonstrates that even at a time when transcription rates from the early histone genes are maximal [9-hour embryos, (18)], high nuclear concentrations of histone mRNAs are not observed. The enormous quantity of histone mRNA in pronuclei (1.3 pg) requires that these transcripts have a very long lifetime. Even if the synthesis rate were as high as in midcleavage [6.25×10^{-4} pg/min·nucleus, (18)], pronuclear concentration would require that the $t_{1/2}$ be at least 100 hours, if indeed all histone mRNA in the egg is pronuclear. If these sequences constitute the pool of early variant mRNAs translated in early cleavage, which seems inevitable, then large changes in the export rate must occur after first cleavage.

A similar accumulation in the pronucleus is not observed for other abundant maternal RNAs. About one-half of maternal RNA molecules are polyadenylated (43, 47), and the distribution of these molecules in unfertilized eggs is revealed by hybridization with [^{3}H]poly U as shown in Figure 4.7. Analysis of many such sections led to the conclusion that levels of polyadenylated RNA in pronuclei are below our limits of detection (30). Similar results have been obtained using a cDNA clone representing a highly abundant egg mRNA (32). Using northern blot analysis of fragments of stratified eggs, Showman et al. (24) have demonstrated that both tubulin and actin mRNAs are not enriched in nucleate egg fragments (see Table 4.2).

Recently Wells et al. (21) have demonstrated that the kinetics of the loading of several early histone mRNAs on polysomes after fertilization are distinct from those for loading of bulk maternal mRNA. They isolated polysomal RNA at various times after fertilization and assayed the concentration of several histone mRNAs by northern blotting. An example of the time course of loading of H3 mRNA and bulk mRNA is shown in Figure 4.8. Bulk mRNA begins to enter polysomes soon after fertilization and gradually loads until about 30% of the mass is being translated after 1.5 hours. In contrast, recruitment of H3 mRNA is delayed until after first cleavage (about 90 minutes after fertilization), after which loading is virtually quantitative. These results agree with those of Woods and Fitschen (20) who used histone cDNA to demonstrate a shift of histone mRNA

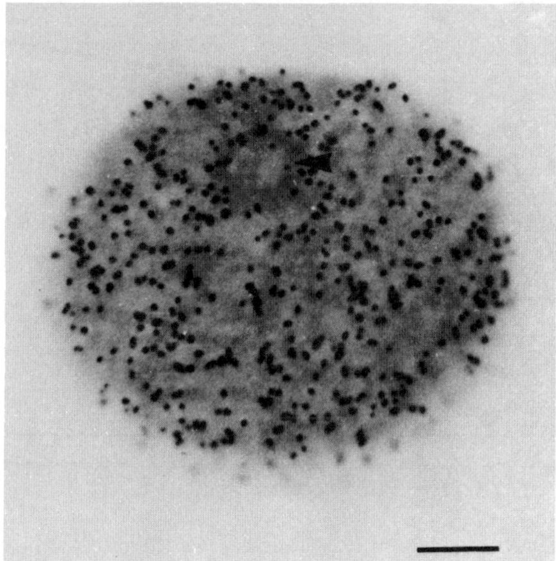

FIGURE 4.7. *In situ* hybridization of ^{3}H poly U (1.7×10^7 dpm/μg) to a section of an unfertilized egg. The exposure time was 12 days. The bar equals 10 μm, and the arrow points to the pronucleus.

into the polysomal compartment at about first cleavage. The differences in translation bear an obvious and striking relationship to the localization of these mRNAs in the egg: The delay in loading until after first cleavage is observed only for early variant histone mRNAs, and only these mRNAs appear to be localized in the egg pronucleus. The quantitative loading of histone mRNAs implies that all of these molecules have equal accessibility to the translation apparatus after first cleavage. All of these results are most simply explained by a mechanism in which the selective control of translation of early variant histone mRNAs is achieved by their sequestration in egg pronuclei and release to the cytoplasm at nuclear membrane breakdown.

The function of such a specialized mechanism for controlling the expression of early histone variants is not currently understood. In this regard, it is of particular interest that Raff and co-workers have found by assaying *in vivo* protein synthesis that mRNAs coding for CS histone variants distribute in proportion to cytoplasmic volume

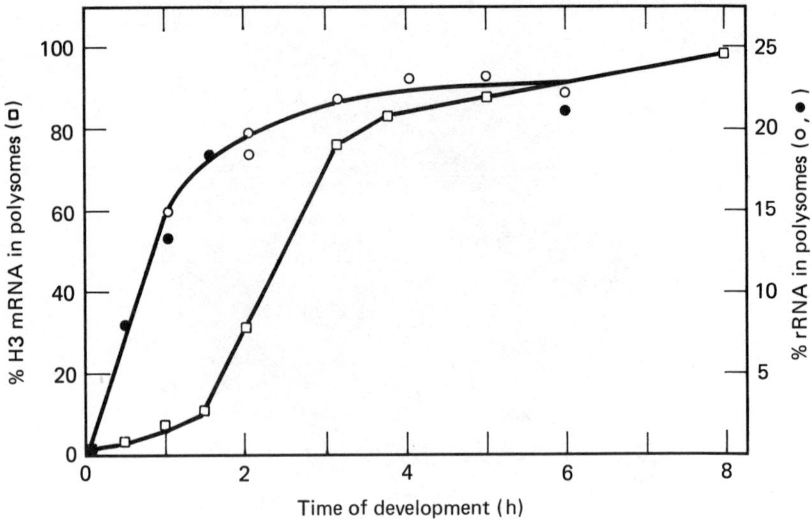

FIGURE 4.8. Time of loading of rRNA and histone H3 mRNA on polysomes of *S. purpuratus* embryos. The postmitochondrial supernatant was sedimented through 5–40% exponential sucrose gradients. RNA from polysomal and nonpolysomal fractions was purified, electrophoresed through formaldehyde gels, transferred to nitrocellulose, and probed with nick-translated recombinant DNAs containing either the genes for histone H3 (□) or rRNA (○). The amount of specific RNA in each fraction was determined by densitometric analysis of autoradiographs. The amount of rRNA in polysomes was also measured by absorbance (●). Further details are provided in (21). Reprinted by permission from *Nature*, Vol. 292, pp. 477–478, Copyright (c) 1981, Macmillan Journals Limited.

of egg fragments, rather than with nuclei (Raff, personal communication). These maternal RNAs are translated in the zygote and during the next several cleavages, and the CS variant proteins both comprise the histone complement of egg pronuclei and displace sperm histones after fertilization (48). It is thus possible that restriction to the pronucleus is a mechanism that prevents "premature" translation of early variant histone proteins and their interference with the initial restructuring of sperm chromatin. The difference in regulation of CS and early variant histone mRNA translation does not necessarily require that the egg pronucleus discriminate between these mRNAs for transport to the cytoplasm although this may be the case. The time of synthesis of CS mRNAs during oogenesis is not yet known.

Thus it is possible that CS genes are transcribed and the mRNAs transported to the cytoplasm before maturation, whereas early variant mRNA transcriptions is initiated only after maturation. Such a hypothesis is consistent with our preliminary observations of very low signals over germinal vesicles hybridized with the early variant histone probe. It is important to note here that there should be no cross-hybridization with our probes and CS mRNAs since these mRNAs are highly divergent (E. Weinberg, personal communication), and the thermal stability of *in situ* hybrids is high (44, 49).

4. *IN SITU* HYBRIDIZATION AS A TOOL FOR STUDYING THE EARLY-LATE HISTONE GENE TRANSCRIPTIONAL SWITCH IN DEVELOPING SEA URCHIN EMBRYOS

The primary utility of the *in situ* hybridization approach is to determine spatial patterns of mRNA expression in individual cells of complex tissues or differentiating systems. This kind of information may be particularly useful in elucidating the function of individual variants in multigene families. As discussed in the introduction, understanding the mechanisms by which the transcriptional switch in expression from the early to the late set is achieved requires knowing whether all cells behave synchronously in this respect. Although early measurements suggested that the transcriptional switch occurred during a rather narrow time interval, recent studies indicate that it is more gradual, since newly synthesized late variant mRNAs can be detected as early as 16-cell stage (4 hours), and early variant mRNAs persist as late as 40 hours (26). Since these studies have been carried out on whole embryos, one plausible explanation for the gradual shift is that different cells make the switch at different times during embryogenesis.

While it is likely that expression of either gene set is not all or none in any cell, quantitative variations in relative expression among cells may exist. There are several factors that could control the timing of this switch. For example, it could occur after a defined number of cell divisions, or be related to the rate of cell division. Both of these parameters differ for different cell lineages and would therefore lead to asynchrony among different cells in the timing of the switch.

Another possibility is that cells switch only when they cease to divide. In this case, a division might yield one daughter cell that continues to divide and produce primarily early histone mRNA and another that ceases to divide and makes predominantly late-variant mRNA. Since the number of nondividing cells increases during development, the effect would be a progressive shift. This interpretation has been suggested by L. Cohen (personal communication) who recently has shown that there is a significant increase in the number of nondividing cells just at the time when the early-late histone transition is maximal. A quite different alternative is that all cells of the embryo may regulate the early and late genes identically and simultaneously. The changing levels of expression of the two sets of genes may then depend on progressive changes in transcription and/or post-transcriptional processing, turnover, and transport rates.

To distinguish among these various possibilities, it will be necessary to measure the concentrations of early and late mRNAs in individual cells of developing embryos. For this purpose, we have developed *in situ* hybridization procedures that allow accurate quantitation of target RNA concentrations in different cells and discrmination between early and late mRNAs through careful adjustment of hybridization conditions. This can be done by exploiting the sequence divergence between these variants. Since solution *Tm* differences between early-early and early-late hybrids (11–13°C) are also observed for duplexes formed *in situ* (44), we can select appropriate discriminating conditions. Thus, by hybridization at permissive criterion and washes at selective criterion, all heterologous hybrids can be dissociated while 80% of homologous hybrids are retained. The other approach, of course, is to utilize probes representing subregions of early and late genes which do not cross-hybridize even at moderate strigency as, for example, the 3′ untranslated sequences. Since such sequences are likely to be quite short, some loss in sensitivity will result.

Preliminary results from *in situ* hybridizations at permissive criterion of the early variant repeat probe to sections of various stage embryos suggest that total histone mRNA concentration (i.e., early and late variant mRNAs) is not uniform in all regions of the embryo. Thus at 12 hours during the peak of early variant gene expression, all cells of the embryo appear to be uniformly heavily labeled (see Figure 4.2), while in the blastula-gastrula period, clusters of labeled or

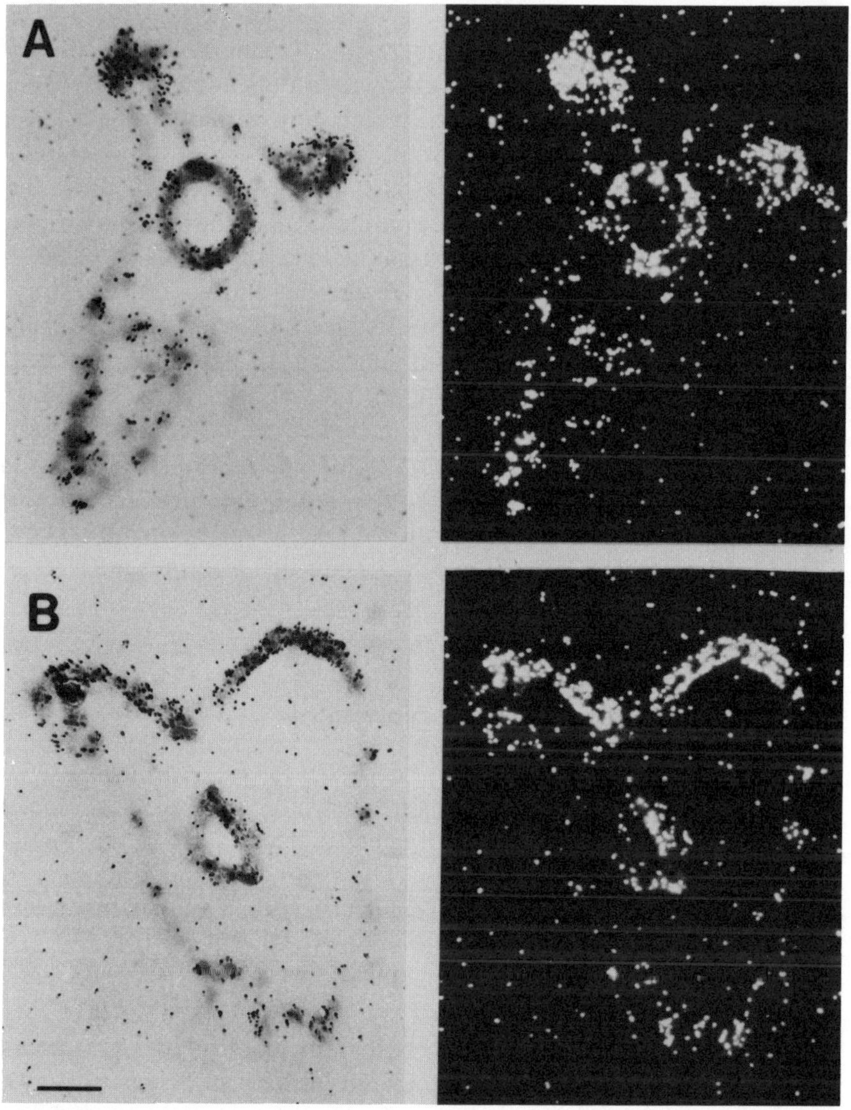

FIGURE 4.9. *In situ* hybridization of pCO2R+ to sections of 68-hour plutei. Hybridization was carried out at Tm-27°C for early-late hybrids. The probe specific activity was 1.8 × 10^7 dpm/μg, and the exposure time was 28 days. (A) Bright-field and dark-field photographs of a section showing heavily labeled anal arms and gut and patches of label in the dorsal ectoderm. (B) Bright-field and dark-field photographs of a section illustrating heavy labeling in the cytoplasm of ventral ectoderm and much lower grain densities in dorsal ectoderm. Bar = 10 μm.

unlabeled cells are observed. One interesting, but unproven, possibility is that this reflects a shift from cell-cycle independent to cell-cycle coupled levels of histone mRNAs, which may or may not be related to the early-late switch. An example of the distribution at pluteus stage, after the switch is complete, is shown in Figure 4.9. Both the magnitude of the signal and the thermal stability of the hybrids formed indicate that the targets are late variant histone mRNAs (44). Clearly, the grain density is much higher over the gut and ventral side (oral face) than in dorsal ectoderm. Studies by L. Cohen and our measurements of the number of dorsal ectoderm cells [marked by their high content of a specific mRNA (35)] indicate that there is little or no cell division in the presumptive dorsal ectoderm cells after mid-blastula. The pattern of distribution of histone mRNAs at pluteus is distinct from that of both polyadenylated and the dorsal ectoderm-specific mRNAs (35), implying that the histone mRNA pattern is not a result of a general difference in accessibility of target mRNAs in these various cell types.

5. CONCLUSIONS AND FUTURE DIRECTIONS

The *in situ* hybridization techniques are now sufficiently well developed to allow measurement of levels of specific RNAs in individual cells of developing sea urchin embryos. The specificity and efficiency of these reactions are good, and the sensitivity is such that we can detect moderately abundant mRNAs that comprise 0.05% of total mRNA nucleotides, and less abundant species if they are localized to subregions of the embryo (35, 44)

This approach led to the discovery that the majority of early variant histone mRNA transcripts are localized in the pronucleus (32, 49). This result was surprising since standard cell fractionation techniques failed to detect such transcripts (15, 16, 20, 24, and our unpublished observations). A considerable amount of circumstantial evidence, namely, the correlation between the loss of pronuclear histone mRNA and the simultaneous increase in its loading on polysomes as well as the fact that this phenomenon appears to be specific for early histone variant messages, suggests that the egg may utilize nuclear storage as one way to regulate selectively the translation of these maternal mRNAs.

Our initial measurements show that the specificity of hybridization *in situ* may be adjusted to allow discrimination between related but divergent mRNAs. This will allow further studies of the expression of different sets of histone genes in individual cells and cell lineages, and the relationship of this switch to other events in the embryo, such as changes in division rates, and differentiation. Such studies will clearly best be carried out by utilizing probes specific for different early and late variant mRNAs. Although localization of an individual RNA is relatively straightforward if that RNA is restricted to a morphologically distinguishable cell type or region of the embryo, this frequently may not be the case, especially at earlier stages of development. If, for example, separate *in situ* hybridizations of early and late probes to sections of midblastulae show hybridization to small clusters of cells, it would be difficult to determine whether individual cells contain one, both, or neither variant mRNA. Such comparisons of the distribution of two different mRNA targets would be greatly facilitated by the development of a double-label technique so that patterns could be directly compared in the same section. This might be achieved by using radiolabeled and chemically modified probes which could be detected sequentially by fluorescent antibody or cytochemical techniques and autoradiography.

ACKNOWLEDGMENTS

We wish to thank Drs. R. Raff and R. Showman for many useful discussions of their work prior to publication and for permission to use some of this data in this article (Figure 4.8 and Table 4.2). Our research was supported by grants from the National Institutes of Health (GM25553) and by an Institutional Biomedical Research Support Grant from the U.S. Public Health Service.

REFERENCES

1. L. H. Kedes, *Ann. Rev. Biochem.* **48,** 837–870 (1979).
2. E. Robbins and T. W. Borun, *Proc. Natl. Acad. Sci. USA* **57,** 409–416 (1967).

3. R. Moll and E. Wintersberger, *Proc. Natl. Acad. Sci. USA* **73,** 1863–1867 (1976).
4. D. M. Prescott, *J. Cell Biol.* **31,** 1–9 (1966).
5. G. S. Stein and T. W. Borun, *J. Cell Biol.* **52,** 292–307 (1972).
6. M. Melli, G. Spinelli, and E. Arnold, *Cell* **12,** 168–174 (1977).
7. L. H. Kedes, P. R. Gross, G. Cognetti, and A. L. Hunter, *J. Mol. Biol.* **45,** 337–351 (1969).
8. G. Stein, J. Stein, E. Shepard, W. Park, and I. Phillips, *Biochem. Biophys. Res. Commun.* **77,** 245–252 (1977).
9. J. V. Ruderman and P. R. Gross, *Dev. Biol.* **36,** 286–298 (1974).
10. S. Detke, A. Lichtler, I. Phillips, J. Stein, and G. Stein, *Proc. Natl. Acad. Sci. USA* **76,** 4995–4999 (1979).
11. L. H. Kedes and P. R. Gross, *Nature* **223,** 1335–1339 (1969).
12. L. M. Hereford and M. A. Osley, *Cell* **24,** 367–375 (1981).
13. M. A. Osley and L. M. Hereford, *Cell* **24,** 377–384 (1981).
14. H. R. Woodland, *FEBS Lett.* **121,** 1–7 (1980).
15. A. Skoultchi and P. R. Gross, *Proc. Natl. Acad. Sci. USA* **70,** 2840–2844 (1973).
16. K. W. Gross, M. Jacobs-Lorena, C. Baglioni, and P. R. Gross, *Proc. Natl. Acad. Sci. USA* **70,** 2614–2618 (1973).
17. T. D. Sargent and R. A. Raff, *Dev. Biol.* **48,** 327–335 (1976).
18. R. E. Maxson and F. H. Wilt, *Dev. Biol.* **83,** 380–386 (1981).
19. B. Moav and M. Nemer, *Biochem.* **10,** 881–888 (1971).
20. D. E. Woods and W. Fitschen, *Cell Diff.* **7,** 103–114 (1978).
21. D. E. Wells, R. M. Showman, W. H. Klein, and R. A. Raff, *Nature* **292,** 477–478 (1981).
22. R. A. Raff and R. M. Showman, in C. B. Metz and A. Monroy, Eds. *The Biology of Fertilization*, Academic Press, in press, 1983.
23. E. H. Davidson, *Gene Activity in Early Development*, Academic, New York, (1976), pp. 94–111.
24. R. M. Showman, D. E. Wells, J. Anstrom, D. A. Hursh, and R. A. Raff, *Proc. Natl. Acad. Sci. USA*, **79,** 5944–5947 (1982).
25. K. M. Newrock, L. H. Cohen, M. B. Hendricks, R. F. Donnelly, and E. S. Weinberg, *Cell* **14,** 327–336, (1978).
26. G. Childs, R. Maxson, and L. H. Kedes, *Dev. Biol.* **73,** 153–173 (1979).
27. P. A. Hieter, M. B. Hendricks, K. Hemminki, and E. S. Weinberg, *Biochemistry* **18,** 2707–2716 (1979).
28. C. C. Hentschel and M. L. Birnstiel, *Cell* **25,** 301 (1980).
29. R. E. Maxson, T. J. Mohun, L. Gormezano, G. Childs, and L. H. Kedes, *Nature* **301,** 120–125 (1983).

30. L. M. Angerer and R. C. Angerer, *Nucleic Acids Res.* **9,** 2819–2840 (1981).
31. C. Godard and J. W. Jones, *Nucleic Acids Res.* **6,** 2849–2861 (1979).
32. D. L. Venezky, L. M. Angerer, and R. C. Angerer, *Cell* **24,** 385–391 (1981).
33. C. Overton and E. S. Weinberg, *Cell* **14,** 247–258 (1978).
34. M. Brahic and A. T. Haase, *Proc. Natl. Acad. Sci. USA* **75,** 6125–6129 (1978).
35. D. A. Lynn, L. M. Angerer, A. M. Bruskin, W. H. Klein, and R. C. Angerer, *Proc. Natl. Acad. Sci. USA*, **80,** 2656–2600 (1983).
36. A. Mauron, L. H. Kedes, B. R. Hough-Evans, and E. H. Davidson, *Dev. Biol.*, in press (1982).
37. A. Mauron, S. Levy, G. Childs, and L. H. Kedes, *Mol. Cell. Biol.* **1,** 661–671 (1981).
38. G. Childs, S. Levy, and L. H. Kedes, *Biochemistry,* **18,** 208–213 (1979).
39. N. S. Kunkel, K. Hemminki, and E. S. Weinberg, *Biochemistry* **17,** 2591–2598 (1978).
40. G. M. Spinelli, M. Melli, E. Arnold, F. Casano, F. Gianguzza, and M. Ciacco, *J. Mol. Biol.* **139,** 111–122 (1980).
41. R. P. Lifton and L. H. Kedes *Dev. Biol.* **48,** 47–55 (1976).
42. R. H. Scheller, F. D. Costantini, M. R. Kozlowski, R. J. Britten, and E. H. Davidson, *Cell* **15,** 189–203 (1978).
43. F. D. Costantini, R. J. Britten, and E. H. Davidson, *Nature* **287,** 111–117 (1980).
44. K. H. Cox, D. V. Deleon, L. M. Angerer, and R. C. Angerer, *Dev. Biol.*, in press (1984).
45. M. M. Salpeter, L. Backmann, and E. E. Salpeter, *J. Cell. Biol.* **41,** 1 (1969).
46. R. J. Arceci, Ph.D. Thesis, University of Rochester, Rochester, NY, 1980.
47. F. H. Wilt, *Cell* **11,** 673–681 (1977).
48. D. Poccia, J. Salik, and G. Krystal, *Dev. Biol.* **82,** 287–296 (1981).
49. D. V. DeLeon, K. H. Cox, L. M. Angerer, and R. C. Angerer, *Dev. Biol.* **100,** 197–207 (1983).

CHAPTER 5

HISTONE GENE EXPRESSION IN *DROSOPHILA* DEVELOPMENT: MULTIPLE LEVELS OF GENE REGULATION

KATHRYN V. ANDERSON AND JUDITH A. LENGYEL
Department of Biology
and Molecular Biology Institute
University of California at Los Angeles
Los Angeles, California

CONTENTS

1. INTRODUCTION
2. DNA REPLICATION IN THE *DROSOPHILA* EMBRYO: THE REQUIREMENT FOR HISTONES
3. CHANGES IN HISTONE mRNA CONCENTRATION DURING EMBRYOGENESIS: LEVELS OF CONTROL
4. RELATIVE CONTRIBUTIONS OF MATERNAL mRNA AND CHANGING RATES OF SYNTHESIS, TURNOVER, AND TRANSLATION TO THE CHANGING LEVEL OF HISTONE GENE EXPRESSION
5. MECHANISMS OF CONTROL: THE DECREASE IN MESSAGE STABILITY
6. MECHANISMS OF CONTROL: TRANSCRIPTION AND GENOMIC ORGANIZATION
7. CONCLUSION

REFERENCES

1. INTRODUCTION

The detailed study of eukaryotic gene regulation, made possible by the tools of recombinant DNA technology, reveals that in many cases gene expression is regulated at multiple levels. Although transcriptional control may be the most general level of eukaryotic gene regulation, it has become clear as more systems have been studied in detail that mechanisms including DNA rearrangement, alternative RNA splicing, changing message stability, and translational control also provide important levels of regulation (for review, Reference 8). To understand how the expression of a particular gene is controlled it is therefore necessary to define quantitatively which levels of control are utilized in the regulation of that gene. This phenomenological description is a necessary prerequisite to any analysis of the mechanism of regulation. Such an analysis requires, in addition, the ability to experimentally manipulate the individual components of the regulatory system either biochemically or genetically.

The high concentration of histone mRNA in early embryos has made the study of embryonic histone gene regulation accessible to more detailed investigation than any other genes expressed in the early embryo. Even with the advent of recombinant DNA technology, which has increased the level of detectibility of specific RNA sequences by orders of magnitude, measurements of rates of transcription, processing, translation, and turnover can be made more accurately for histone mRNAs than for other less abundant RNAs.

Since histones and DNA are required in equal mass in functional chromatin, the rapid rate of DNA replication in early embryos probably places the most stringent demands on the expression of histone genes during the life cycle of higher organisms. Histone gene expression in early embryos is one of the few systems in which gene expression can be quantitatively correlated with an absolute demand for the gene product and where the demand for the gene product changes rapidly as a function of time. How are histone genes regulated during embryogenesis to supply the well-defined, and rapidly changing, amount of histone protein? Are multiple levels of gene regulation employed, and, if so, how are they integrated? Is histone gene expression directly linked to the rate of DNA replication?

The answers to these questions require first the analysis of the levels of regulation of histone gene expression in normal development. In addition, it is necessary to be able to experimentally perturb the components of the regulatory system. *Drosophila* is unique among higher eukaryotes in the relative facility with which defined genetic lesions can be produced and studied *in vivo*. In *Drosophila*, it is potentially possible to disrupt genetically single elements in the regulation of expression of histone genes. Such perturbations could allow a more detailed understanding of the mechanisms of regulation, of how multiple levels of gene regulation are integrated with one another, and of how histone gene expression is coordinated with DNA replication.

2. DNA REPLICATION IN THE *DROSOPHILA* EMBRYO: THE REQUIREMENT FOR HISTONES

The *Drosophila* embryo goes through one of the most rapid cycles of DNA replication of any higher eukaryote, proceeding from a single nucleus to 6000 nuclei as the result of 13 synchronous nuclear divisions in the first 2.5 hours after oviposition (46). During the first nine nuclear divisions, the entire cell cycle time is only 9 min, allowing for little, if any, transcription. In this period, therefore, the embryo is dependent on maternal stores for all of the materials needed to support DNA replication, including histones. At approximately 1.5 hours after oviposition, the nuclei migrate from the central yolky region of the egg to the cortical cytoplasm, at which time their rate of doubling slows from once every 9 min to once every 15 min, and they divide four more times in the next hour. Nuclear divisions then pause for about one hour, during the stage of nuclear elongation and cellularization at blastoderm. Gastrulation begins at 3.5 hours after oviposition, and shortly thereafter mitoses are seen again in all three germ layers (22). Mesodermal cells show the greatest mitotic activity: after a 0.5-hour treatment with colchicine, every mesodermal cell is blocked in mitosis. In the 8-hour-old embryo, mitoses have ceased in the mesoderm, and by 12 hours mitoses have ceased in all cell types but cephalic neuroblasts (22).

These histological observations on DNA synthesis agree with, and have been extended by, biochemical measurements of the rate of DNA synthesis at specific embryonic stages. Synchronously developing embryos were pulse labeled for one hour with ^{14}C-thymidine, and the total counts per minute (cpm) incorporated into DNA as well as the ^{14}C-TTP pool-specific activity during the pulse were determined in order to calculate the absolute rate of DNA synthesis at various times during embryogenesis (4). As shown in Figure 5.1, the measured rate of DNA synthesis during early cleavage stages corresponds well with the rapid nuclear doubling time observed histologically: the biochemical measurements give an average nuclear doubling time of once every 18 min between 1 and 3 hours after oviposition. During gastrulation, the rate of nuclear doubling slows to an average of once per 70–80 min for all cell types. Because the nuclei are increasing in number exponentially, however, the greatest rate of increase in the mass of DNA per embryo occurs during late gastrulation stages, reaching a peak at 6 hours after oviposition. After 6 hours the average rate of DNA replication per nucleus slows dramatically. By 10 hours after oviposition the average doubling time is only once per 11 hours. This decrease in the average rate of nuclear DNA doubling must reflect the decrease in the number of cells undergoing mitosis as well as an overall slowing in the rate of DNA replication per cell.

As an aid to understanding the requirement for changing histone gene expression during embryogenesis, it is useful to arbitrarily divide the time course of DNA replication into three phases. The first phase, the first 2.5 hours after oviposition, is a time of extremely rapid nuclear division during which the embryo depends almost exclusively on maternal stores of histone protein or histone mRNA that is translated immediately after oviposition. The second phase of DNA synthesis occurs during gastrulation and germ band extension. During this phase, the rate of nuclear doubling is slower than during cleavage, but the absolute mass of DNA synthesized per hour is the greatest, and therefore the greatest mass of histone protein is required. In the third phase, after 6 hours, the rate of DNA synthesis, and therefore the need for new histone protein, fall extremely rapidly. How are the rapidly changing requirements for histones met during early embryogenesis?

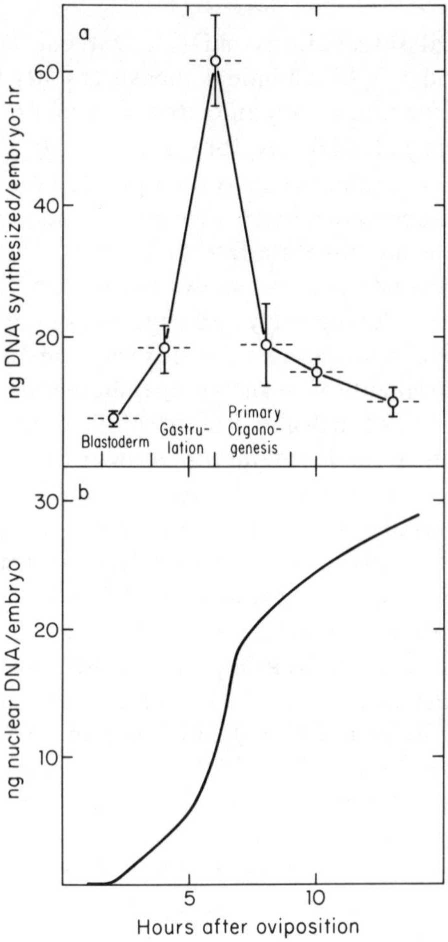

FIGURE 5.1. DNA synthesis during embryogenesis. (*a*) The amount of DNA synthesized per embryo per hour was determined by pulse labeling synchronously developing embryos for 1 hour with ^{14}C-TTP. The pmoles of TTP incorporated were calculated by dividing the ^{14}C incorporated into nuclear DNA by the average specific activity of the ^{14}C-TTP pool during the labeling period (*b*) The total amount of DNA per embryo as a function of developmental stage was calculated by integrating the data shown in (*a*) (4).

3. CHANGES IN HISTONE mRNA CONCENTRATION DURING EMBRYOGENESIS: LEVELS OF CONTROL

Studies of macromolecular synthesis in the *Drosophila* embryo have lagged behind similar studies in other invertebrates such as the sea urchin because the *Drosophila* egg is protected from the external environment by a highly impermeable shell, making it inaccessible to radioactive precursors. This difficulty has been partially eliminated by the discovery of Limbourg and Zalokar (1973) that a brief treatment of dechorionated embryos with 100% octane dissolves enough of the waxy coat on the outside of the vitelline membrane to allow the uptake of small molecules without affecting the course of normal development (21). This has opened the way for the study of RNA and protein synthesis in large numbers of synchronously developing embryos (2, 33, 37, 38).

In situ hybridization to salivary gland polytene chromosomes of *in vivo* labeled RNA from young embryos showed high levels of hybridization to the 39DE region on the left arm of chromosome 2 (1), the site of the histone genes (27). The high level of hybridization *in situ* to the histone locus suggested that the histone mRNAs are among the more abundant transcripts in the early embryo and thus should be accessible for detailed study.

The *Drosophila* histone genes were among the first eukaryotic genes to be cloned and characterized in a recombinant bacterial plasmid (15). *Drosophila* histone genes are present in the genome as tandem repeats with a copy number of about 110. Analysis of the cloned histone gene repeat showed that two of the genes are on one strand of the DNA and three on the other strand (20). The inverted orientation of the H3–H4 pair and the H2A–H2B pair is reminiscent of the organization in yeast (40), but very different from the tandem repeat in sea urchin in which all the genes are on the same strand (16).

The availability of a recombinant DNA probe for all five *Drosophila* histone genes has made it possible to measure accurately the changes in the mass of histone mRNA and of the rate of accumulation of newly synthesized histone mRNA during *Drosophila* embryogenesis (3). The change in the total mass of polysomal histone mRNA during the course of early embryogenesis is shown in Figure 5.2. To determine the percentage of the total histone mRNA associated with polysomes,

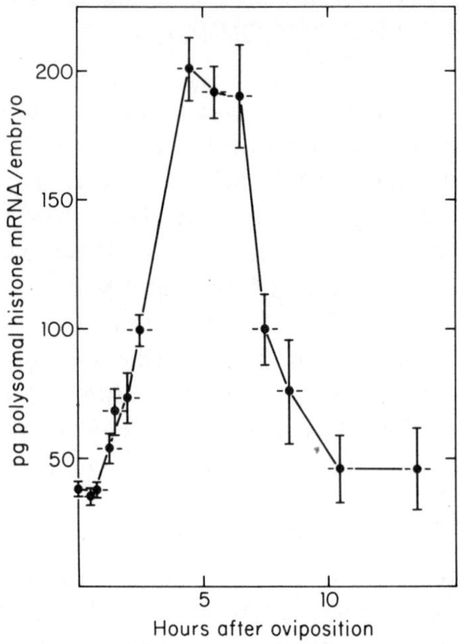

FIGURE 5.2. Mass of polysomal histone mRNA per embryo during embryogenesis. The percentage of total cytoplasmic RNA that is histone mRNA at a specific developmental stage was determined by measuring the percentage of *in vitro* uniformly labeled ^{32}P-RNA hybridizing to filter-bound *Drosophila* histone DNA of the plasmid cDm500 (3). The percentage hybridizing was corrected for hybridization efficiency and converted to total mass of histone mRNA per embryo using the fact that there are 180 ng cytoplasmic RNA per embryo at all stages studied. The zero time point represents RNA prepared form hand-dissected stage 14 oocytes.

polysomes and prepolysomal fractions were prepared, labeled *in vitro* with ^{32}P, and the distribution of histone mRNA in polysomal and prepolysomal fractions was determined (see Figure 5.3*a*). The total amount of histone mRNA per embryo is higher than previously reported (3) because there appears to be a systematic loss of 30% of the histone mRNA relative to total RNA during phenol extraction. This loss is not seen if RNA is prepared without phenol extraction. For the discussion presented here, we consider the amount of histone mRNA associated with polysomes to be proportional to the rate of synthesis of histone protein (14), and thus to indicate the level of expression of the histone genes. As discussed later, the changing level of histone gene expression over the course of embryogenesis is due to regulation at a number of steps in the pathway which leads to the production of translationally active mRNA: the presence of maternal histone mRNA, the rate of synthesis of new histone mRNA, the rate of histone mRNA turnover, and the loading of cytoplasmic histone mRNA onto polysomes.

The mature oocyte (stage 14) contains 134 pg of histone mRNA, or 1×10^8 copies of each of the five types of histone mRNA, 10 times as much as the abundant class of embryonic poly(A)$^+$ mRNAs (19, 4a). Histone mRNA is therefore the most abundant class of mRNA stored in the oocyte. Like the other RNAs found in the egg, it is synthesized in the nurse cells of the ovary and poured into the oocyte at the stage of nurse cell breakdown. The mechanisms that control how much maternal histone mRNA is stored in the egg remain unknown.

Cytoplasmic histone mRNA is under translational control early in embryogenesis. While there is a small increase (from 55% at 1 hour to 70% at 4 hours) in the fraction of total cytoplasmic poly(A)$^+$ RNA associated with polysomes during early embryogenesis, there is a more dramatic recruitment of histone mRNA into polysomes in this same period (Figure 5.3). Immediately after oviposition, 25% of the maternal histone mRNA is associated with polysomes. This increases steadily early in embryogenesis until by 4 hours after oviposition, 90% of the total embryonic histone mRNA is polysomal. This increase in translational activity is similar to the increase in the translational activity of histone mRNA seen in early sea urchin embryogenesis (42, 5), suggesting that similar mechanisms could be involved. Since the proportion in polysomes of pulse labeled ^{3}H-histone mRNA follows the same time course as that of total histone mRNA, we conclude that in *Drosophila*, both maternal and newly synthesized histone mRNA are recruited into polysomes at the same rate.

Synthesis of histone mRNA by zygotic nuclei is detectable at 90 min after oviposition. Histone gene transcription thus begins early in development, before the transcription of ribosomal RNA and of many poly(A)$^+$ sequences (23, 2). The absolute rate of synthesis of histone mRNA at specific embryonic stages has been determined by pulse labeling embryos with ^{3}H-uridine and measuring the cpm incorporated into cytoplasmic histone mRNA and the specific activity

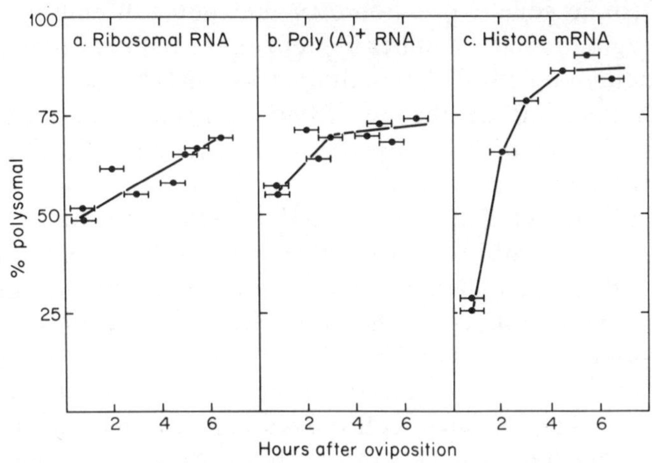

FIGURE 5.3. Loading of RNA on polysomes during embryogenesis. Cytoplasm from staged embryos was sedimented in 5–40% sucrose gradients under conditions that resolve polysomes (9). Disomes and larger polysomes were pooled to constitute a polysomal fraction, and smaller material was pooled as the prepolysomal fraction. RNA was prepared from both fractons and analyzed for rRNA, poly(A), and histone mRNA. The percentage of each class of RNA associated with polysomes as a function of embryonic stage is shown. (*a*) Ribosomal RNA in polysomes, determined from the OD_{254} of the 19S rRNA peak in SDS sucrose gradients of RNA from the polysomal and prepolysomal fractions. (*b*) Poly(A) in polysomes: the poly(A) content of the polysomal and prepolysomal fractions was determined by hybridization with ^{3}H-poly(U) (24). (*c*) Histone mRNA in polysomes, determined by measuring the percentage of the total *in vitro* ^{32}P-labeled RNA which hybridized to filter bound cDm500 DNA. The polysomal histone mRNA at all stages was at least 90% releasable by EDTA.

of the precursor ^{3}H-UTP pool. From the measured change in total mass of cytoplasmic histone mRNA between developmental stages, the behavior of the precursor pool-specific activity during the labeling period and the total cpm incorporated into histone mRNA, it is possible to calculate both k_s, the instantaneous rate of synthesis of histone mRNA per embryo (in this case the rate of appearance of histone mRNA in the cytoplasm), and k_d, the rate of turnover of histone mRNA, at specific developmental stages. The derivation of the equations allowing these calculations has been published elsewhere (3, 4). The changes in the rates of synthesis and turnover of histone mRNA per embryo during the course of early embryogenesis are shown in Figure 5.4 and Table 5.1.

The rate of histone mRNA synthesis per embryo approximately parallels the rate of DNA synthesis per embryo, increasing during the first 6 hours of embryogenesis and falling rapidly thereafter. Not only does the rate of histone mRNA synthesis change substantially during early embryogenesis; more surprisingly, the message stability also changes substantially, decreasing at least 15-fold in the first 6 hours of development. Modification of the stability of specific messages has also been seen in a number of other systems (12, 43, 30), including the histone H2B mRNA of yeast (26).

To understand the control of histone gene expression, it is important to know to what extent the changes in rates of synthesis and turnover are common to all messages and to what extent they are specific to the histone mRNAs. We have therefore analyzed the rates of synthesis and turnover of total cytoplasmic poly(A)$^+$ RNA and of actin mRNA in the same way we have examined histone mRNA synthesis and turnover. The results of this analysis, shown in Table 5.1, indicate that, while the regulation of histone mRNA concentration has characteristics in common with other messages, it also has certain unique properties.

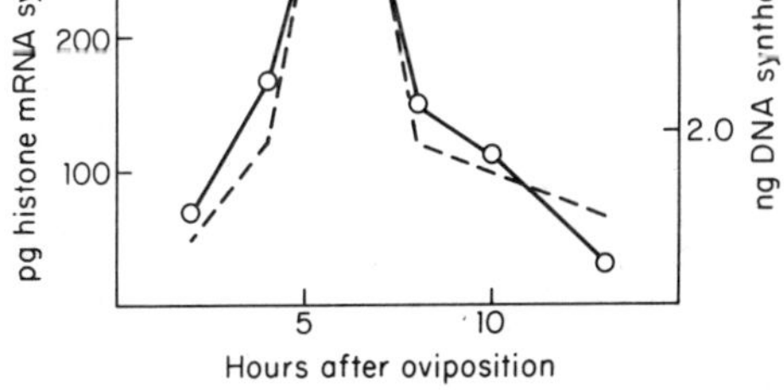

FIGURE 5.4. Rate of synthesis of histone mRNA per embryo during embryogenesis. The rate of synthesis of histone mRNA per embryo (O——O) was calculated as described in the legend to Table 5.1 (note that the k_s in Table 5.1 is per nuclear equivalent, not per embryo, which is shown here). The rates of synthesis of histone mRNA are compared with the rate of DNA synthesis (- - - - -) (from Figure 5.1*a*). The rate of histone mRNA synthesis per embryo is higher than previously reported (3), due to corrections for losses during phenol extraction (see text).

TABLE 5.1. Synthesis and Turnover of Embryonic mRNAs

	k_s: Copies Appearing in the Cytoplasm/Nucleus-Min			Half-Life (hrs)		
Embryonic Age	Histone mRNA	Actin mRNA	Total Cytoplasmic Poly(A)$^+$ RNA	Histone mRNA	Actin mRNA	Total Cytoplasmic Poly(A)$^+$ RNA
1–3 h	370	8 ± 2	1100	2.2[a]	2.8	4.6
3–5 h	210	4 ± 1	520	1.1	0.7	5.0
5–7 h	140	6 ± 1	190	0.4	0.2	5.3
7–9 h	32	5 ± 1	170	0.4	0.2	5.0
9–11 h	16	nd	170	0.3	nd	6.0
12–14 h	6	nd	190	nd	nd	4.0

[a] The histone mRNA has a half-life of 1.1 hour at 2.5 hours after fertilization and appears to be stable (half-life greater than 5 hours) before 1.5 hour; this results in an average half-life of 2.2 hours between 1 and 3 hours after oviposition.

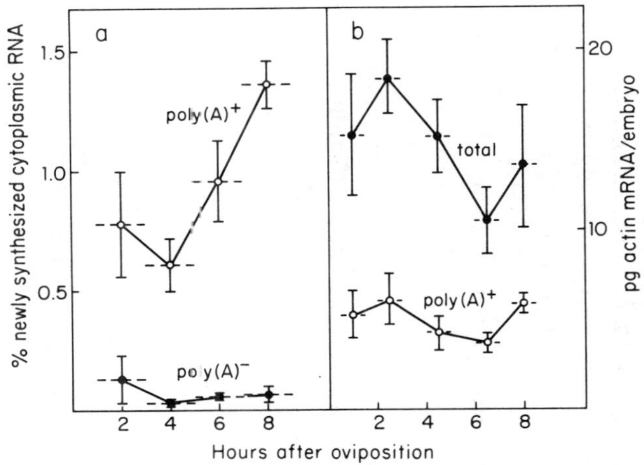

Rates of synthesis and turnover were calcualted as previously described (4) from the change in total message mass per embryo in the given time interval and the amount of radioactivity accumulated in the RNA species during a 1-hour pulse at that stage. Rates of synthesis per embryo were transformed into rates per nucleus, using the previously determined number of nuclear equivalents per embryo at that stage (4). The data on histone mRNA mass are shown in Figure 5.2, while those for histone mRNA accumulation are from Reference 3, but are corrected for the selective loss of both total and newly synthesized histone mRNA during phenol extraction. Values for actin mRNA accumulation and mass are shown below and were obtained, respectively, by hybridization of either (*a*) ^{3}H *in vivo* labeled or (*b*) ^{32}P *in vitro* labeled embryo RNA with a recombinant plasmid (generously supplied by Dr. A. Sodja) containing the conserved portion of the *Drosophila* actin gene, a 0.3 kb Sal I fragment from λDmA2 (10).

Most (>60%) of the pulse labeled actin mRNA is polyadenylated, and actin mRNA makes up 0.7–1.4% of the total A$^+$ mRNA accumulated during labeling in early embryos (Anderson, unpublished data). Data for the rates of accumulation and change in mass for total cytoplasmic poly(A)$^+$ RNA are from References 1, 2 and Anderson, unpublished observations.

The rates of synthesis per nucleus of total poly(A)$^+$ and of histone mRNA fall roughly in parallel during the first 6 hours of embryogenesis. Between 6 and 13 hours, however, the rate of synthesis per nucleus of total cytoplasmic poly(A)$^+$ RNA remains relatively constant while the rate of histone mRNA synthesis per nucleus drops twentyfold. This decrease in the rate of synthesis of histone mRNA parallels the sharp drop in the rate of DNA replication, suggesting that the decrease in the rate of histone mRNA synthesis after 6 hours may be a specific feature of histone gene regulation. It is interesting to note that although the rates of both histone mRNA and total cytoplasmic poly(A)$^+$ RNA synthesis per nucleus both fall during early embryogenesis, the rate of actin mRNA synthesis per nucleus remains roughly constant.

The rate of turnover of total cytoplasmic poly(A)$^+$ RNA appears to be constant throughout early embryogenesis (Table 5.1), implying that the decrease in histone mRNA half-life also represents a specific control. The decrease in stability is not unique to histone mRNA, however, since actin mRNA also decreases in stability in the same time interval, with roughly the same time course (Table 5.1). This suggests that there may exist a subset of all mRNAs, including both actin and histone messages, which become less stable early in embryogenesis.

4. RELATIVE CONTRIBUTIONS OF MATERNAL mRNA AND CHANGING RATES OF SYNTHESIS, TURNOVER, AND TRANSLATION TO THE CHANGING LEVEL OF HISTONE GENE EXPRESSION

The result of the changing rates of synthesis, translation, and turnover of histone mRNA during early embryogenesis is that the embryo has sufficient histone protein to complex with the DNA synthesized at all stages of embryogenesis. Very little histone protein is stored in the mature egg (R. Wiegand, personal communication); thus it is the histone mRNA in the embryo that is critical in the production of normal chromatin. Assuming a translation rate of 3.8 polypeptides per ribosome per minute (the high end of the range of 2.8–3.8/min calculated in Reference 29), we calculate that the total amount of histone protein present in the embryo is in excess of the total mass of DNA until about 7 hours after fertilization, when the total histone:DNA ratio falls to the approximately 1:1 ratio characteristic of normal chromatin (Figure 5.5). The success of the embryo in producing the correct amount of histone protein for the amount of DNA synthesized suggests either that the system is extremely accurately programmed or, alternatively, that there is coordination between the rate of DNA replication and the rate of histone protein synthesis.

The relative contributions of synthesis, translation, and turnover to the regulation of histone gene expression change significantly as the embryo proceeds through the phases of DNA replication. During cleavage, only the amount of maternal histone mRNA present at oviposition and the fraction of that mRNA loaded onto polysomes are significant. During gastrulation, as the DNA mass per embryo increases rapidly, the high rate of synthesis of new zygotic histone

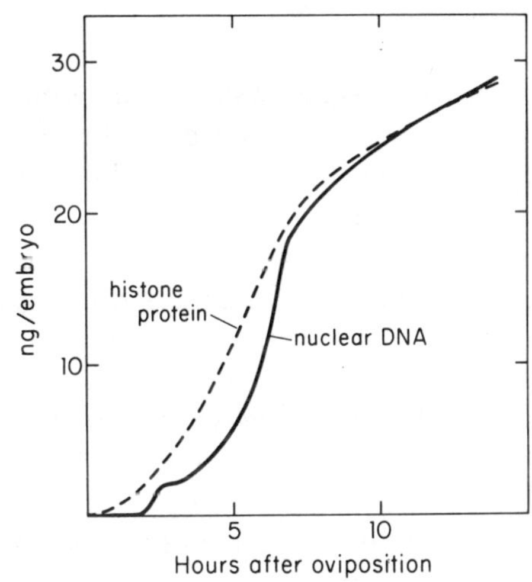

FIGURE 5.5. Comparison of histone protein content and DNA content during embryogenesis. The amount of DNA per embryo as a function of developmental age was calculated by integrating over time the rate of DNA synthesis shown in Figure 5.1*a*. The early rapid rise and plateau in DNA mass during the first 3 hours of embryogenesis was calculated from histological observations on nuclear division during early embryogenesis (46). It was assumed that there are 64 nuclei per embryo at 1 hour after oviposition (46), that there are 0.35 pg DNA per diploid nucleus (32) and that these 64 nuclei undergo three more divisions 9 min apart, followed by four more divisions 15 min apart. The plateau at 3 hours reflects the cytologically observed pause in cell division at the cellular blastoderm and early gastrula (22).

mRNA and the efficient translation of all available histone mRNA are important. After 6 hours, as the rate of DNA replication declines, both the decreasing rate of synthesis and the instability of the histone mRNA are crucial. During all phases of DNA replication, translational control provides only a relatively fine tuning of the system, since the fraction of total histone message associated with polysomes changes only about threefold during early embryogenesis (Figure 5.3*c*). In contrast, histone mRNA stability changes at least 15-fold, and the rate of histone mRNA synthesis per nucleus decreases roughly 60-fold in the first 13 hours of embryogenesis (Table 5.1). Changes in rates of synthesis and turnover thus provide the major levels of modulation of histone gene expression during *Drosophila* embryogenesis.

5. MECHANISMS OF CONTROL: THE DECREASE IN MESSAGE STABILITY

Although the steps at which the rate of histone protein production is regulated have been defined, as described, the larger questions of what the actual regulatory mechanisms are and how the various levels of control are integrated remain unanswered. The change in message stability is, at first glance, more accessible to experimental analysis than the changing rates of synthesis and translation because an obvious hypothesis for its mechanism is suggested by analogy with the switch in histone gene expression in early sea urchin embryos.

It is well documented that the cleavage stage histone mRNAs of sea urchins are transcribed from a set of genes different from those transcribed at the mesenchyme blastula stage. The RNAs from the different genes are readily distinguished by differences in size (11) and by differences in the stability of the hybrids formed between the RNA and cloned histone DNA (18). An analogous gene switch in *Drosophila* could be proposed to explain the observed decrease in half-life: the gradual replacement of moderately stable maternal histone mRNA with less stable zygotic histone mRNA might result in the observed gradual decrease in stability.

To test this hypothesis, the same criteria that have been used in distinguishing sea urchin histone mRNA variants: size and fidelity of hybrids with cloned DNA, have been used to look for histone mRNA variants in the early *Drosophila* embryo. Maternal and zygotic histone mRNAs appear to be the same size in RNA gel blots (Figure 5.6*a*). Since the *Drosophila* histone mRNA species differ from one another in length by 50 nucleotides (Lifton and Goldberg, cited in Reference 16), we estimate that we would detect a change of 20–30 nucleotides in length. In comparison, all five sea urchin histone mRNAs change in size by 15–40 nucleotides at the time of gene switching. We conclude that the *Drosophila* histone mRNAs do not change in size by more than 20–30 nucleotides over the course of embryogenesis.

Hybrids of gastrula type sea urchin histone mRNA with cloned cleavage type histone DNA are sensitive to digestion with RNase. Sixty to 78% of the RNA is released from hybrids by RNase digestion under conditions where well-matched hybrids release only 10–42% of the RNA (18). In contrast to the results obtained with sea urchins, hybrids between either maternal or zygotic *Drosophila* RNA (labeled *in vitro* with ^{32}P) and cloned *Drosophila* histone DNA are equally resistant to RNase digestion. In both cases only 20% of the hybridized cpm are released by treatment with RNase (data are not shown). (The fact that 20% of the presumably perfectly matched hybrids is digestible with RNase is probably a result of the randomly broken, sonicated DNA being smaller than the hybridizing RNA).

Thermal melts of hybrids of histone mRNA with cloned *Drosophila* histone DNA also indicate that there is no detectable difference between the maternal and zygotic histone mRNAs. As shown in Figure 5.6*b* the maternal and zygotic mRNAs hybridized to histone DNA have identical melting curves, with a *Tm* characteristic of perfectly matched hybrids. The RNase digestion and *Tm* characterization experiments indicate that the sequence difference between histone mRNAs present at different times during *Drosophila* embryogenesis does not exceed 1% (6).

The failure to find stage-specific expression of *Drosophila* histone mRNA variants is consistent with analysis of *Drosophila* histone DNA sequences. The histone genes are repeated 110 times per haploid genome, but no restriction site differences within the coding region of cloned repeats have been detected (20; Karp and Hogness, quoted in Reference 16).

Another possible basis for stabilization of the *Drosophila* maternal histone mRNA would be adenylation. This mechanism is suggested by the finding that *Xenopus* maternal histone mRNA appears to be

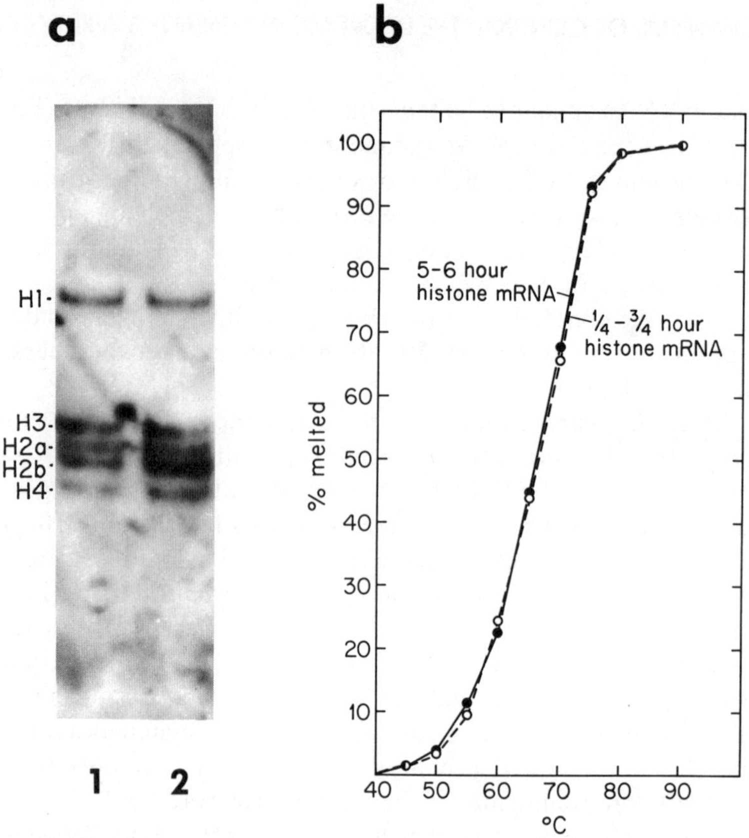

FIGURE 5.6. Comparison of maternal and zygotic histone mRNAs. (*a*) Size of maternal and zygotic histone mRNA. Total cytoplasmic RNA from staged embryos was denatured with glyoxal and electrophoresed in a 2.5% agarose gel. RNA was transferred to nitrocellulose paper and then allowed to hybridize with ^{32}P *in vitro* labeled cDm500 DNA (7×10^6 cpm, 2×10^7 cpm/μg), as described by Thomas (41): (1) 11 μg of cytoplasmic RNA from 0.25–1.25-hour embryos (maternal); (2) 7 μg of cytoplasmic RNA from 6–7-hour embryos (zygotic). Histone mRNAs are identified by the sizes characterized by Lifton and Goldberg (16).

(*b*) Melting characteristics of hybrids of cDm500 DNA with maternal and zygotic histone mRNAs. Cytoplasmic embryo RNA was hybridized in 3X TESS (0.45 M NaCl, 0.015 M EDTA, 0.015 M TES pH 7.4, 0.3% SDS) to filter-bound cDm500 DNA as described previously (3). Unhybridized cpm were removed by repeated washings (without RNase), and total cpm in hybrid were determined by liquid scintillation counting. Scintillant was removed by washing with chloroform, and the hybridized histone mRNA was melted from hybrid by successive increases in temperature in 0.2 × SSC. The eluted cpm were precipitated with TCA and counted. The data are presented as the percentage of the total cpm eluted. Each point is the average of quadruplicate determinations: ○ = 0.25–0.75-hour embryo histone mRNA; ● = 5–6-hour embryo histone mRNA.

stable because it is polyadenylated (13, 34). We have not, however, found any significant difference in the amount of poly(A) associated with the stable maternal histone mRNA and the unstable RNA found in the embryo after 5 hours. Although slightly more maternal than embryonic histone mRNA binds to oligo(dT)-cellulose (7% compared to 1%), this fraction of adenylated histone mRNA is too small to account for the observed change in half-life. Furthermore, no short poly(A) tracts [as assayed by hybridization with ^{3}H-poly(U)] were detected in the maternal histone mRNA, which does not bind to oligo(dT)-cellulose. Thus the change in stability of the *Drosophila* maternal histone mRNA does not appear to be due to a change in its level of polyadenylation.

An additional observation which argues that the decrease in histone mRNA stability is not due to nucleotide sequence changes is that histone mRNA is not the only mRNA that becomes less stable during early embryogenesis. As shown in Table 5.1, actin mRNA also decreases five- to tenfold in stability early in embryogenesis. The similarity in time course of the decreases in actin and histone mRNA stabilities suggests that the change in stability of these two mRNAs is not directed by RNA sequence changes, but by a single mechanism. Since the stability of total cytoplasmic poly(A)$^+$ RNA does not decrease detectably during early embryogenesis (Table 5.1), histone and actin mRNAs must be members of a relatively small subset of the embryonic mRNA population that is selectively destabilized after fertilization. Histone mRNA appears to be stable until 90 min after fertilization (from data in Figures 5.2 and 5.3*c*), the time at which transcription begins in zygotic nuclei. The increased rate of turnover may therefore require the activation of zygotic genes whose products are not stored in the oocyte, resulting in the production of new nucleases or of new molecules that remove proteins protecting the RNA from degradation.

6. MECHANISMS OF CONTROL: TRANSCRIPTION AND GENOMIC ORGANIZATION

Histone mRNA synthesis begins in zygotic nuclei at 1.5 hours after oviposition. This is when the nuclei arrive in the cortical cytoplasm and when transcription first begins, as detected by autoradiography (45). The initial rate of production of histone mRNA is high: 370 copies of each histone mRNA appear in the cytoplasm per nucleus per minute. This rate decreases continuously during embryogenesis, becoming six copies of each mRNA per nucleus per minute at 13 hours (Table 5.1).

These measurements of the rate of appearance of histone mRNA in the cytoplasm reflect rates of processing and nuclear export as well as the rate of transcription. We believe, however, that the rate of transcription, rather than processing or export, primarily controls the rate of histone mRNA appearance in the cytoplasm. The appearance of histone mRNA in the cytoplasm of mouse L cells does not appear to have a lag time (31), implying that the steps of processing and export are very rapid. Although the *in vivo* labeling approach used in the kinetic analysis described here does not allow detailed analysis of separate nuclear events, processing and export appear to be similarly rapid in the *Drosophila* embryos. In a 15-minute label with ^{3}H-uridine, more than 90% of the radioactivity in histone mRNA is found in the cytoplasmic fraction (1). It thus seems likely that virtually all the histone gene transcripts are exported to the cytoplasm. Rates of production of cytoplasmic histone mRNA should therefore be approximately equivalent to rates of synthesis.

At the maximal rate of synthesis, an average of 2 mRNAs are produced from each histone gene per minute (Table 5.1). Since we assume that the efficiency of processing is high, this represents substantially less than the calculated maximal rate of transcription at 25°C of 20 copies per gene per minute (14). This result can be explained by two alternative hypotheses: either all the histone genes may be transcribed at a submaximal rate, or only a fraction of all the histone genes may be transcribed at a maximal rate. The decline in histone mRNA synthesis during embryogenesis could then be explained as due either to a decreased rate of transcription at all copies of the histone genes or to selective transcriptional inactivation of subsets of histone genes.

With these hypotheses in mind, it is reasonable to consider the possibility that different histone genes, all of which are located in the group of bands 39D2-3 through 39E1-2 on the left arm of chromosome 2 (27), might be transcribed differentially at different stages of development. In *Drosophila* there is no evidence for histone genes coding for stage- or tissue-specific histones (47). Analysis of DNA sequences has identified two major classes of histone gene repeat, a 5.0 and a 4.8-kb repeat unit, which differ only by the insertion of an additional 240 nucleotide sequence in the spacer region between H1 and H3 (Figure 5.7; 20, 10a). The 4.8-kb unit constitutes about 25% of the total histone genes in the *Drosophila melanogaster* genome (20). Since no other sequence heterogeneity has been detected by restriction analysis, it seems unlikely that there is a subset of genes selectively transcribed at a high rate on the basis of nucleotide sequence. Analysis of promoters in *in vitro* transcription assays has also failed to reveal any heterogeneity in transcription rate between the 4.8 and 5.0-kb repeat units (C. Parker, personal communication).

The possibility remains that differential transcription of the apparently identical histone genes might occur if these are organized into clusters that could assume different conformations in chromatin. The histone genes at 39DE appear to be grouped into 14 units separated by nonhistone DNA sequences, with an average of five to six repeats per unit (35). Consistent with the idea that there are clusters of histone genes separated by nonhistone DNA is the observation that, in some *in situ* hybridizations with ^{3}H-histone mRNA, the grains over the 39DE region are arranged in a banded pattern (27).

Additional evidence for the existence of separate histone clusters, at least in some *Drosophila* species, comes from studies on *Drosophila virilis*. The cloned histone repeat of *Drosophila melanogaster* hybridizes *in situ* to two separate loci in *Drosophila virilis* polytene

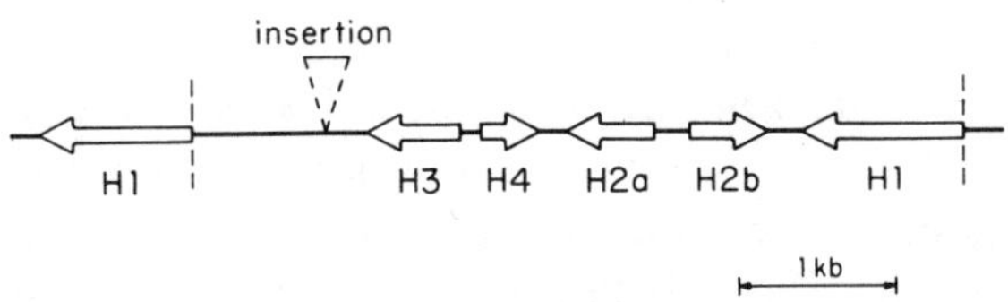

FIGURE 5.7. Structure of the 4.8 and 5.0-kb *Drosophila* histone repeats. The two repeats differ only in that the 5.0-kb repeat contains a 240-nucleotide insertion (10a). The transcribed regions and their directions of transcription are indicated by arrows. The vertical dotted lines indicate the extent of one repeat unit (redrawn from Reference 20).

chromosomes. One of these loci consists of two closely linked bands on either side of band Og on the second chromosome (Figure 5.8*a*,*b*). The other locus is in region K on the fourth chromosome [the homolog to *D. melanogaster* 2L (28)] and hybridizes the probe much less intensely (Figure 5.8*a*). Whether or not these two chromosomal loci are regulated independently during *D. virilis* development is not known at present. Investigation of the organization of these loci could conceivably shed some light on histone gene expression in *D. melanogaster*.

If the various histone clusters in *Drosophila melanogaster* are differentially transcribed because of local chromatin organization, it might be possible to detect this biochemically. Analysis of histone gene chromatin has, however, given no indication of specialized chromatin structures within the repeat units themselves. Apparently all of both the 4.8 and 5.0-kb repeat units show the same chromatin organization in all cell types studied, including the 1-hour-old embryo, tissue culture cells, and polytene nuclei. This chromatin organization consists of phased nucleosomes in the intergenic spacers, nuclease hypersensitive sites at the 5′ ends of all five histone genes, and nonphased nucleosomes within the genes (36).

In *Drosophila* it is possible to ask whether all the histone genes are functionally equivalent by using genetic techniques to delete certain sets of genes. Flies that carry only the haploid number of copies of the histone genes, as well as flies that carry a duplication of the histone gene region are viable (44, 17, 25). This implies that the regulation of histone gene expression does not depend absolutely on the number of copies of histone genes available for transcription. Flies heterozygous for deficiencies that uncover the entire histone locus show the dominant phenotype, Minute. This phenotype, which seems to be characteristic of mutants in which growth is retarded, has been proposed to be the result of deletion of the histone genes themselves rather than the deletion of adjacent loci (39). The level of histone gene expression thus depends, to some extent, on the number of copies of the histone genes present. The observation that stocks carrying histone gene deficiencies are able to increase the number of copies of the histone genes on the homologous chromosome in a process described as "magnification" (7) suggests that the dosage of histone genes may be an important component in the mechanism of histone gene control.

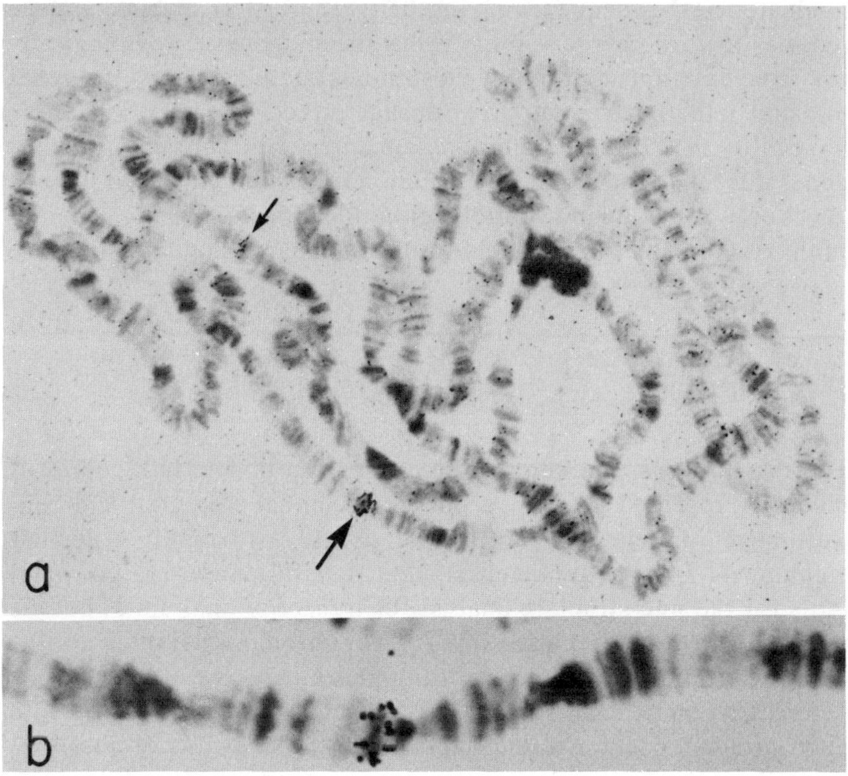

FIGURE 5.8. Hybridization of the *Drosophila melanogaster* histone repeat to *Drosophila virilis* polytene chromosomes. The recombinant plasmid cDm500, containing 1.8 copies of the *Drosophila melanogaster* histone repeat in ColEI, was used as a template for the synthesis of ^{3}H-complementary RNA (cRNA). This ^{3}H-cRNA (50,000 dpm/slide) was hybridized at 67°C in 0.3 M Na^+ to *Drosophila virilis* salivary gland polytene chromosomes *in situ* as described previously (19a). (*a*) Hybridization to two loci in the *Drosophila virilis* genome: O_g (large arrow) on the second chromosome, and K (small arrow) on the fourth chromosome; 707d exposure; 600×. (*b*) Higher magnification of a different field showing hybridization to either side of band O_g on the second chromosome; 124 day exposure, 1900×.

A direct approach to the question of whether there are functional differences between histone gene copies on the same chromosome has been initiated by Siegel (39). By screening for lethal mutations over a deficiency for the distal half of the histone gene region, she was able to identify at least 13 complementing lethal loci within the region. It is possible that these lethal mutations are located in the nonhistone DNA sequences lying between clusters of histone gene repeats (35). Another possibility is that the mutations are cytologically invisible deletions that remove a sufficient number of histone genes to cause lethality from a purely dosage effect. The most intriguing possibility is that these mutations render specific subsets of histone genes nonfunctional, either by deletion of a small cluster of histone genes or by mutation in a control region. Identification of more mutants in this region and further analysis of the mutants should distinguish between these possibilities.

7. CONCLUSION

The mechanisms that control the expression of multiple copies of tandemly repeated genes that possess identical structural and controlling-region sequences are difficult to study with purely molecular techniques. *In vitro* mutagenesis and transformation are powerful tools for the fine structure analysis of transcriptional units, but, as yet, they allow neither analysis of larger chromosomal units nor the identification of separate controlling genes.

Further genetic analysis of the histone gene region, combined with appropriate molecular studies, should answer a number of questions about the transcriptional control and genomic organization of histone genes. Are there nonhistone lethal complementation groups interspersed between the histone genes? If so, do they follow the same pattern of transcription as the histone genes? Is their function in some way related to histone gene function? What is the minimum number of histone genes compatible with viability? Are there noncoding regions that control the activity of a cluster of histone genes? Are all histone gene clusters functionally equivalent?

Most observations suggest that at least some of the control of histone gene expression lies not within the histone gene repeat units, or the RNAs transcribed from them, but in other regulatory systems. The dramatic decrease in histone mRNA stability, for instance, seems not to be a function of changes in the RNA structure but rather a change in availability of a class of RNA sequences to nucleases. This increased accessibility to nucleases, resulting either from changes in ribonucleoprotein structure or from new synthesis of nucleases, is shared by other RNAs, including actin mRNA, at the same developmental stage.

DNA sequence and chromatin structure analysis has so far failed to reveal any structural heterogeneity within the histone gene clusters that can be related to changing rates of transcription during development. It is likely, therefore, that transcriptional regulation is controlled not only by the structure of the histone gene clusters themselves but is also under the direction of elements not included in the tandemly repeated units (i.e., lying either between clusters of histone genes or at genetic loci other than 39DE). The identification of such regulatory elements, either biochemically or genetically, might yield a more satisfying description of the mechanisms of histone gene regulation.

The genetic identification of *trans* regulatory elements for histone genes would not be trivial. A screen for such elements might involve, for example, searching for suppressors of lethality in flies containing several, but a subviable number, of histone gene clusters. The identification of functional relationships between unlinked genes is one of the challenges of modern molecular biology and genetics. Histone gene expression in *Drosophila* embryos might provide a valuable system for such investigations.

REFERENCES

1. K. V. Anderson, Ph.D. thesis, University of California, Los Angeles (1980).
2. K. V. Anderson and J. A Lengyel, *Dev. Biol.* **70,** 217–231 (1979).
3. K. V. Anderson and J. A. Lengyel, *Cell* **21,** 717–727 (1980).
4. K. V. Anderson and J. A. Lengyel, *Dev. Biol.* **82,** 127–138 (1981).
4a. C. G. Arthur, C. M. Weide, and E. S. Goldstein, *Exp. Cell Res.* **121,** 87–94 (1979).
5. E. J. Baker and A. A. Infante, *Proc. Natl. Acad. Sci. USA* **79,** 2455–2459 (1982).
6. T. I. Bonner, D. J. Brenner, B. R. Neufeld, and R. J. Britten, *J. Mol. Biol.* **81,** 123–135 (1973).
7. A. I. Chernyshev, V. N. Bashkiov, B. A. Leibovitch, and R. D. Khesin, *Molec. Gen. Genet.* **178,** 663–668 (1980).
8. J. E. Darnell, *Nature* **297,** 365–371 (1982).
9. S. Falkenthal and J. A. Lengyel, *Biochemistry* **19,** 5842–5850 (1980).

10. E. A. Fyrberg, B. J. Bond, N. D. Hershey, K. S. Mixster, and N. Davidson, *Cell* **24,** 107–116 (1981).
10a. M. L. Goldberg, Ph.D. Thesis, Stanford University (1979).
11. M. Grunstein, *Proc. Natl. Acad. Sci. USA* **75,** 4135–4139 (1979).
12. W. A. Guyette, R. J. Matusik, and J. M. Rosen, *Cell* **17,** 1013–1023 (1979).
13. G. Huez, G. Marbaix, D. Gallwitz, E. Weinberg, R. Devos, E. Hubert, and Y. Cleuter, *Nature* **271,** 572–573 (1978).
14. F. C. Kafatos, *Curr. Topics Dev. Biol.* **7,** 125–191 (1972).
15. R. Karp and D. S. Hogness, *Fed. Proc.* **35,** 1623 (1976).
16. L. Kedes, *Ann. Rev. Biochem.* **48,** 837–870 (1979).
17. R. B. Khesin and B. A. Leibovitch, *Molec. Gen. Genet.* **162,** 323–329 (1978).
18. N. S. Kunkel and E. S. Weinberg, *Cell* **14,** 313–326 (1978).
19. C-H. Kuo and A. Garen, in C. Borek et al., Eds., *Cancer Biology IV. Differentiation and Carcinogenesis*, Stratton Intercontinental Medical Book Corp., New York, 1977, pp. 97–103.
19a. J. A. Lengyel, L. J. Ransom, M. L. Graham, and M. L. Pardue, *Chromosoma (Berl.)* **80,** 237–252 (1980).
20. R. P. Lifton, M. L. Goldberg, R. W. Karp, and D. S. Hogness, *Cold Spring Harbor Symp. Quant. Biol.* **42,** 1047–1051 (1978).
21. G. Limbourg and M. Zalokar, *Dev. Biol.* **35,** 382–387 (1973).
22. M. M. Madhavan and H. A. Schneiderman, *Wilhelm Roux's Arch.* **183,** 269–305 (1977).
23. S. L. McKnight and O. L. Miller, *Cell* **8,** 305–319 (1976).
24. C. Milcarek, R. Price, and S. Penman, *Cell* **3,** 1–10 (1974).
25. G. D. Moore, J. P. Procunier, D. P. Cross, and T. A. Grigliatti, *Nature* **282,** 312–314 (1979).
26. M. A. Osley and L. M. Hereford, *Cell* **24,** 377–384 (1981).
27. M. L. Pardue, L. H. Kedes, E. S. Weinberg, and M. L. Birnstiel, *Chromosoma* **63,** 135–151 (1977).
28. J. T. Patterson and W. S. Stone, *Evolution in the genus Drosophila*, Macmillan, New York, 1952.
29. M. Paul, M. R. Goldsmith, J. R. Hunsley, and F. C. Kafatos, *J. Cell Biol.* **55,** 653–680 (1972).
30. N. J. Pearson, H. M. Fried, and J. R. Warner, (1982). *Cell* **29,** 347–355 (1982).
31. R. P. Perry and D. E. Kelley, *J. Mol. Biol.* **79,** 681–696 (1973).
32. E. M. Rasch, H. J. Barr, and R. W. Rasch, *Chromosoma* **33,** 1–18 (1971).
33. D. B. Roberts and G. Graziosi, *J. Embryol. Exp. Morphol.* **41,** 101–110 (1977).
34. J. V. Ruderman, H. R. Woodland, and E. A. Sturgess, *Dev. Biol.* **71,** 71–82 (1979).

35. K. Saigo, L. Millstein, and L. A. Thomas, *Cold Spring Harbor Symp. Quant. Biol.* **45,** 815–827 (1981).
36. B. Samal, A. Worcel, C. Louis, and P. Schedl, *Cell* **23,** 401–409 (1981).
37. J. B. Santon and M. Pellegrini, *Proc. Natl. Acad. Sci. USA* **77,** 5649–5653 (1980).
38. J. B. Santon and M. Pellegrini, *Dev. Biol.* **85,** 252–257 (1981).
39. J. C. Siegel, *Genetics* **98,** 505–527 (1981).
40. M. M. Smith, see Chapter 1, this volume.
41. P. S. Thomas, *Proc. Natl. Acad. Sci. USA* **77,** 5201–5205 (1980).
42. D. E. Wells, R. M. Showman, W. N. Klein, and R. A. Raff, *Nature* **292,** 477–478 (1981).
43. R. Wiskocil, P. Bensky, W. Dower, R. F. Goldberger, J. I. Gordon, and R. G. Deeley, *Proc. Natl. Acad. Sci. USA* **77,** 4474–4478 (1980).
44. T. R. F. Wright, R. B. Hodgetts, and A. F. Sherald, *Genetics* **84,** 267–285 (1976).
45. M. Zalokar, *Dev. Biol.* **49,** 425–437 (1976).
46. M. Zalokar and I. Erk, *J. Microsc.-Biol. Cell.* **25,** 97–106 (1976).
47. A. Zweidler, in D. J. Prockop and P. C. Champe, Eds., *Gene Families of Collagen and Other Proteins*, Elsevier/North-Holland, New York, 1980, pp. 47–56.

CHAPTER 6

TETRAHYMENA HISTONES: PROTEINS, MESSAGES, AND GENES

GARY A. BANNON AND MARTIN A. GOROVSKY
Department of Biology
University of Rochester
Rochester, New York

CONTENTS

1. INTRODUCTION
2. CHROMATIN STRUCTURE IN MACRO- AND MICRONUCLEI
3. THE MAJOR *TETRAHYMENA* CORE HISTONES
   3.1. H2A
   3.2. H2B
   3.3. H3
   3.4. H4
4. HI TYPE HISTONES
5. MACRONUCLEAR-SPECIFIC VARIANTS
6. SECONDARY MODIFICATIONS
   6.1. Acetylation
   6.2. Phosphorylation
   6.3. Proteolytic Processing
7. NUCLEOSOME HETEROGENEITY IN *TETRAHYMENA*
8. HISTONE MESSAGES
9. HISTONE GENES
   REFERENCES

1. INTRODUCTION

The unicellular eukaryote, *Tetrahymena* is an excellent model system in which to study genes in different structural and functional states. Most species of *Tetrahymena* (like most species of ciliates) possess two types of nuclei: a transcriptionally active macronucleus and a transcriptionally inert micronucleus. The macronucleus controls the phenotype of the cell, is transcriptionally active, divides amitotically during vegetative growth, and is destroyed during the sexual process of conjugation. The micronucleus is responsible for the genetic continuity of succeeding sexual generations; it is analogous to germinal nuclei in multicellular organisms. During vegetative growth, it divides mitotically and is transcriptionally inert. During conjugation, the micronucleus divides meiotically to yield four haploid products, one of which divides mitotically to yield two gametic nuclei. Paired conjugants exchange one gametic nucleus which then fuses with the stationary nucleus to form a zygotic nucleus, which divides to give new macro- and micronuclei.

The common origin of macro- and micronuclei, coupled with their many structural and functional differences, has led us to investigate whether there are differences in the structure and composition of their chromatin. Studies on the DNA sequence organization of these two nuclei are described in detail elsewhere (1, 2). In brief, about 90% of the DNA sequences in the two nuclei are identical. Sequence under-representation (~10% of the micronuclear sequences are not detectable in macronuclei), ribosomal gene amplification, sequence rearrangement, and differences in DNA methylation also accompany macronuclear development. In this chapter we describe the histones of macro- and micronuclei. In particular, we have attempted to relate features of histone composition and histone modification to functions specific to one of the two nuclei: transcription in macronuclei and mitosis in micronuclei. We also describe some preliminary studies on *Tetrahymena* histone mRNAs and histone genes.

2. CHROMATIN STRUCTURE IN MACRO- AND MICRONUCLEI

The histones of both macro- and micronuclei are organized on DNA to produce a typical periodic and particulate (nucleosomal) structure

(3–6). The repeat unit recognized by micrococcal nuclease in micronuclei is shorter than in macronuclei (~175 bp versus ~200 bp). Thus the postulated correlation between transcriptional activity and shorter repeat lengths (7) is not found in ciliated protozoa (3, 8). The internal features of macro- and micronuclear nucleosome structure recognized by extensive (limit) micrococcal nuclease digestion or by DNAse I are similar to those of other organisms (4). As in higher organisms, actively transcribed chromatin in *Tetrahymena* is particularly sensitive to DNAse I, resulting in a rapidly digesting fraction of chromatin which is found only in macronuclei (9). Active genes in macronuclei also are associated with DNAse I hypersensitive sites (10; Pederson, Bannon, and Gorovsky, unpublished observations). We conclude from these observations that the major features of chromatin structure of both active and inactive genes found in higher eukaryotes also occur in *Tetrahymena*.

3. THE MAJOR *TETRAHYMENA* CORE HISTONES

Comparisons of the physical chemical properties (buoyant density, thermal denaturation, staphylococcal nuclease susceptibility) of macro- and micronuclei indicate that DNA from both nuclei is fully complexed with histones (3, 11). Biochemical fractionations (12, 13) indicate that, as in other organisms, there are five major types of histone in macronuclei. Macronuclear core histones have been characterized with regard to their pairwise interactions in solution (14), amino acid composition (13), and electrophoretic variability in different species (15). Some have been partially (16, 17) or completely (18) sequenced. Macro- and micronuclear histones have been compared by one- and two-dimensional electrophoretic analyses (19–21) and by peptide mapping, (20, 21; Bowen and Gorovsky, unpublished observations). Although rigorous quantitative analyses have not been performed, the relative amounts of the major fractions appear to be similar in both nuclei. The following sections summarize the properties of the unmodified forms of the major core histones, followed by a description of H1, of minor variants, and of secondary modifications.

3.1. H2A

H2A exists as two similar but distinct primary sequence variants ($H2A^F$ and $H2A^S$) in both macro- and micronuclei. No differences have been detected in the electrophoretic properties or peptide maps (partial proteolytic or tryptic) between macro- and micronuclear H2A. *Tetrahymena* H2A shows some unusual properties when compared to vertebrate H2A. Its electrophoretic mobility on acid–urea gels is much slower than vertebrate H2A. It does not vary in electrophoretic mobility in different species (a property usually associated with the more evolutionary conservative arginine-rich histones, H3 and H4). Also, *Tetrahymena* H2A could not substitute for calf-thymus H2A in nucleosome reconstitution experiments (4), even though it was able to form a strong complex with calf H2B (14). An understanding of the molecular basis of these observations awaits sequence data not currently available.

3.2. H2B

Tetrahymena appears to contain a single primary sequence form of H2B that is indistinguishable in macro- and micronuclei. H2B from different species of *Tetrahymena* shows electrophoretic variability similar to that observed for different vertebrate species. The sequence of *Tetrahymena* H2B has recently been determined (18). It is 119 residues long (the shortest H2B yet described) and differs from calf H2B at 45 positions. Although many of these substitutions are not conservative, and some occur in the hydrophobic, structured portion of the molecule, *Tetrahymena* H2B still can interact strongly in solution with calf H2A and H4 (14). It also can substitute for calf H2B in nucleosome reconstitution experiments (4).

3.3. H3

The majority of *Tetrahymena* macronuclear H3 consists of a primary sequence form that differs in electrophoretic mobility from calf thymus H3 but shows conservatism in electrophoretic mobility in different species of *Tetrahymena*. Although it differs slightly from vertebrate

H3, *Tetrahymena* H3 can complex with calf H4 and can substitute for calf H3 in nucleosome reconstitution. Preliminary sequence analysis of macronuclear H3 (17) indicates that it is unique among all H3 proteins sequenced to date in having alanine instead of glycine at position 13. Except for differences caused by secondary modifications (see later discussion), macro- and micronuclear H3 proteins are indistinguishable electrophoretically and by peptide mapping (20).

3.4. H4

The unmodified forms of macro- and micronuclear H4 also are indistinguishable electrophoretically and by peptide mapping. *Tetrahymena* H4 is almost indistinguishable electrophoretically from calf H4, and shares with its vertebrate homologue electrophoretic constancy in related species. However, *Tetrahymena* H4 and calf H4 were demonstrably different in their abilities to complex with (homologous or heterologous) H3 or H2b in solution, and *Tetrahymena* H4 failed to substitute for calf H4 in nucleosome reconstitutions. These differences in what is generally thought to be the most evolutionarily conserved histone (22) almost certainly are due to a surprisingly large difference in the sequences of calf and *Tetrahymena* H4 (16): of 66 (of a total of 102) residues sequenced, 15 differ from calf H4. Many of the replacements are not conservative.

In summary, we can draw the following conclusions about *Tetrahymena* core histones: (1) Where available, sequence comparisons indicate that *Tetrahymena* histones show more divergence than those of any other phylogenetic group, consistent with an extremely early evolutionary origin of ciliates. (2) In some cases *Tetrahymena* histones are sufficiently different from vertebrate histones to show altered properties in functional assays (pairwise histone-histone interactions, nucleosome reconstitutions). Since these functional differences are associated with multiple amino acid replacements, it is not possible to attribute them to particular changes in sequence. However, in cases such as H2B, where a number of sequence alterations (18) fail to result in changes in the functional assays, it is possible to conclude that the particular nonconservative replacements that have occurred have no detectable effect on the properties measured. (3) There are no significant differences in the electrophoretic properties or in the peptide maps of the unmodified forms of the major core histones in macro- and micronuclei (Bowen and Gorovsky, unpublished observations). Thus none of the functional differences between macro- and micronuclei can be associated with differences in the primary sequences of the major core histones, and none of the sequence differences between the major *Tetrahymena* macronuclear core histones and those of other organisms can be ascribed to the peculiar amitotic division of the ciliate macronucleus.

4. H1 TYPE HISTONES

Tetrahymena macronuclear H1 appears to consist of a single primary sequence with two unusual properties. It migrates almost as rapidly as H4 on low pH acrylamide gels, even though it is the largest *Tetrahymena* histone, consistent with its having an unusually large number of positive residues (23). *Tetrahymena* H1 also shows a phosphorylation-induced change in mobility on SDS gels (24), a property it shares with Drosophila H1 (25). The significance of these properties is not clear. In all other respects examined to date, macronuclear H1 is typical.

Micronuclei do not contain any detectable macronuclear-type H1 as assayed either by biochemical analyses (11, 19) or by *in situ* staining with a fluorescently labeled antibody to macronuclear H1 (26). Instead, three peptides (α, β, γ) appear to be associated with linker DNA in micronuclear chromatin. Compositional analyses of these proteins have not yet been performed. It should be noted, however, that they do not show the perchloric acid solubility typical of H1, or can they be dissociated from chromatin with 0.5M NaCl. Thus micronuclei share with other transcriptionally inert nuclei such as red blood cells (27) and some histone-containing sperm (28) the property of having an altered complement of H1 type histones.

5. MACRONUCLEAR-SPECIFIC VARIANTS

Two histone variants (hv1 and hv2), which are present in macro- but not in micronuclei, can be demonstrated using two-dimensional gel electrophoresis (21). Each is present in about ⅕–⅙ the concentration of a major core histone. Partial proteolytic and tryptic peptide mapping indicated that hv2 was only slightly different from H3. hv1 was not strikingly similar to any other major histone, although it had a few peptides in common with H2A and has triton-binding properties and secondary modifications most like H2A (21). Recently we have raised a polyclonal antiserum to hv1. Immuno-blotting experiments indicate that this antiserum reacts only with hv1 in *Tetrahymena* histone, indicating that hv1 and H2A are distinct immunologically as well as by peptide mapping (29).

As expected, the antiserum to hv1 stains macronuclei but not micronuclei using *in situ* immunofluorescence techniques (29), confirming the biochemical observation of macronuclear specificity. Surprisingly, however, anti-hv1 also stains small, punctate regions in the nucleoli of several mammalian cell lines (29). Preliminary experiments (Ziegler, Allis, Olmsted, and Gorovsky, manuscript in preparation) using immuno-blotting techniques, indicate that a protein with similar electrophoretic properties and with immunological cross-reactivity to hv1 exists in plants (wheat), yeast, and vertebrates (mouse). These observations suggest that *Tetrahymena* variant hv1 is an evolutionarily conserved histone, distinct from H2A. Restriction of hv1 to the transcriptionally active macronucleus, coupled with its high concentration in transcriptionally active ribosomal gene-containing chromatin of mammalian cells, suggest that this conserved variant may be preferentially associated with transcribed DNA sequences.

6. SECONDARY MODIFICATIONS

Tetrahymena macro- and micronuclei provide a favorable system for testing proposed associations between particular histone modifications and either chromosome condensation or transcriptional activity, since each of these functions is found in only one of the two nuclei. To date, three types of secondary modifications of *Tetrahymena* histones have been studied: postsynthetic (presumarly ϵ-amino lysine) acetylation, phosphorylation, and proteolysis.

6.1. Acetylation

Tetrahymena macronuclear histones H2A, H2B, H3, H4, hv1, and hv2 are highly acetylated; H1 is not. Micronuclear histones are acetylated little, if at all (30). Thus the acetylation patterns of macro- and micronuclear histones provide a striking example of the frequently noted correlation (see References 31 and 32 for review) between histone acetylation and transcriptional activity. *Tetrahymena* also contains an active histone acetylase activity and a typical, butyrate-sensitive deacetylase activity. Since micronuclear histones incorporate little or no acetate when cells are treated with high concentrations of butyrate (30), it seems likely that the absence of acetylated histones in micronuclei is due to a lack of acetylase activity rather than to a highly active deacetylase. It should be noted, however, that even in macronuclei, the acetylation of individual histones is extremely complex. The steady-state level of acetate on each core histone appears to be regulated independently. Each histone shows different rates of acetate turnover and different fractions of molecules that are not acetylated, even after prolonged treatment with high concentration of butyrate (30).

6.2. Phosphorylation

Macronuclear histones H2A, hv1, and H1 are phosphorylated. In micronuclei, the H1-type histones (α, β, γ) and H3 are phosphorylated. Phosphorylation of the H1-type histones (macronuclear H1, micronuclear α, β, and γ) and of micronuclear H3 is much higher in actively growing than in nongrowing cells, suggesting that these events may be related to progression through the cell cycle. Macronuclear H2A is phosphorylated whether cells are growing or not (33).

These observations have interesting implications for the function of histone phosphorylation (see Reference 33 for discussion). Finding H1 phosphorylation in the amitotically dividing macronucleus argues strongly that H1 phosphorylation is not involved exclusively in mitotic chromosome condensation, although the growth-dependent phosphorylation of α, β, and γ suggest that one function of phosphorylation of H1-type histone could be in mitosis. A nonmitotic function for

H1 phosphorylation in *Tetrahymena* is also suggested by the observation that heat shock induces extensive macronuclear H1 phosphorylation in starved nongrowing cells (24). The finding that H3 is phosphorylated exclusively in the micronucleus (albeit on a proteolytically modified form of the parent molecule; see later discussion) argues strongly for a uniquely mitotic function for H3 phosphorylation, especially when the spatial segregation of H3 phosphorylation in the *Tetrahymena* micronucleus is added to the temporal correlation between this event and mitosis in mammalian cells (34). Unpublished observations (Glover, Allis, Williams, Abraham, and Gorovsky, manuscript in preparation) indicate that phosphorylation of micronuclear H3 occurs at serine-10 (which becomes serine-4 in the proteolytically processed form of micronuclear H3), just as it does in higher organisms (35). The absence of H2A phosphorylation in micronuclei is surprising, since H2A phosphorylation is constitutive in all other nuclei examined. Clearly, this finding precludes an essential role for H2A phosphorylation in chromatid replication, segregation, or condensation (all of which occur in micronuclei). Macronuclear specificity of H2A phosphorylation suggests that it may be involved in transcription.

Finally, although its significance is unclear, it should be noted that trace amounts of radioactive phosphate also can be incorporated into micronuclear, but not macronuclear, H4 and H2B.

6.3. Proteolytic Processing

Two forms of H3 ($H3^F$ and $H3^S$) are observed in micronuclei (19, 20). Only one form, whose electrophoretic mobility and peptide map are indistinguishable from $H3^S$, is found in macronuclei. Pulse-chase experiments with radioactive amino acids indicated that a growth-dependent, precursor-product relationship exists between $H3^S$ and $H3^F$. Partial proteolytic peptide mapping and partial sequencing of the amino terminus of $H3^F$ and $H3^S$ clearly demonstrated that $H3^F$ was derived from $H3^S$ by a proteolytic cleavage of six amino acid residues from the amino terminus (20). Interestingly, only the proteolytically processed form of micronuclear H3 ($H3^F$) is phosphorylated (33).

The proteolytic processing of micronuclear H3 is the only example reported so far of physiologically regulated proteolysis of histones.

Its function remains unclear, but it could be related to the unusual requirement of *Tetrahymena* to phosphorylate micronuclear H3 while leaving macronuclear H3 unmodified.

7. NUCLEOSOME HETEROGENEITY IN *TETRAHYMENA*

Table 6.1 summarizes the primary sequence forms of macro- and micronuclear histones and the secondary modifications studied to date. We think it unlikely that a significant number of additional primary sequence variants exist in either nucleus. However, additional modified subspecies probably do exist, since other modifications known to occur on histones (ADP-ribosylation, methylation, complexing with nonhistone proteins) have not been extensively studied. It seems clear, even from the minimal estimates provided by the information in Table 6.1, that there is enormous potential for nucleosome heterogeneity in *Tetrahymena*. Because nucleosome heterogeneity due to both primary sequence heterogeneity (hv1, hv2) and to secondary modification (acetylation, H2A phosphorylation, H3 phosphorylation) show nuclear specificity in *Tetrahymena*, we think it likely that it results in structural and functional differences in nucleosomes that are of physiological significance.

8. HISTONE MESSAGES

The diversity of histone primary sequence types in macro- and micronuclei (Table 6.1) implies there must be a remarkable degree of specificity in the regulation of histone synthesis and deposition in *Tetrahymena*. For example, histones hv1 and hv2 must be deposited only in macronuclei and in only one-fifth to one-sixth the amounts of a major core histone. H1 must be restricted to macronuclei, while α, β, and γ are micronuclear specific. The mechanisms by which this compartmentalization is achieved, and amounts are regulated, are unknown. It seems likely that regulation of the synthesis of specific histones during the cell cycle plays a major role since macro- and micronuclei have completely nonoverlapping S-periods (36). It is also worth emphasizing that the major core histones ($H2A^F$, $H2A^S$, H2B, H3, H4) of macro- and micronuclei are indistinguishable, raising

TABLE 6.1. Primary sequence types and Secondary Modifications of *Tetrahymena* Macro- and Micronuclear Histones

Histone Type	Primary Sequence Types		Secondary Modifications	
	Macronucleus	Micronucleus	Macronucleus	Micronucleus
H1	H1[a]	α, β, γ	P	P
H2A	$H2A^F$, $H2A^S$	$H2A^F$, $H2A^S$	Ac, P	—[c]
H2B	H2B	H2B	Ac	(P)
H3	H3, hv2	H3[b]	Ac	P[d]
H4	H4	H4	Ac	(P)
?	hv1	—	Ac, P	—

[a] Electrophoretic variants are demonstrable in SDS-containing gels but appear to be due to phosphate-induced alteration of electrophoretic mobility.
[b] Electrophoretic variants ($H3^F$, $H3^S$) of micronuclear H3 are due to proteolytic processing of $H3^S$ to $H3^F$ and are therefore not true primary sequence variants.
[c] Small amounts of phosphate found in micronuclear H2A are probably due to macronuclear contamination but the existence of low levels of phosphorylated H2A in micronuclei cannot be completely ruled out.
[d] On micronuclear-specific, processed form only.

the possibility that a single group of core histones genes is expressed coordinately at different times in the cell cycle either with macronuclear H1 or with the micronuclear linker proteins (α, β, γ). These considerations have led us to initiate studies on the regulation of histone gene expression in *Tetrahymena*.

Using probes isolated from cloned yeast histone genes (kindly provided by Dr. M. Mitchell Smith and by Dr. Lynna Hereford), we have found that most of the mRNAs for histones H3 and H4 are polyadenylated. Preliminary experiments indicate that H2A and H2B mRNAs are also polyadenylated. Thus *Tetrahymena*, like yeast (37) and frog oocytes (38), are unusual in having polyadenylated histone mRNAs. On a per cell basis, the amounts of H4 and H3 mRNAs in log phase *Tetrahymena* is thirty- to fortyfold higher than in nongrowing (starved) cells. In addition, histone H3 and H4 mRNAs are loaded on polysomes in the same concentrations as the general message population in both log and starved cells [~65% loaded in log cells and ~4% loaded in starved cells (39, 40)] In both cases these mRNAs are polyadenylated, suggesting that polyadenylation bears no obvious relationship to message storage, as has been suggested for mRNAs in *Xenopus* oogenesis (41), or to mRNA loading on polysomes.

Northern gel analyses have revealed the surprising result that, for both H3 and H4, the number of different sized mRNAs exceeds the number of known proteins by one. Thus there are two H4 messages, but only a single H4 protein has been detected. Similarly, there are three H3 mRNAs, but only two H3 proteins (H3 and hv2) have been detected. All of the multiple mRNAs are found on polysomes. Interestingly, the steady-state concentrations of the three H3 and the two H4 mRNAs are not coordinately regulated. When cells are placed in starvation conditions, both forms of H4 mRNA are found on polysomes, but in relative amounts different from log phase cells. Even more strikingly, only one of the three H3 mRNAs is found on starved cell polysomes. Recently, Wu and Bonner (42) reported that 10% of all histone synthesis in CHO cells is due to the basal (noncell cycle dependent) synthesis of specific H2A variants (Z, X), a specific form of H3 (H3.3.), H2B, and H4. We have found that hv2 is preferentially synthesized and deposited in starved *Tetrahymena* nuclei (40), probably representing basal histone synthesis in this organism. Thus it is likely that the H3 mRNA found on polysomes of starved cells is for hv2, and that the differential synthesis of basal histones and replication histones is regulated at the level of mRNA abundance.

We have also observed that in *Tetrahymena*, as in other organisms (43), the steady-state concentration of histone mRNAs is regulated independently of the general message population. Thus, the thirty- to fortyfold reduction in H4 and H3 mRNAs per cell which occurs upon starvation is accompanied by only a fourfold reduction in the total content of mRNA per cell. Although the relative abundance of histone mRNAs can vary, the fraction of total H4 and H3 mRNA loaded onto polysomes is the same as for bulk mRNAs. Also, the relative amounts of different H4 and H3 messages loaded onto polysomes in both starved and growing cells is indistinguishable from the amount found in total cellular RNA. Thus there appears to be little, if any, selective translational control of *Tetrahymena* histone message; regulation of histone synthesis in *Tetrahymena* occurs largely, if not entirely, at the level of mRNA abundance.

9. HISTONE GENES

Southern blot analyses using yeast histone gene sequences as probes indicate that there are two kinds of H4 genes and 3–4 kinds of H3 genes, suggesting that each message is derived from a different gene. The H4-I gene has been cloned on an 8.6 kb Eco RI restriction fragment from a micronuclear library. A portion of this clone has been subcloned into M13, progressively shortened by the method of Hong (44) and sequenced using the Sanger dideoxy method, yielding the first histone gene sequences from a protozoan and the first protein coding sequence from *Tetrahymena*. The coding sequence is 46% G + C, in contrast to the low (<25%) G + C content of the total genome and of the flanking sequence. Yeast and *Tetrahymena* H4 genes show remarkable (~80%) conservation of nucleotide sequence in the coding regions. No evidence was found for linkage of H4-I to H4-II or to H3, but the H4-II gene appears to be linked to an H3 sequence. Utilizing nullisomic strains (lacking both copies of one or more of the 5 micronuclear chromosomes but having a normal macronuclear DNA complement; 45) the two types of H4 genes were shown to be on different chromosomes. Thus, *Tetrahymena* H4 genes are similar to yeast H4 genes in number of types, in polyadenylation of their mRNAs, in coding sequence, and in dispersion. Unlike yeast, *Tetrahymena* H4 genes are not always associated with H3 genes.

Comparisons of Southern blots of macro- and micronuclear DNA indicate that there is no detectable amplification or rearrangement of the two H4 genes. Since these sequences account for all of the H4 genes, the extensive sequence rearrangement (2) that accompanies macronuclear development need not occur in close proximity to these genes for them to be transcriptionally active. The existence of DNAse I-hypersensitive sites (unpublished observations) specific to macronuclear chromatin flanking the H4-I gene suggests that chromatin differentiation rather than gene arrangement is responsible for transcriptional activation of histone genes during macronuclear development.

These studies, though still in their early stages, clearly indicate the feasibility of isolating and characterizing the entire complement of *Tetrahymena* histone genes. Such analyses should not only provide the complete sequences of the proteins themselves, but will also provide insights into the process of gene activation during macronuclear development and the mechanisms regulating differential gene expression in *Tetrahymena*.

NOTE ADDED IN PROOF

The complete amino acid sequence of histone H4 (46) and of the two major forms of H2A (47) of *Tetrahymena pyriformis* have recently been published.

ACKNOWLEDGMENT

The unpublished work described here was supported by research grants from the National Institutes of Health and by an American Cancer Society postdoctoral fellowship to Gary A. Bannon. We are grateful to our colleagues, both past and present, whose published and unpublished results provided much of the substance for this review.

REFERENCES

1. M.-C. Yao and J. G. Gall, *J. Protozool.* **26,** 10–13 (1979).
2. M. A. Gorovsky, *Ann. Rev. Genet.* **14,** 203–239 (1980).
3. M. A. Gorovsky and J. B. Keevert, *Proc. Natl. Acad. Sci. USA* **72,** 3536–3540 (1975).
4. M. A. Gorovsky, C. Glover, C. A. Johmann, J. B. Keevert, D. J. Mathis, and M. Samuelson, *Cold Spring Harbor Symp. Quant. Biol.* **42,** 493–503 (1977).
5. L.-L. Frado, A. T. Annunziato, and C. L. F. Woodcock, *Biochim. Biophys. Acta* **475,** 514–520 (1977).
6. C. P. Giri and M. A. Gorovsky, *Nucl. Acids Res.* **8,** 197–214 (1980).
7. J. O. Thomas and R. J. Thompson, *Cell* **10,** 633–640 (1977).
8. H. J. Lipps and N. R. Morris, *Biochem. Biophys. Res. Commun.* **74,** 230–234 (1977).
9. K. J. Vavra, D. S. Pederson, and M. A. Gorovsky, *Nucl. Acids Res.* **9,** 5825–5943 (1981).
10. S. Borchsenius, B. Bonven, J. C. Leer, and O. Westergaard, *Eur. J. Biochem.* **117,** 245–250 (1981).
11. M. A. Gorovsky and J. B. Keevert, *Proc. Natl. Acad. Sci. USA* **72,** 2672–2676 (1975).
12. K. Hamana and K. Iwai, *J. Biochem.* **69,** 1097–1111 (1971).
13. C. A. Johmann and M. A. Gorovsky, *Biochemistry* **15,** 1249–1256 (1976).
14. C. V. C. Glover and M. A. Gorovsky, *Biochemistry* **17,** 5705–5713 (1978).
15. C. A. Johmann and M. A. Gorovsky, *Arch. Biochem. Biophys.* **175,** 694–699 (1976).
16. C. V. C. Glover and M. A. Gorovsky, *Proc. Natl. Acad. Sci. USA* **76,** 585–589 (1979).
17. C. V. C. Glover, Ph.D. Thesis, University of Rochester, Rochester, N.Y. (1979).
18. M. Nomoto, H. Hayashi, and K. Iwai, *J. Biochem.* **91,** 897–904 (1982).
19. C. D. Allis, C. V. C. Glover, and M. A. Gorovsky, *Proc. Natl. Acad. Sci. USA* **76,** 4857–4861 (1979).
20. C. D. Allis, J. K. Bowen, G. N. Abraham, C. V. C. Glover, and M. A. Gorovsky, *Cell* **20,** 55–64 (1980).
21. C. D. Allis, C. V. C. Glover, J. K. Bowen, and M. A. Gorovsky, *Cell* **20,** 609–617 (1980).
22. I Isenberg, *Ann. Rev. Biochem.* **48,** 159–191 (1979).
23. M. A. Gorovsky, J. B . Keevert, and G. L. Pleger, *J. Cell Biol.* **61,** 134–145 (1974).

24. C. V. C. Glover, K. J. Vavra, S. D. Guttman, and M. A. Gorovsky, *Cell* **23,** 73–77 (1981).
25. P. C. Billings, J. W. Orf, D. K. Palmer, D. A. Talmage, C. G. Pan, and M. Blumenfeld, *Nucl. Acids Res.* **6,** 2151–2164 (1979).
26. C. A. Johmann and M. A. Gorovsky, *J. Cell Biol.* **71,** 89–95 (1976).
27. A. Ruiz-Carrillo, L. J. Wangh, V. C. Littau, and V. G. Allfrey, *J. Biol. Chem* **249,** 7358–7368 (1974).
28. D. Alder and M. A. Gorovsky, *J. Cell Biol.* **64,** 389–397 (1975).
29. C. D. Allis, Y. S. Ziegler, M. A. Gorovsky, and J. B. Olmsted, *Cell* **31,** 131–136 (1982).
30. K. J. Vavra, C. D. Allis, and M. A. Gorovsky, *J. Biol. Chem.* **257,** 2591–2598 (1982).
31. V. G. Allfrey, *Chromatin and Chromosome Structure*, Academic, New York, 1977, pp. 167–191.
32. D. Mathis, P. Oudet, and P. Chambon, *Prog. Nucl. Acid Res. Mol. Biol.* **24,** 1–55 (1980).
33. C. D. Allis and M. A. Gorovsky, *Biochemistry* **20,** 3828–3833 (1981).
34. L. R. Gurley, R. A. Tobey, R. A. Walters, C. E. Hildebrand, P. G. Hohmann, J. A. D'Anna, S. S. Burham, and L. L. Deaven, *Cell Cycle Regulation*, Academic, New York, 1978, pp. 37–60.
35. J. R. Paulson and S. S. Taylor, *J. Biol. Chem.* **257,** 6064–6072 (1982).
36. J. Woodard, E. Kaneshiro, and M. A. Gorovsky, *Genetics* **70,** 251–260 (1972).
37. K. Fahrner, J. Yarger, and L. Hereford, *Nucl. Acids Res.* **8,** 5725–5738 (1980).
38. J. V. Ruderman and M. L. Pardue, *J. Biol. Chem.* **253,** 2018–2025 (1978).
39. F. J. Calzone, R. C. Angerer, and M. A. Gorovsky, submitted to *J. Biol. Chem.* (1982).
40. G. A. Bannon, F. J. Calzone, J. K. Bowen, C. D. Allis, and M. A. Gorovsky, *Nucl. Acids Res.* **11,** 3903–3917 (1983).
41. J. V. Ruderman, H. R. Woodland, and E. A. Stargess, *Dev. Biol.* **71,** 71–82 (1979).
42. R. S. Wu and W. M. Bonner, *Cell* **27,** 321–330 (1981).
43. D. E. Wells, R. M. Showman, W. H. Klein, and R. A. Raff, *Nature* **292,** 477–478 (1981).
44. G. F. Hong, *J. Mol. Biol.* **158,** 539–549 (1982).
45. P. J. Bruns and T. E. B. Brussard, *Science* **213,** 549–551 (1981).
46. H. Hayashi, M. Nomota, and K. Iwai, *Proc. Japan Acad.* **56,** Ser. B, 579–584 (1980).
47. Y. Fusauchi and K. Iwai. *Proc. Japan Acad.* **93,** 1487–1497 (1983).

CHAPTER 7

ORGANIZATION AND EXPRESSION OF HISTONE GENES IN THE BRINE SHRIMP, *ARTEMIA*

JOSEPH C. BAGSHAW,[1] MATTHEW T. ANDREWS, AND BRIAN A. PERRY
Department of Biochemistry
Wayne State University School of Medicine
Detroit, Michigan

[1] Present address: Department of Biology and Biotechnology, Worcester Polytechnic Institute, Worcester, MA 01609.

CONTENTS

1. INTRODUCTION
2. CLONING AND ISOLATION OF *ARTEMIA* HISTONE GENES
3. ORGANIZATION OF HISTONE GENES IN *ARTEMIA*
4. HISTONE GENE EXPRESSION IN EARLY LARVAE
5. 5S-RELATED SEQUENCES INTERSPERSED WITH HISTONE GENES IN *ARTEMIA*
6. CONCLUSION: PRESENT STATUS AND FUTURE GOALS

REFERENCES

1. INTRODUCTION

Postgastrula embryonic and early larval development of the brine shrimp, *Artemia*, provide a useful system for studies of gene expression in a developing eucaryote. *Artemia* is a small crustacean native to hypersaline habitats throughout the world. Embryos develop within the ovisac to early gastrula, then development is reversibly arrested, and the now dormant embryos are encysted in a chitinous shell and expelled into the environment. The dormant embryos accumulate along the shoreline in large windrows, where they are desiccated by wind and sun. The encysted and dessicated embryos are completely dormant but will resume development immediately upon rehydration at a suitable temperature. The rehydrated embryos develop synchronously inside the shell, producing prenauplius larvae which emerge at a predictable time. Shortly thereafter, free-swimming nauplius larvae hatch. Encysted gastrulae, commonly but erroneously called brine shrimp "eggs," are collected and sold to aquarium supply shops and commercial hatcheries and fish farms; thus the dormant embryos are available at any time in large quantity. They can be stored for years, and synchronously developing embryos and early larvae can be obtained at the investigator's convenience. We routinely raise 60 g of embryos or 30–40 g of synchronously hatched larvae in one batch. These practical advantages have enabled many research groups to study various aspects of *Artemia* development (1, 2).

We are exploiting the inherent biological advantages of *Artemia* to identify and isolate specific genes of interest, and a particularly attractive model system is the expression of histone genes. In *Artemia*, DNA synthesis does not occur between the resumption of development of dormant embryos and the hatching of prenauplius larvae (3). In early larvae, DNA synthesis is restricted to a period immediately after hatching (4). Thus the system is naturally synchronous with respect to DNA synthesis. Since histones are synthesized in nauplius larvae in concert with DNA replication, but not in preemergence embyros (5), we can easily produce tens of grams of synchronously developing organisms that are or are not (our choice) expressing their histone genes. This is not possible with other animals commonly used in developmental studies.

The organization of histone genes in *Artemia* is also a subject of more than passing interest, owing to the phylogenetic position of this organism. Other chapters in this book show that the sea urchins, *Drosophila*, and several vertebrates exhibit a variety of patterns of histone gene organization. Among these only *Drosophila* is a protostome, a member of one of the two major groups of animal phyla that diverged more than 600 million years ago. *Artemia* is a primitive crustacean, thus a protostome, and determining the order and transcriptional polarity of the histone genes in *Artemia* could help to elucidate the possible pathways of histone gene evolution.

2. CLONING AND ISOLATION OF *ARTEMIA* HISTONE GENES

The first step in the isolation of the *Artemia* histone genes was construction of a "library" (6, 7) of *Artemia* DNA sequences cloned in the bacteriophage vector lambda Charon 4A. To screen the library for histone genes, we used cloned *Drosophila* genes from the plasmid cDm500 (8), a gift of Dr. David Hogness. A DNA fragment containing one repeat unit of histone genes was excised from cDm500 by cleavage with Hind III, isolated by agarose gel electrophoresis, and radiolabeled by nick translation. Approximately 10,000 phage from the *Artemia* gene library were plated, and the resulting plaques were transferred to nitrocellulose, as described by Benton and Davis (9), and hybridized with the histone gene probe. Four probe-positive recombinants were found, and each of these was plaque purified by rescreening at lower phage densities. The four isolates, designated λBSh1, λBSh2, λBSh3, and λBSh4 were then propagated in bulk for preparation and analysis of DNA.

Restriction maps of the inserts of all four isolates, oriented as they exist in the recombinant phage genomes, are shown in Figure 7.1. It is immediately evident that these four isolates arose from four independent cloning events. Recombinants λBSh1 and λBSh3 each have the same set of inserted Eco RI fragments, with lengths of approximately 3.4, 3.0, and 2.1 kb, but these fragments are arranged in opposite orientations in these two isolates. In λBSh4, the 2.1-kb fragment is divided by an additional Eco RI site into fragments of 1.2 and 0.9 kb. The Eco RI fragments of λBSh2, which measure

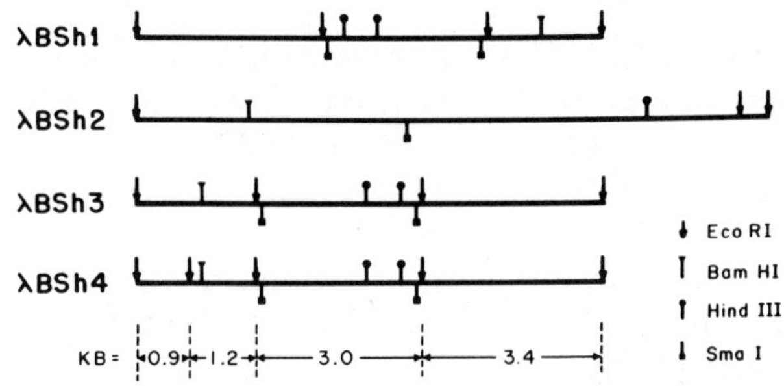

FIGURE 7.1. Restriction maps of recombinant bacteriophage DNAs. Only the insert regions of the phage DNAs are shown, oriented as they exist in the bacteriophage genomes, that is, with the left cloning arm of the vector to the left, and so on. Lengths of the Eco RI fragments of λBSh4 are given in kilobase pairs (kb).

10.9 and 0.6 kb, are very different from the other three, but hybridization studies have shown that the inserted DNA sequences in λBSh2 are homologous to specific fragments of the other isolates. The restriction maps of the inserted *Artemia* DNA in λBSh1, λBSh3, and λBSh4 are identical except for the additional Eco RI site in λBSh4. The restriction maps suggest that the structures of λBSh1, 3, and 4 represent the most common form of histone gene repeat unit in *Artemia*, and this has been confirmed by hybridization of histone gene subclones to restriction fragments of *Artemia* genomic DNA (see Section 3). The insert in λBSh2 appears to represent a rare variant of histone gene structure.

3. ORGANIZATION OF HISTONE GENES IN *ARTEMIA*

Restriction fragments derived from λBSh2 DNA were hybridized with a probe containing one complete set of *Drosophila* histone genes by the method of Southern (10) to determine the location of histone genes in this recombinant. The results revealed two regions of cross-hybridization: the Eco RI to Bam HI fragment at the left end of the insert, and a region at the right end including the small Eco RI fragment. The Eco RI to Bam HI fragment was subcloned into pBR322, and the resulting plasmid (pPS11) was used for more detailed mapping of this fragment. We also obtained from Dr. Carl Parker a set of recombinant plasmids, each containing portions of one or two *Drosophila* histone genes. From these we isolated probes specific for single histone genes, which we then hybridized with Hae III fragments of the insert region isolated from plasmid pPS11. The results of these experiments are illustrated in Figure 7.2. Hybridization of probes for histone genes H2B, H3, and H4 to specific Hae III fragments indicated that these three genes lie within the Eco RI to Bam HI fragment in the order shown. Placement of the H2A gene is based on the observation that a probe for this gene hybridizes weakly with the left end of the pPS11 insert and strongly with the small Eco RI fragment at the right end of the λBSh2 insert. We have not detected any cross-hybridization between a cloned *Drosophila* H1 gene and any *Artemia* DNA sequence, probably owing to greater sequence divergence in this gene than in other histone genes.

Preliminary experiments have allowed a less precise assignment of histone genes to the Eco RI fragments of λBSh4, as shown in Figure 7.2. Tandem repetition of a structure similar to the insert in λBSh4 (see later discussion) permits a gene order of H2A-H4-H3-H2b, the same order found in λBSh2. However, because λBSh2 is a rare variant, it will be essential to establish independently the order of histone genes in λBSh4. Note that the order found in λBSh2 is different from the order of histone genes in *Drosophila* as well as in sea urchins (see appropriate chapters in this book).

As noted previously (Section 2), the similarity in structure among λBSh1, 3, and 4 suggested that these three recombinants contained examples of the most common histone gene repeat unit in the *Artemia* genome. The presence of a single Bam HI site and closely spaced

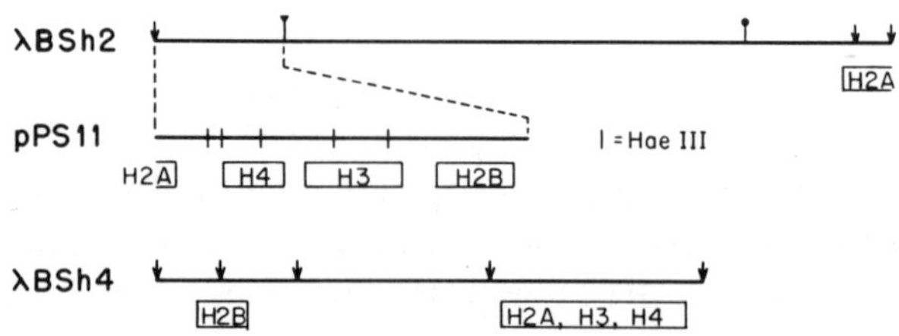

FIGURE 7.2. Histone gene order in λBSh2 and λBSh4. Plasmid pPS11 was constructed by inserting the indicated Eco RI to Bam HI fragment into pBR322. Histone genes were located within this insert by hybridizing gene-specific probes with Hae III fragments as described in the text.

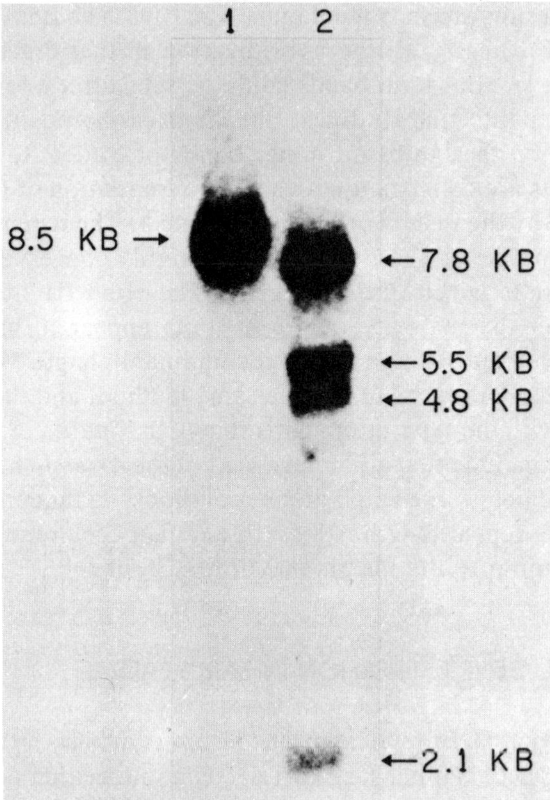

FIGURE 7.3. Southern blot analysis of *Artemia* genomic DNA. *Artemia* DNA was digested with either Bam HI (lane 1) or Hind III (lane 2). The resulting fragments were separated on agarose gels, blotted onto "Gene Screen" (New England Nuclear), and hybridized with nick-translated plasmid pPS11 DNA. Washed filters were exposed to Kodak XAR film with one Du Pont Cronex Lightning Plus intensifying screen. The indicated fragment sizes were estimated graphically, using as standards restriction fragments of phage lambda DNA run on the same gel.

Hind III sites in these three inserts enabled us to test the hypothesis that the *Artemia* histone genes are arranged in a tandem repetition of this structure. First we determined that plasmid pPS11 hybridizes with the 2.1 and 3.4 kb Eco RI fragments of λBSh3 and with the 0.9, 1.2, and 3.4 kb fragments of λBSh4, as predicted from the structures shown in Figure 7.2. We then used pPS11 as a probe in Southern blot analysis of Bam HI and Hind III fragments of *Artemia*

DNA. The results are shown in Figure 7.3. An 8.5 kb Bam HI fragment accounted for nearly all the hybridization in that digest, although two or three smaller faint bands could be seen after a long exposure of the film. In the Hind III digest, the major probe-positive fragment was 7.8 kb, with significant minor bands of 5.5, 4.8, and 2.1 kb. These results are consistant with a tandem repetition of the structure represented by the inserts of λBSh 1, 3, and 4. The repeat unit length is 8.5 kb, and the single Bam HI site is highly conserved, probably because it falls within the H2B gene. The Hind III sites are more variable, but the majority of repeat units appear to have the two closely spaced sites found in our recombinant phage. We have also examined Eco RI fragments by the same method, and the results are consistent with the restriction maps shown in Figure 7.1. From these results we conclude that all or nearly all of the *Artemia* histone genes are arranged in one or a small number of blocks of tandemly repeated sets, that the repeat length is 8.5 kb, and that each repeat unit has a structure similar to the inserts in λBSh 1, 3, and 4.

4. HISTONE GENE EXPRESSION IN EARLY LARVAE

As noted before, DNA replication and histone synthesis are coordinated in newly hatched nauplius larvae of *Artemia*, but neither occurs during preemergence development of encysted embryos (3–5). Amaldi et al. (11) detected histone mRNAs by translation *in vitro* of poly(A)-minus RNAs isolated from hatched larvae, but found no histone mRNAs in dormant embryos by the same test. We have confirmed these results, using mRNA preparations that were not fractionated according to poly(A) content. The absence of functional histone mRNAs, as judged by translation *in vitro*, can not rule out the possible presence of "masked" histone mRNAs in dormant or developing embryos. This possibility can be tested by employing molecular hybridization with a suitable probe, which will react with the putative mRNAs regardless of their functional status.

We used the recombinant λBSh2 to determine whether the dormant embryos of *Artemia* contain cryptic or "masked" histone mRNAs. Whole-cell RNAs were extracted from dormant embryos and hatched larvae, fractionated by agarose gel electrophoresis in the presence of formaldehyde (12), and blotted onto nitrocellulose (13). These RNAs were then hybridized with λBSh2 DNA radiolabeled by nick translation. The results of one such experiment are shown in Figure 7.4. The λBSh2 probe hybridized with a cluster of larval RNAs which, in a separate experiment, were shown to have the same electrophoretic mobility as *Drosophila* core histone mRNAs (5). No such hybridization signal appeared in the lane containing RNAs from dor-

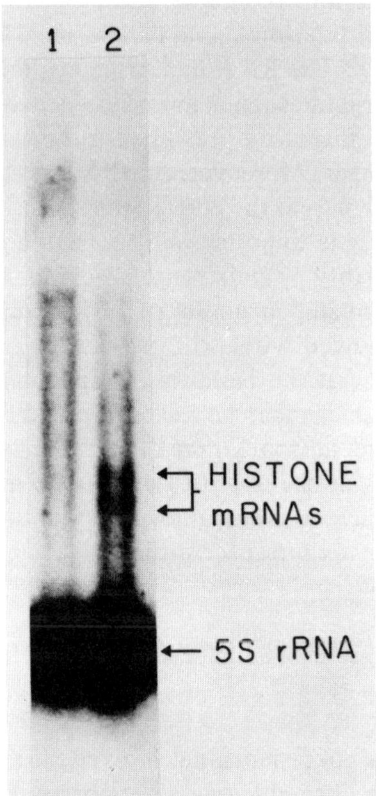

FIGURE 7.4. Northern blot analysis of *Artemia* RNAs. Whole-cell RNA was prepared from dormant embryos (lane 1) or hatched larvae harvested 48 hours after immersion of dormant embryos (lane 2). RNAs fractionated on agarose gels in 2.2 M formaldehyde were blotted onto nitrocellulose and hybridized with nick-translated λBSh2 DNA. The x-ray film shown here was exposed for 10 days. A short (4-hour) exposure showed a single sharp signal at the position labeled 5S rRNA.

mant embryos. We have not yet determined the lower limit of detection in these experiments, and we can not say with certainty that histone mRNAs are completely absent in dormant embryos. However, it is clear from these results that the vast majority of histone mRNAs present in hatched larvae must arise from transcription of the histone genes.

It is apparent from Figure 7.4 that λBSh2 hybridizes with RNA molecules other than the core histone mRNAs. The small, abundant RNA near the bottom of the gel is discussed in detail in Section 5. An additional hybridization signal can be seen some distance above the histone mRNAs. The RNA that produces this signal has not yet been identified. It migrates much more slowly than *Drosophila* histone H1 mRNA, and is therefore presumed to be larger than a putative *Artemia* H1 messenger. Moreover, this RNA is substantially depleted in hatched larvae, whereas the core histone messengers become moderately abundant. It is hypothetically possible that this RNA is an oversized and possibly polycistronic precursor to the larval histone mRNAs, but the unusual structure of λBSh2 suggests that this possibility must be viewed with extreme caution. A direct test of this possibility must await the isolation of specific subcloned *Artemia* histone genes, which can then be used as defined hybridization probes. In any case, it is evident from Figure 7.4 that the hypothetical precursor could not account for the quantity of histone mRNAs found in the hatched larvae. Thus the great majority of larval histone messengers must arise from *de novo* transcription.

5. 5S-RELATED SEQUENCES INTERSPERSED WITH HISTONE GENES IN *ARTEMIA*

The hybridization experiments that showed the absence of detectable histone mRNAs in dormant *Artemia* embryos also provided an unexpected and interesting surprise. The recombinant DNA λBSh2 hybridizes to a small, highly abundant RNA found in both embryos and larvae. Shorter exposures of the same blots show that this hybridization signal emanates from a single RNA band. The hybrids formed between this RNA and λBSh2 are quite strong, surviving the most stringent washing criteria we use (3 mM Tris base at room temperature). The size of this RNA, about 120 nucleotides, and its presence in polysomal but not in cytoplasmic RNA, suggested that it might be 5S ribosomal RNA. This was confirmed by the observation that the RNA was found exclusively in the 60S subunit of *Artemia* ribosomes. The possibility that the small RNA might be 5.8S rRNA could be excluded by a lack of restriction site homology between λBSh2 and another recombinant, λBSrl, that contains the *Artemia* rRNA (18S + 5.8S + 28S) genes (15). More recently we have confirmed this identification by hybridization with a *Drosophila* 5S RNA gene probe excised from the plasmid 12D1 (15). This probe hybridizes to the same small abundant RNA as does λBSh2, and to specific restriction fragments of λBSh1, 2, 3, and 4. The Eco RI fragments from all four recombinants were fractionated and blotted, then hybridized either with *Drosophila* históne genes from a plasmid containing one full repeat unit (pSK10), or with *Drosophila* 5S RNA genes. The locations of histone and 5S genes in these four recombinant phage are shown in Figure 7.5. As described, histone genes are located in the Eco RI fragments at the ends of the inserts of λBSh1, 3, and 4, and the analogous regions of λBSh2. The 12D1 probe hybridized only to a region of λBSh1, 3, and 4 in the central Eco RI fragment and bounded by one of the Hind III sites and one of the

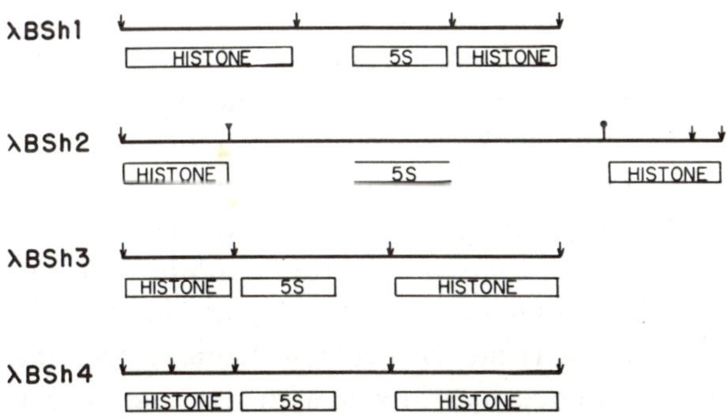

FIGURE 7.5. Arrangement of histone genes and 5S-related sequences in recombinant phage DNAs. Restriction fragments were analyzed by the method of Southern (10). The probe for histone genes was pSK10, a plasmid constructed by inserting one repeat unit of *Drosophila* histone genes from cDm500 into the Bam HI site of pBR322. No attempt was made to map individual histone genes. The probe for 5S-related sequences was 12D1, a plasmid containing *Drosophila* 5S RNA genes.

Sma I sites in that fragment. The exact location of 5S-related sequences in λBSh2 is less clearly defined; it lies between the Bam HI and Hind III sites, and on both sides of the single Sma I site. The latter observation is particularly significant because the *Artemia* 5S RNA sequence (16) does not include a Sma I site. If the 5S-related sequences in λBSh2 are authentic or nearly authentic 5S genes, then there must be at least two copies of the gene, one on each side of the Sma I site.

The *Drosophila* genomic DNA sequences that hybridize with 12D1, presumably comprising the authentic 5S RNA genes, are tandemly linked, apparently in a single large block (15). We used the 12D1 insert as a probe to explore the organization of the cross-hybridizing sequences in the *Artemia* genome. *Artemia* DNA was digested with several different restriction enzymes, and the fragments were analyzed by Southern blots with 12D1 as the probe. The enzymes Eco RI, Bam HI, and Hind III, which do not have recognition sites in the *Artemia* 5S sequence (16), each produced a large, probe-positive fragment that barely entered a 0.8% agarose gel. In addition, each enzyme produced a broad spectrum of probe-positive fragments ranging in size from greater than 20 kb to less than 2 kb. Sau 3A, which has one recognition site in the *Artemia* 5S sequence, also produced a very heterogeneous pattern of fragments that hybridized with 12D1. From these results, it appears that the *Artemia* genomic sequences that cross-hybridize with *Drosophila* 5S RNA genes (and their associated spacers) are highly heterogeneous in their organization. Some of these sequences are arranged in a large block of tandem repeats, but many others are either very heterogeneous in sequence, and thus in restriction sites, or are scattered throughout the *Artemia* genome. From the structures of the four recombinant phage, it is apparent that some of these 5S-related sequences are interspersed with histone genes in *Artemia*, and this hypothesis is supported by southern blot analysis of genomic DNA. A plasmid (pPS13) containing the central Bam HI to Hind III fragment from λBSh2 and including the 5S-related sequence hybridizes with the same 8.5-kb Bam HI fragment and 7.8-kb Hind III fragment shown in Figure 7.3. In addition, pPS13 hybridizes with a 3-kb Eco RI fragment and a 2-kb Hind III to Eco RI fragment. These are the results expected if the 5S-related sequences are interspersed between sets of histone genes, the combined repeat unit having the structure shown for λBSh1, 3, and 4 in Figures 7.1 and 7.5. From these results we conclude that some 5S genes (or pseudogenes) are interspersed with histone genes in the *Artemia* genome.

6. CONCLUSION: PRESENT STATUS AND FUTURE GOALS

The studies presented here have answered certain questions about the organization and expression of histone genes in *Artemia*, and they have also defined some interesting new questions for further investigation. Our initial objective in taking up this project was to determine how—indeed, whether—histone gene expression is controlled during postgastrula embryonic and early larval development of *Artemia*. For this question, we now have a first approximation to an answer. Histones are synthesized in newly hatched larvae at the time of DNA replication (5). The majority, and quite possibly all of the histone mRNAs present in the larvae appear to arise by *de novo* transcription of histone genes, rather than by recruitment from a preexisting pool of "masked" messengers. It remains to be directly demonstrated that histone genes are actively transcribed in the nuclei of hatched larvae, but we now have the tools necessary for this task. Assuming that this is the case, we have the ability to produce tens of grams—hundreds if necessary—of synchronously developing organisms in which the structural components that dictate transcription of the histone genes are in place. Equally important, we can produce a like amount of organisms in which the same genes are transcriptionally inactive. This contrast could be of great value in defining the molecular parameters that control transcription of the histone genes in a biologically relevent setting, that is, the living organism.

We have addressed two questions regarding the organization of the *Artemia* histone genes: Are the genes clustered in a tandem array? What is the order of the individual histone genes?

The results of Southern blot analysis of genomic DNA fragments clearly indicate that the vast majority of *Artemia* histone genes are indeed clustered, possibly in a single block of tandemly repeated units. This majority repeat unit is experimentally defined by the 8.5-kb Bam HI fragment shown in Figure 7.3. There are, however, a small number of DNA fragments that hybridize with the histone gene probe but do not have the structure of the major repeat unit. Considering that a typical *Artemia* DNA preparation is derived from hundreds of thousands of individual organisms, these minor structures are much too abundant to be "orphons" of the type found in other species. They may simply represent a few sequence variants scattered throughout the histone gene block, or possibly the borders of more than one block of repeats. A more interesting possibility is that the minor structures encode developmental stage-specific histones. At present we have no evidence that any such stage-specific histone variants exist in *Artemia*. On the other hand, we have no direct evidence that the genes we have isolated actually encode the histone messengers found in early larvae. The possibility exists that the major repeat units encode early, pregastrula histones (compare sea urchins). To test this possibility, we will need to examine very carefully the extent of base sequence homology between the larval mRNAs and the genes we have isolated.

We have established a tentative order for the *Artemia* core histone genes of H2A-H2B-H3-H4 (Figure 7.2). This order is distinct from that of the major repeat unit in *Drosophila* (8), and interconversion of the *Artemia* and *Drosophila* patterns would require the complete transposition of at least one gene. Even more rearrangement might be required to achieve the same transcriptional polarities, which we have yet to determine for the *Artemia* histone genes. However, major organizational differences between *Artemia* and *Drosophila* are not too surprising, given the evolutionary distance between the two. Current paleontological, morphological, and embryological evidence divides the arthropods into three taxonomically and evolutionarily distinct phyla: the Chelicerata, the Crustacea, and the Uniramia (17, 18). The presence of Crustacea in the mid-Cambrian period implies that the early Crustacea and Uniramia (the latter include the modern insects) must have diverged over 500 million years ago. Thus, in terms of evolutionary distance, *Artemia* is scarcely more related to *Drosophila* than to sea urchins and vertebrates.

The presence of sequences related to 5S ribosomal RNA interspersed with the *Artemia* histone genes came as a complete surprise. Individual isolates of cloned *Drosophila* histone genes have been found to contain contiguous or inserted nonhistone sequences, including the mobile genetic element named 297 (19, 20). However, interspersion of histone genes and a defined nonhistone sequence in a regularly repeated structure has not been previously reported. At present, all we know about the 5S-related sequence in our recombinants is that it occurs in a constant position with respect to the histone genes and that it hybridizes exclusively with 5S RNA. Whether this sequence is a "true" gene or a pseudogene remains to be determined by direct DNA sequencing. How this curious juxtaposition arose is completely open to question, but the apparent presence of the 5S-related sequence in each copy of the repeated structure suggests that the construction occurred prior to duplication of the histone gene set during evolution of the ancestors of *Artemia*. We are presently refining techniques for transcription *in vitro*, using a whole cell extract from *Artemia*, and we will be very eager to learn whether the 5S-related sequence subcloned from λBSh4 will support RNA polymerase-III-dependent transcription *in vitro*. If this proves to be true, it will be even more interesting—and challenging—to determine whether these sequences are transcribed *in vivo*.

ACKNOWLEDGMENTS

The authors wish to express their thanks to Drs. David Hogness, Carl Parker, Walter Gehring, and Sarah Elgin for gifts of recombinant plasmids, and to Ms. Paula Smolenski for indispensible technical assistance. This work was supported by National Science Foundation Research Grant No. PCM 81-11485 to Joseph C. Bagshaw.

REFERENCES

1. J. C. Bagshaw and A. H. Warner, Eds., *Biochemistry of Artemia Development*, University Microfilms International, Ann Arbor, Michigan, 1979.
2. G. Persoone, P. Sorgeloos, O. Roels, and E. Jaspers, Eds. *The Brine Shrimp Artemia*, Universa Press, Wetteren, Belgium, 1980.
3. Y. H. Nakanishi, T. Okigaki, H. Kato, and T. Iwasaki, *Annot. Zool. Japon* **35,** 223–228 (1962).
4. D. K. McClean and A. H. Warner, *Dev. Biol.* **24,** 88–105 (1971).
5. J. C. Bagshaw, *Dev. Genet.* **3,** 41–51 (1982).

6. T. Maniatis, R. C. Hardison, E. Lacy, J. Lauer, C. O'Connell, D. Quon, G. K. Sim, and A. Efstratiadis, *Cell* **15,** 687–701 (1978).
7. B. Hohn and K. Murray, *Proc. Nat. Acad. Sci. USA* **74,** 3259–3263 (1977).
8. R. P. Lifton, M. L. Goldberg, R. W. Karp, and D. S. Hogness, *Cold Spring Harbor Symp. Quant. Biol.* **42,** 1047–1051 (1978).
9. W. D. Benton and R. W. Davis, *Science* **196,** 180–182 (1977).
10. E. M. Southern, *J. Mol. Biol.* **98,** 503–517 (1975).
11. P. O. Amaldi, L. Felicetti, and N. Campioni, *Biochim. Biophys. Acta* **518,** 518–524 (1978).
12. N. Rave, R. Crkvenjakov, and J. Boedtker, *Nucl. Acids Res.* **6,** 3559–3567 (1979).
13. P. S. Thomas, *Proc. Natl. Acad. Sci. USA* **77,** 5201–5205 (1980).
14. J. C. Vaughn, D. J. Whitman, J. C. Bagshaw, and J. C. Helder, *Biochim. Biophys. Acta* **697,** 156–161 (1982).
15. S. Artavanis-Tsakonas, P. Scheld, C. Tschudi, V. Pirrotta, R. Steward, and W. J. Gehring, *Cell* **12,** 1057–1067 (1977).
16. L. Diels, R. DeBaere, A. Vandenbergheand, and R. Wachter, *Nucl. Acids Res.* **9,** 19 (1981).
17. S. M. Manton and D. T. Anderson, in M. R. House, Ed., *The Origin of Major Invertebrate Groups*, Academic, New York, 1979, pp. 269–321.
18. H. B. Whittington, *ibid.*, pp. 253–268.
19. R. W. Karp, doctoral dissertation, Stanford University, 1979.
20. H. Ikenaga and K. Saigo, *Proc. Natl. Acad. Sci. USA* **79,** 4143–4147 (1982).

PART 2

INTERMEDIATE EUCARYOTES

CHAPTER 8

DNA ORGANIZATION AND EXPRESSION OF *XENOPUS* HISTONE GENES

W. M. A. M. VAN DONGEN, A. F. M. MOORMAN,
AND O. H. J. DESTRÉE
Anatomisch-Embryologisch Laboratorium
Universiteit van Amsterdam
Amsterdam, The Netherlands

CONTENTS

1. INTRODUCTION

2. SEQUENCE ORGANIZATION OF A CLONED *XENOPUS LAEVIS* HISTONE GENE CLUSTER

3. GENOMIC ORGANIZATION OF THE HISTONE GENES IN *XENOPUS LAEVIS LAEVIS*
   3.1. Genomic Organization of Histone Genes in Other Species of the Genus *Xenopus*

4. EXPRESSION OF HISTONE GENES DURING OOGENESIS AND EARLY EMBRYOGENESIS IN *XENOPUS LAEVIS*
   4.1. No Evidence for Major Changes in the mRNA Population Coding for the Nucleosomal Histones During Early Development in *Xenopus laevis*

5. DIFFERENT SPECIES USE DIFFERENT STRATEGIES FOR THE SYNTHESIS OF HISTONE mRNA AND HISTONE PROTEINS DURING EARLY DEVELOPMENT

REFERENCES

1. INTRODUCTION

Differential expression of histone genes during development of *Xenopus* has been known for a decade (1, 2, 3). Changes in the histone protein complement in *Xenopus* development appear to be rather subtle in comparison with those found in sea urchins (see 4, 5). Although the study of histone gene expression in early sea urchin embryos set the ideas, it is clear now that *Xenopus* uses a different strategy in histone synthesis, at least in early development. Whereas in sea urchins most of the histone mRNA is made in early embryogenesis for immediate use, in *Xenopus* histone mRNAs are stored along with histone proteins in the oocyte for later use (see 4, 5). These developmental aspects and the fact that histone genes are reiterated only 45–50-fold in *Xenopus* (6, 7) make it a very interesting system for the study of histone gene expression.

Here we review recent studies on the molecular organization and expression of histone genes in *Xenopus*. In Section 2 we describe some features of a cloned histone DNA fragment that enabled us to study the organization of histone genes in genomic DNA of *Xenopus* and that was used to select other histone DNA clones (Section 3).

The study of the expression of histone genes in *Xenopus* was initiated with emphasis on oogenesis and early development as described in Section 4. As no evidence for major changes in the mRNA population of nucleosomal histone genes was found in early development of *Xenopus* (4.1), a short survey is given of the different strategies that different species use to supply their needs in histones and histone mRNAs during early development (Section 5).

2. SEQUENCE ORGANIZATION OF A CLONED *XENOPUS LAEVIS* HISTONE GENE CLUSTER

To be able to study the organization of the histone genes in the *Xenopus* genome and to study their expression in development by different hybridization techniques, homologous gene probes were isolated. Using a cloned histone gene repeat unit of the sea urchin *Psammechinus miliaris* an EcoRI fragment was detected and cloned from genomic DNA of *Xenopus laevis* that contains genes coding for histones H3, H4, H2A, and H2B (Figure 8.1) (8). The orientation of the H2A gene in this histone gene cluster (X1-hi-1) is antiparallel with respect to that of the H3, H4, and H2B genes (9). Such an orientation rules out that these genes are transcribed as a polycistronic pre-mRNA. The nucleotide sequences of the genes and most of the surrounding DNA in this cluster have been determined (9, 10). The amino acid sequences for histones H3 and H4, derived from these nucleotide sequences, are only slightly divergent from those of H3 and H4 histones in other organisms, confirming the evolutionary conserved nature of these proteins (11). On the other hand, parts of the derived amino acid sequences of histones H2A and H2B are less conserved. This fits in with the idea that H2A and H2B genes may be evolutionary hybrids (11), consisting of evolutionary conserved and more variable parts. However, it is remarkable that, for all histone genes, the sequence organization is less conserved on the DNA level: nucleotide sequences of the histone genes of *Xenopus* (9, 10, 12) and sea urchin (13, 14, 15) are highly divergent at the wobble positions of the codons. This implies that evolutionary restraints are imposed primarily on the amino acid sequence of the proteins, but that the nucleotide sequence of the genes may vary within the limitations set by fixed amino acid sequences of the gene products.

The nucleotide sequences surrounding the histone genes in X1-hi-1 DNA contain short elements of homology with sequences surrounding histone genes in other organisms (reviewed in Reference 4). The 3′ flanking sequences of all four histone genes in X1-hi-1 DNA contain the dyad symmetry motif GGCTCTTTCAGAGCC.

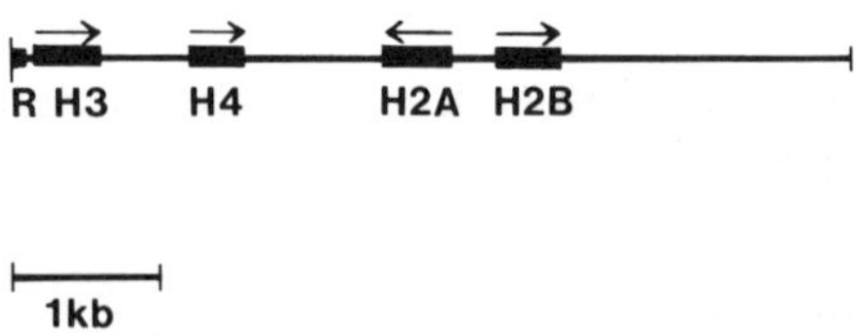

FIGURE 8.1. Schematic representation of a cloned 5.8-kb EcoRI fragment (X1-hi-1) containing histone genes from the *Xenopus laevis laevis* genome. The position of individual histone genes on this cloned fragment is indicated. Arrows show the direction of transcription of the genes. The solid block marked R, in front of the H3 gene, indicates a sequence in X1-hi-1 DNA that is highly repeated in genomic DNA.

Such symmetry motifs have been demonstrated in all histone genes sequenced until now (4) and probably define the 3′ termini of the histone mRNA (12, 16). Hentschel and Birnstiel (4) described that the 5′ flanking sequences of histone genes from various organisms also contain conserved regions or "homology blocks" characteristic for histone genes. Such "consensus sequences" can, however, not yet be unequivocally assigned in the 5′ flanking sequences of the histone genes in X1-hi-1 DNA, except for the "TATA-boxes" that play a role in promoting specific initiation of transcription by RNA polymerase II (17).

Interestingly, the cloned histone gene fragment X1-hi-1 contains just upstream of the H3 gene a sequence that is highly repeated in genomic DNA (18). This sequence contains internal short direct and inverted repeats (see Figure 8.1 and Reference 10). Related sequences were found in all of 16 cloned histone gene clusters analyzed by hybridization with a DNA fragment containing only the repeated sequence (R, see Figure 8.1). Under relatively stringent conditions of hybridization some of these hybrids are unstable. This indicates that these repeated sequences belong to families of repeated elements. The sequenced repeat has a stretch of 13 nucleotides with a high degree of homology with sequences that define the start position for polymerase III transcription of Alu family sequences in human DNA (19) and with sequences defining the start of a transposonlike repeat in rat DNA (20). What the exact nature and function of this highly repeated sequence is, remains unknown for the moment. An interesting but not yet tested possibility seems that it might act as a kind of transposable element and that it might be involved in the polymorphic organization of the 45–50 histone gene clusters in the genome of different *Xenopus* individuals (see Section 3).

3. GENOMIC ORGANIZATION OF THE HISTONE GENES IN *XENOPUS LAEVIS LAEVIS*

The genome of *X. laevis laevis* contains 45–40 genes for the nucleosomal histones (6, 7, 21) while the exact number of H1 genes remains to be determined. This gene number is intermediate between those found in the genomes of different sea urchin species [(several hundred copies (22, 23)], the fruit fly *Drosophila* [(110 copies 24)], and the newt *Notophthalmus* [(600–800 copies (25)] on the one hand, and those found in the genomes of mouse [(10–20 copies (26)], chicken [(16 copies (27)], and man [(10 copies (28)]. It has been suggested that the different numbers of genes in the genomes of different animals may be related to the demand for histone mRNA synthesis during early development of these species (29, 30, and see next sections).

It appears from restriction analysis of genes coding for nucleosomal histones both in genomic *Xenopus* DNA (7) and in cloned genomic DNA fragments (8, 12, 18, and O. H. J. Destrée, unpublished results) that most of the histone genes are clustered in the genome. Many, if not all, clusters contain one gene for each of the nucleosomal histones and an H1 gene. This type of clustered organization has also been found for the majority of the histone genes in the genomes of sea urchins (31, 32), *Drosophila* (24), and *Notophthalmus* (25). It has been suggested (30) that this type of clustering might be functional in the expression of equimolar amounts of the four-core histone classes. However, clear-cut indications for coordinated expression have not been deduced from the sequence organization yet.

We have studied the organization of histone gene clusters in the *Xenopus* genome by hybridization of digested, size-fractionated genomal DNA with radioactive probes derived from the cloned histone DNA fragment (X1-hi-1 DNA) (7, 18). In the DNA of one particular animal we found that approximately 30 of the histone gene clusters have identical EcoRI digestion patterns (7), indicating that they are repeated in an identical way (but not necessarily *in tandem*) in the *Xenopus* genome. However, about 20 other clusters appear to have unique restriction patterns in genomic DNA of the same animal. The structure of the repeated cluster, 14 kb in length, has been determined (Figure 8.2). Interestingly, the order of the H3 and H4 genes in the repeated cluster is reversed with respect to that found in the DNA of clone X1-hi-1. An *Eco*RI fragment with a length equal to that of cloned X1-hi-1 DNA (5.8 kb) belongs to the minor fragments detected in genomic DNA and may even be absent in the DNA of some individual animals. Clones λX1-hi-101, −103, −105, and −107 have been purified from a library of partially digested (*Alu*I × *Hae*III) genomic *Xenopus* DNA cloned in λ charon-4 DNA. This library was

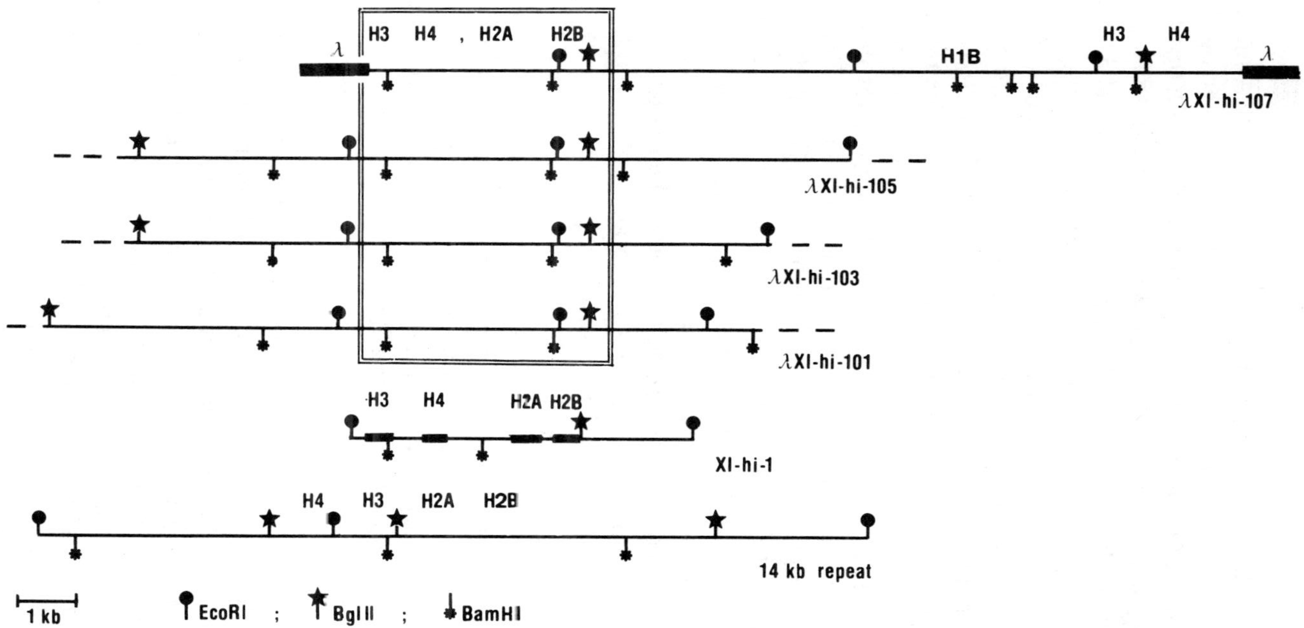

FIGURE 8.2. Restriction maps of the major 14-kb repeat in genomic *Xenopus* DNA and of the cloned *Xenopus* histone DNA fragments X1-hi-1, λX1-hi-101, λX1-hi-103, λX1-hi-105, and λX1-hi-107 for restriction endonucleases EcoRI, BamHI and BglII. The restriction maps of the major 14-kb repeat and the cloned fragment X1-hi-1 have been discussed in References 7 and 8.

obtained from Wahli and Dawid (33). For λX1-hi-107 DNA the entire map is depicted: black bars represent the ends of the vector DNA. Note the presence of two genes for histones H3 and H4 on this DNA; this cloned fragment might be identical to the one described by Zernik et al. (12). From the cloned DNA fragments λ X1-hi-103, −105, and −107 are only those parts shown that contain the genes for the nucleosomal histones and the surrounding restriction sites, none of these sites is located in the vector DNA. The DNA of clone λ X1-hi-105 may partially overlap with the DNA in λ X1-hi-107. Those parts of the cloned fragments where the genes and intergenic spacer DNA are located are shown boxed; in these parts the restriction sites are highly conserved; they differ, however, from the sites in X1-hi-1 DNA and the 14-kb repeat detected in genomic DNA. The exact order of the H4 and H2A genes in the charon clones is not yet known.

The existence of an extensive polymorphism in the organization of the histone gene clusters within the *Xenopus* population, as judged by restriction mapping, is not restricted to unique fragments but also occurs in the repeated clusters. Figure 8.3 shows the *Eco*RI restriction patterns for the histone gene clusters in the DNA of two individual animals and their offspring. The restriction profile of the repeated cluster in a male animal shows hybridizing bands at 8.9 kb (H3, H4, H2A, and H2B genes) and at 5.1 kb (H4 genes), consistent with the structure of the 14-kb repeat unit in genomic DNA shown in Figure 8.2. More major bands are visible when the digested DNA of a female animal is hybridized with the same probes (panel B): there are at least four major hybridizing bands between 8.9 and 20 kb. Furthermore, the H4 probe does not hybridize with a major 5.1-kb genomic DNA fragment, as in the male animal, but with a major 2.9-kb fragment. The restriction patterns of histone genes in the DNA of the offspring are the sum of those of the male and female animals (panel C). This indicates that more than one type of repeated gene cluster exists in the animal population. These results can be explained by processes like continuous amplification and elimination of histone gene clusters, together with processes that change the structures of these clusters, such as point mutations, inversions, additions or deletions of spacer DNA, gene conversions, or transposition of gene clusters to different parts of the genome.

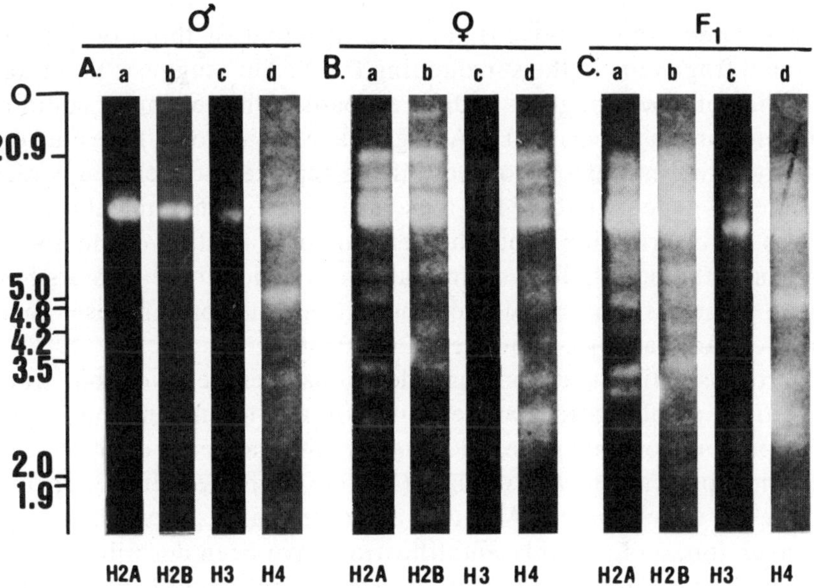

FIGURE 8.3. Hybridization patterns of the genes coding for the nucleosomal histones in the DNA of embryos and their parents after digestion with *Eco*RI. DNA was isolated from the blood of a male (panel A) and female (panel B) animal and from embryos from a mating of male and female (panel C). The DNA was digested with EcoRI, electrophoresed on 0.8% agarose gels, transferred to nitrocellulose sheets, and hybridized with ^{32}P-labeled fragments of X1-hi-1 DNA containing genes for histones H2A (lanes a), H2B (lanes b), H3 (lanes c), and H4 (lanes d). Probes used for hybridization and hybridization conditions are described in Reference 7. Bacteriophage λDNA was included in the digestions to monitor completeness of digestion. The arrows indicate the major 8.9-kb EcoRI fragment. Asterisks indicate the 5.1-kb EcoRI fragments in embryonic and paternal DNA and the 2.9-kb fragments in embryonic and maternal DNA (lanes d). Owing to the relatively short exposure time of the film, only the major repeated histone gene clusters in genomic DNA are clearly visible in this picture. λDNA digested with EcoRI and HindIII served as a size marker (left).

Indeed, there are indications that different histone gene clusters may be in a different environment of genomic DNA. These indications stem from the restriction analysis of four isolated histone gene clusters, λX1-hi-101, −103, −105, and −107 (Figure 8.2). The maps indicate that the restriction sites for the endonucleases *Eco*RI, *Bam*HI and *Bgl*II are conserved in those regions of the clusters where the genes are located (shown boxed), but are different in three of the four cloned fragments in the surrounding DNA. This suggests that transposition of blocks of genes to different parts of the genome or additions or deletions in the spacer DNA might have occurred in these clusters. It would be interesting to know if the repeated sequence upstream of the H3 gene is involved in these processes. Unfortunately, we do not have information about the genomic organization of the histone genes in the populations of animals from which these clusters have been cloned, so that we do not know if one or more of these cloned clusters are part of a major repeated cluster.

Processes like additions and deletions, gene conversion, transposition, amplification, and elimination are not uncommon in gene families. Ribosomal genes in *Xenopus* (34) and genes coding for the chorion proteins in *Drosophila* (35) are amplified during specific stages of development. Addition and deletion of spacer DNA, gene conversion, and possibly amplification have been described for the tandemly repeated chorion gene clusters in silkmoths (36). Addition or deletion of spacer DNA has been observed in the histone gene clusters of sea urchins (31) and *Drosophila* (24, 37); subsequent amplification of these clusters may cause the differences in the restriction patterns of the tandemly repeated histone gene clusters in individual sea urchins (31). There are also indications that histone genes in these species can move and insert into different regions of the genome (38). Many of these changes in the structure of these clusters may be the result of fortuitous events. However, in some instances they may serve a functional purpose. For example, amplification of the clusters with the genes coding for the early histone variants in sea urchin may allow rapid synthesis of high amounts of histone mRNA and proteins during early development in this species (39). It may well be that the amplification of histone gene clusters in *Xenopus* DNA serves a similar purpose (discussed later).

An interesting question is whether all 45–50 nucleosomal histone genes code for the same type of histone protein variants, or if some of them code for variants with a different primary structure, as has been found in sea urchin (40–43). At this moment we do not have direct evidence for different protein variants of the nucleosomal histones in *Xenopus* oocytes and embryos in different stages of development (see next section), but variants of the H1 proteins have found in different adult tissues (1, 2, 44–46) and possibly exist in cleavage stages (47). However, there is evidence that the genes for the nucleosomal histones are not all identical. For example, the nucleotide sequences of two *Xenopus* H4 genes show extensive divergence, however, without any impact on the amino acid sequences of the proteins they code for (10, 12). Also, heterogeneity in cleavage sites for restriction endonucleases is found, for example, in the H2B gene (see Figure 8.2). Furthermore, S1 mapping of hybrids between histone genes in X1-hi-1 DNA or λX1-hi-103 DNA and mRNA from oocytes or embryos indicates that the H3 and H2A genes in these cloned DNAs are not fully complementary to the mRNAs expressed in these developmental stages (see Section 4; 7, 10). The differences are located mainly in the 3′ and 5′ nontranslated sequences that may be involved in the regulation of the expression of the genes. These results also imply the existence of various types of genes with different primary structures.

3.1. Genomic Organization of Histone Genes in Other Species of the Genus *Xenopus*

The organization of the histone genes in genomic DNA of individuals of several species and subspecies of the genus *Xenopus* was analyzed by hybridization with specific gene probes for the nucleosomal histone genes and for the H1B gene. In all species that were investigated major bands of hybridization were found together with bands of much less intensity (Figure 8.4). Here only the hybridization patterns found with the H3 gene probe are shown.

In some species (*X. borealis*, *X. fraseri*, *X. l. victorianus*), a relatively high percentage of hybridization was found in a few major bands; for example, in *X. borealis* three major fragments of 15 kb, 10 kb, and 6 kb were found. Both the 15-kb and 10-kb fragments contain all five histone genes, while the 6-kb fragment did not hybridize with an H4 probe but contains all other histone genes. However, in most species and subspecies a rather heterogeneous array of fragments is found. At present we do not know if the differences in restriction patterns found are due to differences between individuals, as was found for *X. l. laevis* (7). It is clear, though, that no obvious conservation of particular size fragments has taken place within the genus *Xenopus* or within the species *X. laevis* (compare Figure 8.2).

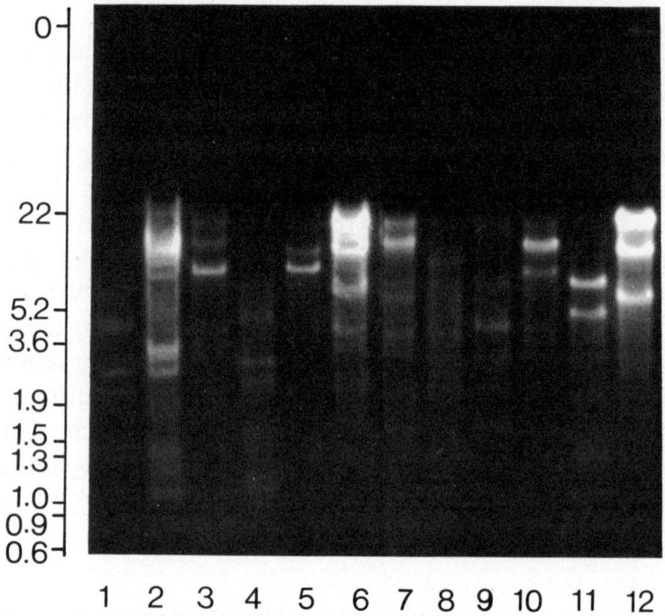

FIGURE 8.4. Hybridization patterns of EcoRI digested DNA from individuals of different species and subspecies of *Xenopus*. Genomic DNA from erythrocytes was digested with EcoRI, electrophoresed on an 0.8% agarose gel, blotted to nitrocellulose paper, and hybridized with a 450-bp XhoI-BamHI fragment containing most of the H3 gene from X1-hi-1, but not the repeated element R. The DNA in lanes 1–12 is from: (1) *X. tropicalis*; (2) *X. vestitus*; (3) *X. laevis bunyoniensis*; (4) *X. ruwenzoriensis*; (5) *X. l. petersi*; (6) *X. muelleri*; (7) *X. sp. n. I*; (8) *X. l. laevis*; (9) *X. wittei*; (10) *X. l. victorianus*; (11) *X. fraseri*; (12) *X. borealis*. Sizes of EcoRI × Hind III digested bacteriophage fragments are shown at the left side of the figure; o represents the origin. Bacteriophage λDNA was used to monitor the EcoRI digestion.

4. EXPRESSION OF HISTONE GENES DURING OOGENESIS AND EARLY EMBRYOGENESIS IN *XENOPUS LAEVIS*

It has been noted that the synthesis of histone proteins in oogenesis (48) and early embryogenesis (3) of *Xenopus* is not strictly linked with DNA replication, as it is in somatic cells. *Xenopus* oocytes accumulate a large store of preformed histone proteins that are used predominantly after fertilization, in early embryonic development, when cell division and DNA replication rates are high (48). A store of 40–50 ng of each of the nucleosomal histones has been measured in full-grown oocytes (48). This store is accumulated at low rates during oogenesis (49). The amount of stored histones is sufficient to complex the newly synthesized DNA in embryos until approximately the 30,000 cell stage (early blastula stage). We have determined that at least one of the H1 proteins, H1A, is also stored in oocytes: 8–10 ng H1A was found in full-grown oocytes (50). Remarkably, this store of H1A protein is made very early in oogenesis (Figure 8.5): approximately 50% of the final amount found in full-grown oocytes is already present in oocytes in Dumont's (52) stage II of development (50), and 80% is already present in stage III oocytes, approximately

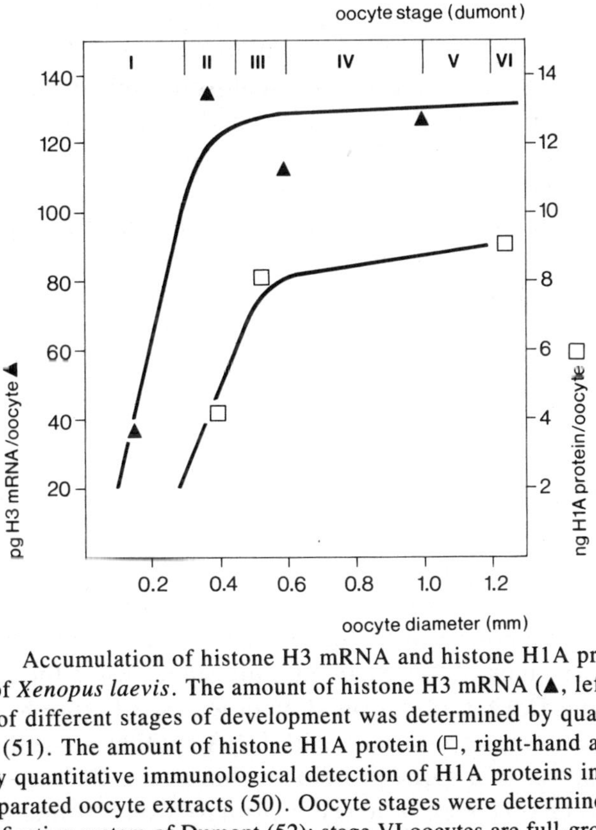

FIGURE 8.5. Accumulation of histone H3 mRNA and histone H1A protein during oogenesis of *Xenopus laevis*. The amount of histone H3 mRNA (▲, left-hand axis) in oocytes of different stages of development was determined by quantitative hybridization (51). The amount of histone H1A protein (□, right-hand axis) was determined by quantitative immunological detection of H1A proteins in electrophoretically separated oocyte extracts (50). Oocyte stages were determined according to the classification system of Dumont (52); stage VI oocytes are full-grown. Average duration of each stage of development is, according to Callen et al. (53), stage I: 7–8.5 months; stage II: 2–2.5 months; stage III: 0.5 months; stage IV: 1 month; stages V + VI: 12 months.

one year before the oocyte becomes fertilizable (46). As histone synthesis can still be detected in mature oocytes [although at low rates (49)] and as the total amount of histone protein does not considerably increase after stage III of oocyte development, the pool of stored histones is probably continually and slowly turned over.

Xenopus oocytes also contain a large stock of histone mRNA. This was first discovered when it appeared that fertilized eggs of *Xenopus* were able to synthesize histone proteins when mRNA synthesis was inhibited by Actinomycin D treatment or by enucleation of the oocytes (49). Furthermore, histones belong to the most prominent products after cell-free translation of oocyte mRNA (54, 55, 56). Direct quantitation revealed that approximately 130 pg, or 5×10^8 copies of H3 mRNA are present in full-grown oocytes (51). Similar amounts of the messengers for the other nucleosomal histones are present. The store of histone mRNA is also made very early in oogenesis, and accumulation is completed in stage II (52) oocytes, more than one year before the oocyte becomes fertilizable. It has been demonstrated that total poly(A)$^+$ RNA (57) and several unspecified mRNAs (58) accumulate during the same stage of oogenesis (Figure 8.5). This store of histone mRNA seems to be translated predominantly after oocyte maturation: histone synthesis rates increase dramatically then from 40 pg/hour in the oocyte to 2.5 ng/hour in the mature egg and 55 ng/hour in the blastula embryo (3). The histone mRNA stock accumulated in the oocyte is translated in the embryos until late blastula/early gastrula stage of embryogenesis (59); around this time it is replaced by newly made transcripts synthesized by the embryo. Histones made of the store of histone mRNA that the embryo inherits from the oocyte complement the store of histone proteins made during early embryogenesis, when, due to the very high rates of DNA replication, high amounts of histones must be available for the embryos within short times. After early gastrula stage, when the rates of cell division and DNA replication slow down (see Figure 8.5), histone synthesis becomes coordinate with DNA synthesis. At this time the inherited stores of histone protein and histone mRNA are exhausted, and from then on the embryo relies primarily on histones and histone mRNA made by itself (51).

An intriguing question is whether all or only part of the 45–50 genes in the *Xenopus* genome are involved in the accumulation of the histone mRNA pool in *Xenopus* oocytes. From literature data on

RNA synthesis rates in *Xenopus* oocytes, we can estimate the minimum amount of genes required for the accumulation of 5×10^8 copies of H3 mRNA within 7–8 months, the time that the oocyte spends in stage I of oogenesis (53). Assuming that stored mRNA is stable after synthesis, as shown by Ford et al. (68), that there is a (high) RNA chain-elongation rate of 20 bases per second (69), that the RNA packing interval has the same high value as observed for ribosomal genes in *Xenopus* oocytes [1 RNA polymerase molecule per 100 bases (70)], and that the oocytes have a 4 C genome (71), it can be calculated that synthesis of 5×10^8 copies of H3 mRNA will take several years if only one gene per haploid genome is transcribed. For accumulation of this amount of mRNA within 7–8 months under the previously mentioned conditions, at least 30–35 histone genes per haploid genome must be active in histone mRNA production. This number represents the majority of the 45–50 histone genes in the *Xenopus* genome. Amplification of the *Xenopus* histone genes may thus be required to enable the high production of histone mRNA in the oocytes.

An attractive, although not yet tested, idea is that the genes in the major repeated cluster are the ones that are expressed during oogenesis. If this appears true, the situation will bear some analogy with that in sea urchins where the tandemly repeated sets of histone gene clusters are active in the developmental stage that requires synthesis of high amounts of histones in short times, that is, early embryogenesis up to midblastula stage (72, 73). Results of S1 mapping experiments indicate that histone mRNA fully complementary to the H3 and H2A genes in clones X1-hi-1 and λX1-hi-103 is not, or only in very low amounts, present in *Xenopus* oocytes (18). Genes in these two clusters are apparently involved in the synthesis of, at most, a very small portion of the histone mRNA in oocytes. We have not yet tested if the majority of oocyte histone mRNA is fully complementary to genes in the major repeated clusters.

4.1. No Evidence for Major Changes in the mRNA Population Coding for the Nucleosomal Histones during Early Development in *Xenopus laevis*

The developmental changes in the histone complement in early sea urchin embryos are well described (40, 41, 74–76). It has been demonstrated that early and late histone variants are produced by different sets of histone genes (58, 72). Histone mRNAs coding for the different variants have different nucleotide sequences (73), run with different mobilities on polyacrylamide gels (58, 72), and differ in their homology to cloned histone gene clusters (58, 72). Structural changes in the sea urchin chromatin have been observed in parallel with the appearance of the different sets of histones on the DNA (77, 78), especially with respect to the nucleosome repeat length.

We have studied whether such histone gene switches also occur in *Xenopus* early development. Therefore we have investigated the transcription products of histone genes in different stages of early development by three different methods.

1. We have electrophoresed RNAs from oocytes and embryos in different stages of development on high-resolution polyacrylamide gels (79). Except for a lower mobility of oocyte nucleosomal histone mRNAs on these gels, no other changes in the electrophoretic mobilities were detected, at least not before tailbud stage of embryogenesis. Although we did not thoroughly investigate why oocyte histone mRNAs have lower mobilities than embryonic mRNAs, an obvious reason for this difference may be the presence of a short (±20 bp long) poly-A tail on oocyte mRNA (55, 56) that is released after fertilization (80). No other developmental changes were detected between histone RNAs from the different stages. However, it appeared that in all stages tested two types of H4 messengers and four types of H2B messengers could be observed, with different mobilities. This indicates that during oogenesis and early embryogenesis more than one type of gene set might be expressed. Heterogeneity of the H2B mRNA population was confirmed by S1 mapping experiments (79).

2. Although H3 mRNA runs as a homogeneous band on polyacrylamide gels, S1 mapping of hybrids between the histone H3 gene in X1-hi-1 DNA and H3 mRNA revealed that also histone H3 mRNA, both in oocytes and in embryos, may consist of a population of different types of messenger molecules, each of them with a different degree of homology to the H3 gene in X1-hi-1 DNA (18). However, qualitative developmental changes in the H3 mRNA population were not observed using this method.

3. Cell-free translation products of total oocyte and tailbud-stage embryonic RNA and of purified histone mRNA from oocytes and blastula and neurula embryos (purified by hybridization to immobilized X1-hi-1 DNA) were analyzed on SDS–polyacrylamide gels and on acid urea Triton X-100 gels with two different concentrations of urea (79). From these experiments we have no evidence for developmental changes in the translation products of mRNAs isolated from the different stages of embryonic development. In addition, RNA of all stages tested gave an identical series of minor translation products in the H2B/H4 region of acid urea Triton X-100 gels, but it remains to be assessed whether these minor products represent translation artifacts.

From these experiments we conclude: (1) No major qualitative developmental changes in the transcription products of the nucleosomal histone genes occur during oogenesis and early embryogenesis up to tailbud stage, that is, the same types of nucleosomal histone genes seem to be expressed in this period. (2) The presence of two types of H4 mRNA and four types of H2B mRNA in all developmental stages tested indicates that more than one type of gene for these histone classes is expressed during early development. However, we are not certain if these different types of genes code for different variants of these histone classes.

As we have only recently obtained probes for histone H1 genes, we have not yet studied extensively what developmental changes in H1 gene expression occur. Flynn and Woodland (45) have reported that the same H1 protein variants (H1A, H1B, and H1C) are present in the chromatin during oogenesis and early embryogenesis. Koster et al. (47), on the other hand, observed around gastrula stage developmental changes in the electrophoretic mobility of H1 proteins on acid urea Triton X-100 gels. It remains difficult to assess, however, which of these changes are due to the expression of different H1 protein variants and which to changes in secondary modification of identical H1 variants.

The expression of H1 variant genes is now under study. Hybridization-translation experiments were performed using a subcloned H1B coding DNA fragment from λX1-hi-107 (see Figure 8.2) with RNA from full-grown oocytes and erythroblasts under conditions that allow cross-hybridization between the H1B probe and H1A mRNA. It appears that mature oocytes have stored mainly H1A and relatively little H1B mRNA. In erythroblasts the proportions of mRNAs for H1A and H1B are approximately equal in accordance with results from direct translation of mRNAs (J. G. Koster, manuscript in preparation).

The change in relative concentrations of the mRNAs for H1A and H1B in development appears to parallel the changes found in the protein contents of the lysine-rich histones. Besides these quantitative changes in the H1 variants and their mRNAs, additional lysine-rich histones have been described in *Xenopus* that are specific for adult tissues: in erythrocytes (1, 2, 44) and sperm cells (81), as well as other adult tissues (81). The erythrocyte lysine-rich histone [H5 (44)] is probably related to the mammalian H1°, and its concentration steadily increases during maturation of erythrocytes in animals recovering from induced anemia (J. G. Koster, unpublished results).

5. DIFFERENT SPECIES USE DIFFERENT STRATEGIES FOR THE SYNTHESIS OF HISTONE mRNA AND HISTONE PROTEINS DURING EARLY DEVELOPMENT

The chromatin in higher eukaryotes contains DNA and histones in a 1:1 ratio on weight base. In somatic cells, histone and DNA synthesis are closely linked: histone synthesis mainly occurs when the DNA is replicating (82). During each cell cycle the cell synthesizes sufficient new histones to complex the newly formed DNA. During early embryogenesis, however, cell division and DNA replication rates are extremely high in several species (see Figure 8.6); this implies that

FIGURE 8.6. Growth curves for early embryos of the sea urchin (*a*) *S. purpuratus* at 15°C, (*b*) *X. laevis* at 21°C, and (*c*) *D. melanogaster* at 25°C, derived from data in References 60 and 61 (*a*), 62, 63, and 64 (*b*), and 65 and 66 (*c*). Both accumulation of DNA and increase in number of cells are shown in the figures. As chromatin contains DNA and histones in a 1:1 ratio, DNA accumulation curves also represent the minimal amounts of histones required. Time between fertilization and first cleavage in *Xenopus* is omitted. For comparison: rabbit embryos of only 32 cells after approximately 40 hours of development, mouse embryos of eight cells (67). Different stages of embryonic development are depicted at the top of the figure: b: blastula/blastoderm; g: gastrula; o: primary organogenesis.

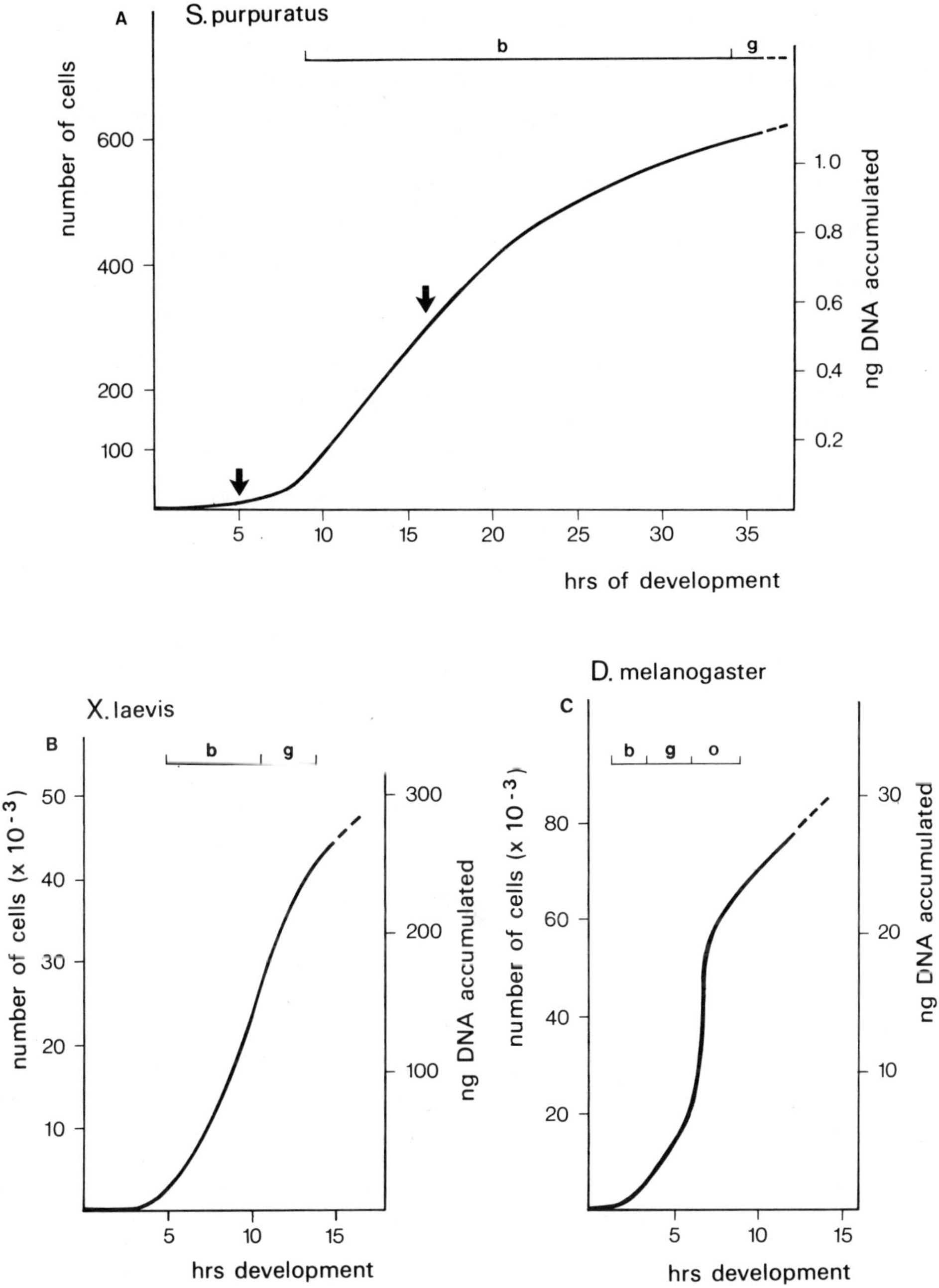
A
S. purpuratus
b
g
number of cells
600
400
200
100
ng DNA accumulated
1.0
0.8
0.6
0.4
0.2
5
10
15
20
25
30
35
hrs of development
B
X. laevis
b
g
number of cells (x 10 -3)
50
40
30
20
10
ng DNA accumulated
300
200
100
5
10
15
hrs development
C
D. melanogaster
b
g
o
number of cells (x 10 -3)
80
60
40
20
ng DNA accumulated
30
20
10
5
10
15
hrs development

high amounts of histones must become available for embryos of these species within very short periods. Comparison of the data on histone and histone mRNA synthesis in *Xenopus* early development with those data obtained by other groups for mammals, sea urchins, and *Drosophila* may give an indication of the different strategies that may be used by these species for the synthesis of sufficient histones during early development. The growth curves for embryos of *Xenopus*, the sea urchin *S. purpuratus*, and the fruit fly *D. melanogaster* (obtained from literature data) are shown in Figure 8.6. Cell division rates during early embryogenesis of these species are high compared to rates in mammals: after 40 hours of development, mammalian embryos consist only of 8–32 cells (58). The amounts of DNA accumulated in embryos of the three species in similar time intervals differ widely (see Figure 8.6), both because of differences in the nuclear DNA content of cells between these species and because of differences in the DNA replication rates. As chromatin consists of DNA histones in a 1:1 ratio, the amounts of DNA will also represent the amounts of histones required for chromatin formation. What are the different strategies used for histone synthesis during early development?

1. In early mammalian embryos, the cell division rate is rather low—each cell divides on average once every 8–12 hours (67). Histone synthesis in mammalian embryos appears to be coordinated with DNA replication (83, 84), and there is no evidence for a store of histone protein inherited from the oocyte. Probably, sufficient histones can be produced in each cell cycle by the 10–20 histone gene sets present in the mammalian genome.
2. In early sea urchin embryos cell division and DNA replication proceed at much higher rates. Each cell divides on average every 1.5 hours (60 i.e.). Sea urchins use different strategies to produce sufficient histones in this period of rapid DNA replication:
    a. It uses, only during the first 2–3 cell divisions after fertilization, a small amount of preformed histones inherited from the oocyte (75, 85).
    b. Embryos also inherit from the oocyte a small store of histone RNA [2–3 pg (86)] that is translated only after fertilization.

c. Most of the histone mRNA is synthesized (39) and translated (87) during embryonic development itself. A defined set of histone genes that is tandemly repeated hundreds of times in the genome (22) is transcribed specifically during early development (72, 73). Maximum rates of histone mRNA synthesis are reached very early in embryogenesis, in the 128-cell stage. The high reiteration frequency of these sets of histone genes enables the transcription of sufficient amounts of histone mRNA during this stage of development. At midblastula stage, when the cell division rate slows down, transcription of these highly repeated sets of genes is abolished, and transcription proceeds from other, less reiterated, sets of genes.

3. During early development of *Drosophila* DNA replication proceeds even faster than in sea urchins. Each cell divides every 10–20 min (65). A large amount of the histone mRNA in early *Drosophila* embryos is synthesized in and inherited from the oocyte (88), the rest is made during embryogenesis itself (88). The high histone mRNA content decreases toward the end of gastrulation, when the high DNA replication rate decreases also. There is no evidence for a large maternal store of preformed histone proteins inherited by *Drosophila* embryos (89).

4. *Xenopus* has a slower DNA replication rate compared to *Drosophila*, but during early embryogenesis requires histones at an even faster rate than does *Drosophila* owing to the almost twenty-fold higher diploid DNA content per cell. To provide the newly made DNA in early embryogenesis with sufficient histones, *Xenopus* uses both a large store of histone proteins inherited from the oocyte [approximately 170 ng (50, 52)] and a large store of histone mRNA [approximately 600 pg (51)] that is made by the oocyte and translated after fertilization (49). These inherited stores of histones and histone mRNA are sufficient for development to end-blastula/early stage of embryogenesis (59). Around this period DNA synthesis rates decrease, and histone synthesis becomes coordinated with DNA replication.

It was calculated that, if histone and histone mRNA synthesis in *Xenopus* embryos would be coordinated with DNA replication in the stage of rapid cell proliferation, even under the most favorable transcription and translation conditions more than 1000 histone gene sets have to be active in the *Xenopus* genome to support the synthesis of sufficient histone mRNA and protein in cleavage stages of development. This means that, given the high rates of DNA replication in *Xenopus* and *Drosophila* embryos, the inheritance by the embryos of preformed stores of histone mRNA and histone proteins made in the oocytes may drastically reduce the number of genes required in the genomes of these species. Or, alternatively, that the inheritance of histones and histone mRNA (together with other necessary products, such as DNA polymerase and ribosomes) allows the high rate of cell division observed during early embryogenesis.

ACKNOWLEDGMENTS

Part of the studies described were supported by grants from the Netherlands Organization for the Advancement of Pure Research (ZWO), grant numbers 13-50-09 and 14-97-15. The selection of variant histone genes from a *Xenopus laevis* genomic library was performed in the Institut für Molekularbiologie II, University of Zürich (head, Prof. Dr. M. L. Birnstiel) in collaboration with Prof. Dr. W. Schaffner supported by a short-term fellowship to O. H. J. Destrée from the European Molecular Biology organization. We acknowledge Dr. W. Wahli and Dr. I. B. Dawid for their generous gift of the *Xenopus laevis* genomic library. We are grateful to Dr. M. Bendig and Mr. R. Weiss for their help in the selection and preparation of clones. We greatly appreciate Prof. Dr. M. Fischberg's gift of different species of *Xenopus*. We are much indebted to Lia de Laaf and Piet de Boer for their excellent technical assistance. We thank Mrs. J. Husslage for help with the preparation of the manuscript.

REFERENCES

1. O. H. J. Destrée, H. A. d'Adelhart-Toorop, and R. Charles, *Acta Morphol. Neerl.-Scand.* **10,** 233–248 (1972).

2. O. H. J. Destrée, H. A. d'Adelhart-Toorop, and R. Charles, *Cell Diff.* **2,** 229–242 (1973).
3. E. D. Adamson and H. R. Woodland, *J. Mol. Biol.* **88,** 263–285 (1974).
4. C. C. Hentschel and M. L. Birnstiel, *Cell* **25,** 301–313 (1981).
5. H. R. Woodland, *FEBS Lett.* **121,** 1–7 (1980).
6. E. Jacob, G. Malacinsky, and M. L. Birnstiel, *Eur. J. Biochem.* **69,** 45–54 (1976).
7. W. van Dongen, L. de Laaf, R. Zaal, A. Moorman, and O. Destrée, *Nucl. Acids Res.* **9,** 2297–2311 (1981).
8. A. F. M. Moorman, R. T. M. de Laaf, O. H. J. Destrée, J. Telford, and M. L. Birnstiel, *Gene* **10,** 185–193 (1980).
9. A. F. M. Moorman, P. A. J. de Boer, R. T. M. de Laaf, and O. H. J. Destrée, *FEBS Lett.*, in press (1982).
10. A. F. M. Moorman, P. A. J. de Boer, R. T. M. de Laaf, W. M. A . M. van Dongen, and O. H. J. Destrée, *FEBS Lett.* **136,** 45–52 (1981).
11. I. Isenberg, *Ann. Rev. Biochem.* **48,** 159–191 (1979).
12. M. Zernik, M. Heintz, I. Boime, and R. G. Roeder, *Cell* **22,** 807–815 (1980).
13. W. Schaffner, G. Kunz, M. Daetwyler, J. Telford, H. O. Smith, and M. L. Birnstiel, *Cell* **14,** 655–671 (1978).
14. M. Busslinger, R. Portmann, J. C. Irminger, and M. L. Birnstiel, *Nucl. Acids Res.* **8,** 957–977 (1980).
15. I. Sures, J. Lowry, and L. H. Kedes, *Cell* **15,** 1033–1044 (1978).
16. C. Hentschel, J. C. Irminger, P. Bucher, and M. L. Birnstiel, *Nature (London)* **285,** 147–151 (1980).
17. R. Grosschedl and M. L. Birnstiel, *Proc. Natl. Acad. Sci. USA* **77,** 1432–1436 (1980).
18. W. M. A. M. van Dongen, academic thesis, University of Amsterdam, ed. Rodopi, Amsterdam.
19. C. H. Duncan, P. Jagadeeswaran, R. C. T. Wang, and S. R. Weisman, *Gene* **13,** 185–196 (1981).
20. G. S. Page, S. Smith, and H. M. Goodman, *Nucl. Acids Res.* **9,** 2087–2104 (1981).
21. V. A. Hilder, R. N. Livesey, P. C. Turner, and M. T. Vlad, *Nucl. Acids Res.* **9,** 5737–5746 (1981).
22. L. H. Kedes and M. L. Birnstiel, *Nature New Biol.* **230,** 165–169 (1971).
23. E. S. Weinberg, M. L. Birnstiel, I. F. Purdom, and R. Williamson, *Nature (London)* **240,** 225–228 (1972).
24. R. P. Lifton, M. L. Goldberg, R. W. Karp, and D. S. Hogness, *Cold Spring Harbor Symp. Quant. Biol.* **42,** 1047–1051 (1977).
25. E. C. Stephenson, H. P. Erba, and J. G. Gall, *Cell* **24,** 639–647 (1981).

26. E. Jacob, *Eur. J. Biochem.* **65,** 275–282 (1976).
27. R. J. Crawford, P. Krieg, R. P. Harvey, D. A. Hewish, and J. R. E. Wells, *Nature (London)* **279,** 132–137 (1979).
28. M. C. Wilson, M. Melli, and M. L. Birnstiel, *Biochem. Biophys. Res. Comm.* **61,** 404–410 (1974).
29. J. D. Engel and J. B. Dodgson, *Proc. Natl. Acad. Sci. USA* **78,** 2856–2860 (1981).
30. D. B. Sittman, I. M. Chiu, C. J. Pan, R. H. Cohn, L. H. Kedes, and W. F. Marzluff, *Proc. Natl. Acad. Sci. USA* **78,** 4078–4082 (1981).
31. G. C. Overton and E. S. Weinberg, *Cell* **14,** 247–257 (1978).
32. R. H. Cohn and L. H. Kedes, *Cell* **18,** 855–864 (1979).
33. W. Wahli and I. B. Dawid, *Proc. Natl. Acad. Sci USA* **77,** 1437–1441 (1980).
34. L. Watson-Coggins and J. G. Gall, *J. Cell Biol.* **52,** 569–576 (1972).
35. A. C. Spradling and A. P. Mahowald, *Proc. Natl. Acad. Sci. USA* **77,** 1096–1100 (1980).
36. C. W. Jones and F. C. Kafatos, *Mol. Cell. Biol.* **1,** 814–828 (1981).
37. K. Saigo, L. Millstein, and C. A. Thomas, *Cold Spring Harbor Symp. Quant. Biol.* **45,** 815–827 (1980).
38. G. Childs, R. Maxson, R. H. Cohn, and L. H. Kedes, *Cell* **23,** 651–663 (1981).
39. R. E. Maxson and F. H. Wilt, *Dev. Biol.* **83,** 380–386 (1981).
40. L. H. Cohen, K. M. Newrock, and A. Zweidler, *Science* **190,** 994–997 (1975).
41. K. M. Newrock, C. R. Alfagemene, R. V. Nardi, and L. H. Cohen, *Cold Spring Harbor Symp. Quant. Biol.* **42,** 421–431 (1977).
42. J. V. Ruderman and P. R. Gross, *Dev. Biol.* **36,** 286–298 (1974).
43. W. F. Brandt, W. N. Strickland, M. Strickland, L. Carlisle, D. Woods, and C. Von Holt, *Eur. J. Biochem.* **94,** 1–10 (1979).
44. O. H. J. Destrée, H. J. Hoenders, A. F. M. Moorman, and R. Charles, *Biochim. Biophys. Acta* **577,** 61–70 (1979).
45. J. M. Flynn and H. R. Woodland, *Dev. Biol.* **75,** 222–230 (1980).
46. M. S. Risley and R. A. Eckhardt, *Dev. Biol.* **84,** 79–87 (1981).
47. J. G. Koster, M. E. Kasinsky, and O. H. J. Destrée, *Cell Differ.* **8,** 93–104 (1979).
48. H. R. Woodland and E. D. Adamson, *Dev. Biol.* **57,** 118–135 (1977).
49. E. D. Adamson and H. R. Woodland, *Dev. Biol.* **57,** 136–149 (1977).
50. W. M. A. M. van Dongen, A. F. M. Moorman, and O. H. J. Destrée, *Cell Differentiation* **12,** 257–264 (1983).
51. W. Van Dongen, R. Zaal, A. Moorman, and O. Destrée, *Dev. Biol.* **86,** 303–314 (1981).
52. J. N. Dumont, *J. Morphol.* **136,** 153–164 (1972).

53. J. C. Callen, N. Dennebouy, and J. C. Mounolou, *Dev. Growth Differ.* **22,** 831–840 (1980).
54. O. H. J. Destrée, A. L. Haenni, and M. L. Birnstiel, *Nucl. Acids Res.* **4,** 801–811 (1977).
55. J. V. Ruderman and M. L. Pardue, *Dev. Biol.* **60,** 48–68 (1977).
56. R. G. Levenson and K. B. Marcu, *Cell* **9,** 311–322 (1976).
57. M. Rosbash and P. J. Ford, *J. Mol. Biol.* **85,** 87–101 (1974).
58. P. A. Hieter, M. B. Hendricks, K. Hemminki, and E. S. Weinberg, *Biochemistry* **18,** 2707–2716 (1979).
59. H. R. Woodland, J. M. Flynn, and A. J. Wyllie, *Cell* **18,** 165–172 (1979).
60. R. T. Hinegardner, in F. H. Wilt and N. K. Wessels, Eds., *Methods of Developmental Biology*, Crowell, New York, 1967, p. 139.
61. R. T. Hinegardner, *Comp. Biochem. Physiol. B., Comp. Biochem.* **49,** 219–231 (1974).
62. P. D. Nieuwkoop and J. Faber, *Normal Table of Xenopus laevis*, North-Holland, Amsterdam, 1956.
63. H. R. Woodland and J. B. Gurdon, *J. Embryol. Exp. Morphol.* **19,** 363–385 (1968).
64. I. B. Dawid, *J. Mol. Biol.* **12,** 581–599 (1965).
65. K. V. Anderson and J. A. Lengyel, *Dev. Biol.* **82,** 127–138 (1981).
66. E. M. Rasch, H. J. Barr, and R. W. Rasch, *Chromosoma* **33,** 1–18 (1971).
67. L. W. Browder, *Developmental Biology*, Saunders College, Philadelphia, 1980.
68. P. J. Ford, T. Mathieson, and M. Rosbash, *Dev. Biol.* **57,** 517–526 (1977).
69. D. M. Anderson and L. D. Smith, *Dev. Biol.* **67,** 274–285 (1978).,
70. O. L. Miller and A. Bakken, in *Fifth Karolinska Symposium on Research Methods in Reproductive Endocrinology*, 1972, pp. 155–173.
71. E. H. Davidson, *Gene Activity in Early Development*, 2nd ed., Academic, New York, 1976, p. 322.
72. G. Childs, R. Maxson, and L. H. Kedes, *Dev. Biol.* **73,** 153–173 (1979).
73. M. Grunstein, K. E. Diamond, E. Knoppel, and J. E. Grunstein, *Biochemistry* **20,** 1216–1223 (1981).
74. D. Poccia, J. Salik, and G. Krystal, *Dev. Biol.* **82,** 287–296 (1981).
75. B. Ramsay Shaw, G. Cognetti, W. M. Sholes, and R. G. Richards, *Biochemistry* **20,** 4971–4978 (1981).
76. L. H. Kedes, *Ann. Rev. Biochem.* **48,** 837–870 (1979).
77. R. J. Arceci and P. R. Gross, *Dev. Biol.* **80,** 186–209 (1980).
78. A. Savić, P. Richman, P. Williamson, and D. Poccia, *Proc. Natl. Acad. Sci. USA* **78,** 3706–3710 (1981).
79. W. M. A. M. van Dongen, A. F. M. Moorman, and O. H. J. Destrée, *Differentiation* **24,** 226–233 (1983).

80. J. V. Ruderman, H. R. Woodland, and E. A. Sturgess, *Dev. Biol.* **71,** 71–82 (1979).

81. M. S. Risley and R. A. Eckhardt, *Dev. Biol.* **84,** 79–87 (1981).

82. L. Golden, U. Schafer, and M. Rosbash, *Cell* **22,** 835–844 (1980).

83. R. C. Matheson and G. A. Schultz, *J. Exp. Zool.* **213,** 337–349 (1980).

84. P. L. Kaye and R. G. Wales, *J. Exp. Zool.* **216,** 453–459 (1981).

85. J. Salik, L. Herlands, H. P. Hoffman, and D. Poccia, *J. Cell Biol.* **90,** 385–395 (1981).

86. A. Skoultchi and P. R. Gross, *Proc. Natl. Acad. Sci. USA* **70,** 2840–2844 (1973).

87. A. S. Goustin, *Dev. Biol.* **87,** 163–175 (1981).

88. K. V. Anderson and J. A. Lengyel, *Cell* **21,** 717–727 (1980).

89. R. Wiegand, ibid.

CHAPTER 9

ORGANIZATION AND EXPRESSION OF NEWT HISTONE GENES

EDWIN C. STEPHENSON
Department of Biology
Indiana University
Bloomington, Indiana

CONTENTS

1. INTRODUCTION
2. STRUCTURE AND ORGANIZATION OF NEWT HISTONE GENES
3. EXPRESSION OF NEWT HISTONE GENES
REFERENCES

1. INTRODUCTION

Newts and other urodele amphibians have been of interest to molecular biologists for two reasons:

1. Lampbrush chromosomes, the decondensed, transcriptionally active chromosomes found in diplotene oocytes, offer the possibility of correlating morphological and molecular aspects of nuclear structure and function.
2. Molecular analysis of urodeles, which have some of the largest genomes known, may help in understanding the *C*-value paradox, that is, why eukaryotic genomes are much larger than necessary to encode the expected number of genes, and why eukaryotic genomes vary in size over several orders of magnitude.

Study of newt histone genes was undertaken for both of these reasons and has provided information that bears upon both points. In this Chapter, we review the structure and genomic organization of newt histone genes. Newt histone genes differ in several respects from those of other vertebrates studied so far, and we speculate on the functional and evolutionary reasons for these differences. Data concerning the transcription of these genes from lampbrush chromosomes are also reviewed, and a model for lampbrush chromosome structure and function is discussed.

2. STRUCTURE AND ORGANIZATION OF NEWT HISTONE GENES

Histone gene organization is known in detail for only one urodele, the American spotted newt *Notophthalmus viridescens* (1). Molecular clones were isolated on the basis of their hybridization to a previously identified newt H4 pseudogene. Several clones containing identical 9.0 kb newt DNA segments were obtained; one was analyzed in detail and found to contain one gene for each of the five histones. The positions of the H2A, H2B, H3, and H4 genes were determined by hybridization with cloned sea urchin histone genes; the H1 gene was assumed to correspond to the position of the largest mRNA encoded in the cloned segment. Partial sequencing confirmed these assignments and allowed determination of the direction of transcription. The order of genes in this cloned segment is H1-H3-H2B-H2A-H4, with each gene except H2B being transcribed in the same direction. These gene mapping data are summarized in Figure 9.1.

The reiteration and heterogeneity of histone genes in the newt genome were examined by Southern hybridization (2). The cloned histone gene cluster appears to be a faithful representative of a large, relatively homogeneous family of sequences. For each restriction enzyme tested, histone gene probes hybridize to major bands of the sizes predicted by the spacing of these sites in the cloned cluster, as illustrated in Figure 9.1. Major bands of 9.0 kb in the Bgl II digest (lane b), 3.25 kb in the Pvu II digest (lane c) and 2.7 kb in the Sac I digest (lane d) exactly reflect the spacing of these sites in the cloned DNA. In most of the digests, minor bands are also visible, most of which were interpreted to reflect heterogeneity in the position or occurrence of single sites. For instance, the second most prominent band in the Bgl II digest (lane b), a 4.2-kb fragment, was found to represent a fraction of newt histone gene clusters that contain an extra Bgl II site near the middle of the cluster. Histone genes are reiterated about 600–800 times per haploid genome; this value was determined by comparing the hybridization of histone gene probes to total genomic DNA digests and to known quantities of cloned DNA in an adjacent lane. The conclusion from these experiments was that newt histone genes comprise a large gene family, the members of which are identical with respect to structure and the positions of most restriction enzyme sites.

Not all restriction enzymes produce easily interpretable hybridization patterns in whole genome Southern blots. For instance, in Figure 9.1, lane a, it is evident that Bam HI fails to produce discrete hybridizing fragments within the resolving range of this gel; instead, hybridizing fragments are larger than 25 kb. Stephenson et al. (2) concluded from this and other evidence that 9.0-kb clusters could not be arranged with other 9.0-kb clusters in a simple tandemly reiterated array. If clusters were tandemly arranged, hybridization to a 4.5-kb fragment should be observed; such a fragment would represent the leftmost 2.0 kb and the rightmost 2.5 kb of adjacent gene clusters, flanked on each end by Bam HI sites. Analogous experiments, using low percentage agarose gels, allowed these authors to conclude that most of the Bam HI sites within histone gene clusters

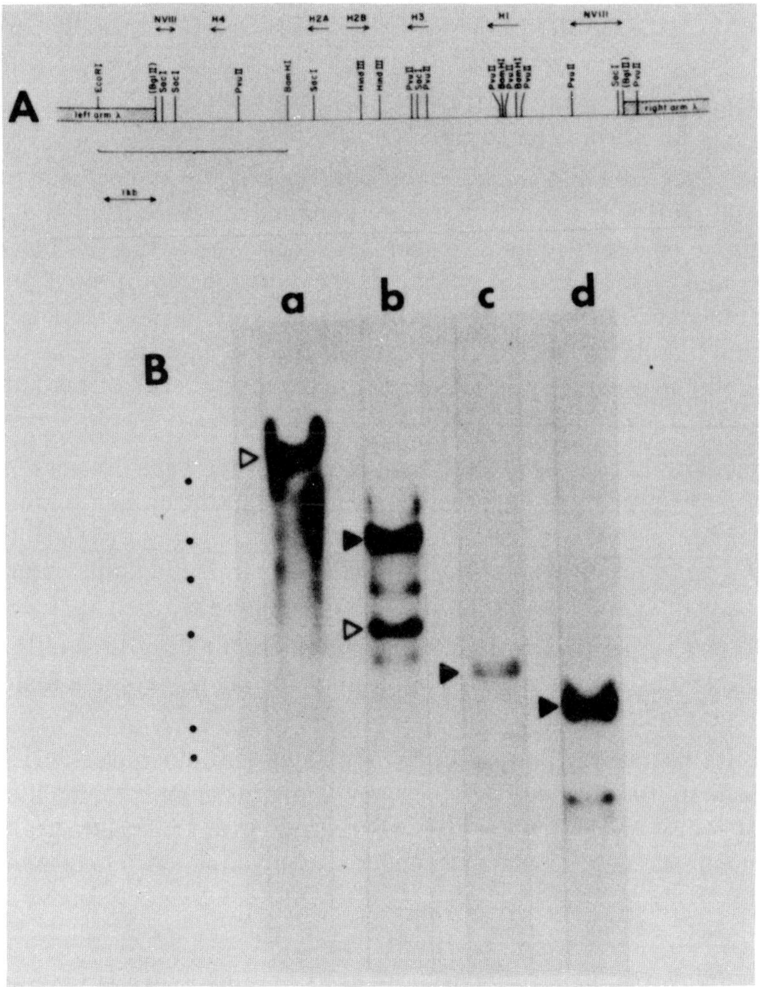

FIGURE 9.1. (*A*) Restriction and transcription map of a cloned histone gene cluster of the newt *Notophthalmus viridescens*. The clone consists of a 9.0 kb Bgl II genomic fragment inserted in the lambda vector ES6. Restriction enzyme sites were mapped by conventional means, and the positions of the genes were determined by heterologous hybridization and sequencing (1). Brackets indicate a fragment used as a hybridization probe in part B. *B*. Southern hybridization using a newt histone gene probe. Restriction enzyme digested genomic DNA was electrophoresed in an agarose gel, blotted to a nitrocellulose filter, and hybridized with the fragment indicated in (*A*). The filter was washed stringently to remove satellite DNA hybridization (2). Dots are the positions of Hind III fragments of λDNA, run in an adjacent lane, the sizes of which are 23.8, 9.6, 6.6, 4.3, 2.4 and 2.1 kb. Filled arrowheads point to bands of sizes predicted by the spacing of restriction enzyme sites in the cloned cluster. Open arrowheads point to other bands discussed in the text. Digests are (a) Bam H1, (b) Bgl II, (c) Pvu II, (d) Sac I.

must be separated by over 50 kb in the newt genome, with some separated by 100 kb or more.

The sequences that separate neighboring histone gene clusters are predominantly, or entirely, satellite 1 sequences. Satellite 1 is a 222 nucleotide tandemly reiterated satellite, which comprises a few percent of the newt genome and is found at the centromeres of most or all newt chromosomes in addition to its (relatively minor) association with the histone genes (3). The interspersion of satellite 1 DNA and histone gene clusters was inferred from several pieces of evidence:

1. Satellite 1 sequences flank the cloned histone gene clusters, one or a few repeats being present at each end of the cloned segment. Sequences with the same occurrence and spacing of restriction enzyme sites as in satellite 1 were demonstrated to flank virtually all of the gene clusters in the newt genome (2).
2. Restriction enzyme sites known to be uncommon in satellite 1 sequences are uncommon in the sequences that separate histone gene clusters (2).
3. Both sequences are present at the same chromosomal loci, and both are transcribed from the same lampbrush chromosome loops. The pattern of satellite 1 transcription from lampbrush chromosomes suggests that these sequences are arranged in long tandem arrays (3, 4).

A model for the organization of newt histone genes is presented in Figure 9.2. Each of 600–800 clusters has the same fine structure as shown for the cluster in Figure 9.2*a*; that is, gene order, the positions of most restriction enzyme sites examined, and presumably, the DNA sequences themselves, are relatively homogeneous. Every cluster is separated from every other cluster by satellite 1 sequences, as shown in Figure 9.2*b*. It is clear that no gene clusters are immediately adjacent in the genome and that all of the clusters are directly flanked by satellite 1 sequences. However, the notion that the only sequences that separate clusters are satellite 1 is based on indirect evidence, all of which is consistent with this idea. The distances that separate neighboring clusters are probably heterogeneous; the available evidence, also indirect, points to distances of 50 kb or above for most clusters.

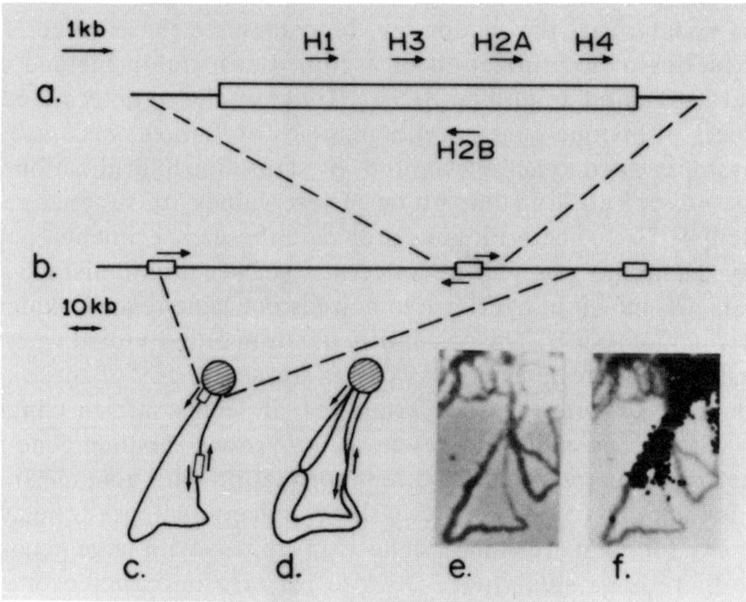

FIGURE 9.2. Genomic and chromosomal organization of newt histone genes. In parts *a*, *b*, and *c*, histone gene clusters are indicated by boxes and satellite 1 sequences by single lines. Parts *b*, *c*, and *d* represent an interpretation of the loop shown in *e* and *f*. The bright circular object in *e*, drawn as a cross-hatched circle in *c* and *d*, is an axial sphere. The spheres, of unknown structure or function, are convenient morphological markers for the histone gene loci in this and other newt species (3). The scale for *a* is the 1-kb arrow, and for the rest of the figure the 10-kb double-headed arrow. (*a*) Detailed view of a single histone gene cluster. The cluster is flanked by satellite 1 sequences. (*b*) Expanded view of the histone gene locus. Histone gene clusters are separated by long tracts of satellite DNA. Arrows indicate the initiation and direction of transcription within some of these clusters. (*c*) Interpretation of the folding of this chromosomal region into the loop shown in *e* and *f*. (*d*) Accumulation of ribonucleoprotein matrix on this loop. (*e*) Phase micrograph of part of a lampbrush chromosome, before hybridization. (*f*) The same chromosome after hybridization of a single strand of satellite 1 to nascent RNA. The probe used hybridizes to readthrough transcripts of the H1, H3, H2A, and H4 promotors, but not the H2B promotor.

The number of *Notophthalmus* histone genes, while large, is approximately proportional to the large genome size of this animal. The reiteration of histone genes in other urodeles has been measured, and in each case histone genes appear to comprise a large family. Sommerville (5) determined the reiteration of histone genes in the genomes of *Triturus cristatus*, the European crested newt, and *Nec-*

turus masculosus, the mudpuppy; he calculated these values from the kinetics of hybridization of a cloned sea urchin histone gene repeat to sheared amphibian DNA. Hilder et al. (6) determined the numbers of histone genes in the genomes of *Triturus cristatus* and *Ambystoma mexicanum*, the axolotl, by saturation hybridization with a *Xenopus* H4 cDNA clone probe. These values are summarized in Table 9.1. Use of heterologous probes in these experiments, particularly sea urchin genes, is of concern, since sea urchin histone gene spacers are known to hybridize to non-histone gene related sequences in newt genomes (9). The possibility of strain or individual variation in histone gene reiteration was not addressed in any of these measurements. Despite these uncertainties, it seems safe to conclude that, with the possible exception of *Ambystoma*, histone gene reiteration shows a more or less direct correlation with genome size.

It is tempting to speculate that this rough equivalence reflects the necessity for greater histone synthesis in animals with large genomes, since there is so much more DNA to package into nucleosomes at each division. Kedes (10, 11) pointed out that the requirements for histone mRNA synthesis in early development could explain the observed reiteration of sea urchin and *Xenopus* histone genes. Various sea urchin species have histone genes reiterated in hundreds of copies per haploid genome, while *Xenopus* has only 50 or so; Kedes correlated these values with the different mechanisms these organisms use in meeting histone protein synthesis during early development. *Xenopus* embryos utilize maternally stored histone and histone mRNA for the early rapid cleavages and evidently require little new message synthesis during the early rapid cleavages, whereas sea urchin embryos rely heavily upon histones translated from newly transcribed mRNAS. It is difficult to extend this model to explain urodele histone gene reiteration, since so little is known about the requirements for and mechanisms of histone synthesis in urodele embryogenesis. There are many conceivable ways to increase the level of a gene product without increasing the reiteration of its gene, in this system, for example, by storage of larger quantities of histone protein or mRNA in the oocyte, or by increased mRNA stability. In other words, the large size of urodele genomes would not seem to require large numbers of histone genes. It is pertinent to note here that histone genes are not the only repetitive sequences to be highly reiterated in animals

TABLE 9.1. Histone Gene Reiteration and Genome Sizes for *Xenopus* and Four Urodeles

	Haploid Genome Size (Picograms)[a]	Histone Genes per Haploid Genome	Histone Genes per Picogram DNA[b]
Xenopus laevis	2.75–3.15	45–50 (8)	14.3–18.2
Notophthalmus viridescens	31.5–49.0	600–800 (2)	12.2–25.4
Triturus cristatus	29.0–33.5	350–570 (5), 636 ± 21 (6)	10.4–21.9
Necturus masculosus	78.0–95.0	1200–2000 (5)	12.6–25.6
Ambystoma mexicanum	21.9–48.0	2685 ± 349 (6)	55.9–122.6

[a] Haploid genome sizes were taken from the review by Olmo (7) and represent the range of values obtained by various authors.
[b] The extremes of histone gene reiteration in relation to genome size, obtained by dividing the lowest and highest reported values for histone gene reiteration by, respectively, the highest and lowest reported genome sizes.

with large genomes; genes for ribosomal RNAs and 5s RNAs are also reiterated in proportion to genome size in urodeles (12). It is not intuitively clear why a large genome should mandate a large number of genes for ribosomal components, either; inevitably, one comes back to the function of these large genomes themselves, and why eukaryotes, with presumably similar coding requirements, should have genomes that vary widely in size—the *C*-value paradox. Comparing histone gene reiteration may just be a more molecularly oriented way of restating the *C*-value paradox; it is clear nonetheless that few answers will be forthcoming without a deeper understanding of the requirements for and mechanisms of histone synthesis in development.

Another feature distinguishing *Notophthalmus* histone genes from those of other vertebrates studied to date is their homogeneity. Newt histone genes comprise a large homogeneous family; histone genes of other vertebrates, including those of another amphibian, *Xenopus*, are heterogeneous, without a common repeating unit or pattern. Stephenson et al. (2) suggested that the homogeneity of newt histone genes and the heterogeneity of other vertebrate histone genes are consequences of their degrees of reiteration. Theoretical models have shown that unequal crossing over can explain the homogeneity of satellite DNAs and repeated genes (13, 14). Newt histone genes, highly reiterated and interspersed with large blocks of satellite DNA, comprise a sizable proportion of the newt genome and thus provide large targets for the exchanges envisioned in this model. The same model can be extended to the histone genes of invertebrates. Both *Drosophila* and sea urchin early histone genes are highly reiterated and relatively homogeneous (11), whereas yeast histone genes, which are not highly reiterated, are heterogeneous with respect to structure and surrounding sequences (15, 16). In short, the simplest interpretation of histone gene (and other repeated sequence) homogeneity is that it bears some relation to the reiteration of these sequences in the genome.

The interspersion of *Notophthalmus* histone gene clusters and satellite 1 sequences represents a novel form of gene organization in eukaryotes. Repeated gene spacers containing simple repeated sequences have been described in other animals (17–20); however, in none of these cases is the simple sequence highly reiterated elsewhere in the genome. Satellite 1, in contrast, is a major satellite sequence in newts, representing a few percent of the genome (4). Like a typical satellite DNA, satellite 1 is found primarily in the centromeric heterochromatin (3), probably no more than 10% being associated with the histone genes.

What is the functional significance of this association? There is little information, so any models will be speculative and require one or more *ad hoc* assumptions. The first possibility worth considering is that the proper and/or efficient expression of the histone genes requires this arrangement. One could suppose, for instance, that proper expression requires that clusters not be immediately adjacent in the genome; the nature of the sequences separating the clusters would presumably be unimportant, the only relevant feature being that, for whatever reason, gene clusters not be tandemly arranged. Another possibility involves the opposite assumption, that the satellite DNA *is* itself important, for instance is a regulatory sequence that directs polymerases to the histone genes or keeps the genomic region in an appropriate configuration for transcription. These models are weakened by the fact that other animals express their histone genes without interpersion with long sequences, satellite or otherwise; the secondary acquisition of an interspersion requirement for newt histone gene expression seems unlikely.

A second class of models supposes that the interspersion of histone genes and satellite DNA is related to the other features that distinguish newt histone genes from those of other vertebrates—their high reiteration and homogeneity. One possibility is that satellite sequences are required to maintain the homogeneity of newt histone genes and thus prevent the accumulation of deleterious quantities of mutant genes. Satellite sequences presumably increase the effective size of the histone gene locus and the unequal crossovers envisioned necessary for the maintenance of homogeneity (13, 14). Another possibility is that the satellite DNA is an evolutionary relic of the unknown process by which urodele genome sizes and histone gene reiteration increased in the course of evolving from a form with a smaller genome and fewer histone genes. The lack of data makes it impossible to judge the merits of these or other models.

These recent studies raise a number of interesting questions concerning the evolution of repeated gene families. Future research directed toward determining the organization of histone genes in other animals with large genomes should be helpful in assessing the significance of the unusual organizational pattern of newt histone genes. In addition to urodeles, some apodan and anuran amphibians possess the large genomes which would make interesting studies, as do some bony fishes. Species closely related to newts, as well as those that have undergone independent increases in genome size, should provide critical information in this regard.

3. EXPRESSION OF NEWT HISTONE GENES

Most studies of urodele histone gene expression have dealt with their transcription from lampbrush chromosomes during oogenesis. Unlike the situation in sea urchins or *Xenopus*, there are few data concerning the quantities and pattern of histone mRNA and protein synthesis in oogenesis and early development. The information available for newts, then, offers few clues as to the developmental significance of histone gene expression. The data do, however, allow several important conclusions concerning the structure of lampbrush chromosomes. This section reviews the structure of lampbrush chromosomes, a model for the transcription of histone genes from them, and the data upon which this model is based. It concludes with a brief discussion of the significance of this unusual pattern of transcription with regard to the presumed role of lampbrush chromosomes and histone mRNAs in development.

For the sake of brevity, we offer here only a short review of lampbrush chromosome structure and refer the reader to more extensive treatments elsewhere (21, 22). Lampbrush chromosomes are found in the oocytes of many animals, but species with large yolky eggs and large genomes provide the most useful material. A lampbrush chromosome consists of an axis, which contains most of the DNA packaged into a compact form, and thousands of lateral loops, containing the rest of the DNA stretched out to its full, or nearly full length. The loops are active in RNA synthesis and, in fact, are composed largely of nascent RNA and protein, while the axis is transcriptionally nearly inert. The ribonucleoprotein matrix of the loops is never uniform in thickness but tapers from thin to thick, one or several thin-to-thick matrices occuring on a single loop. Electron microscope studies have shown that each thin-to-thick matrix unit corresponds to a transcription unit (TU), with the thin end of the matrix representing the origin of transcription, and the thickening of the matrix representing the accumulation of long nascent transcripts and associated protein. Within a loop, multiple transcription units may be arranged tandemly, with transcription proceeding in the same direction for each TU, or with transcription in both directions, so that the thin or thick ends of adjacent TUs meet at a common point on the loop.

In situ hybridization of newt satellite DNA to the nascent RNA of lampbrush chromosome loops provided an opportunity to examine the structure of these transcription units (4). For clarity, we present the model suggested by these studies and then some of the data upon which it is based. Transcription from histone gene loops initiates at promotors upstream from one of the five histone genes and proceeds through the gene, the downstream spacer sequences, any downstream genes, and finally into the satellite DNA which flanks each cluster. Transcription continues until the TU encounters a TU transcribing with the opposite polarity (TUs meet finish to finish), or until the TU comes to the initiation point of the adjacent TU, transcribing in the same direction (adjacent TUs are tandemly arranged).

This model is illustrated in Figure 9.2. The probe used in Figure 9.2*f* was one strand of satellite 1; it has been hybridized to the nascent RNA of the chromosome loop. This probe is known to hybridize to readthrough transcripts that initiate at the H1, H3, H2A, or H4 promotor, but not to transcripts from the H2B promotor. Of the three TUs on this loop, two are labeled, having evidently initiated at one of the H1, H3, H2B, or H4 promoters, and one is not labeled, having initiated at the H2B promotor. A phase micrograph of this loop, taken before hybridization, is shown in Figure 9.2*e*, and a diagram of the TUs in Figure 9.2*d*. An interpretation of the configuration of histone gene clusters and satellite 1 sequences in this loop is shown in Figure 9.2*c*.

The original data upon which this model was based, hybridization of satellite 1 single strands to nascent RNA on lampbrush chromosome loops, provided only indirect support. The most suggestive feature of the data with regard to the specificity of transcription initiation at histone genes was the asymmetry of transcription; transcription occurs in one direction several times more frequently than in the other. TUs that transcribe in the "major" direction contain the satellite read-through expected from initiation at an H1, H3, H2A, or H4 promotor (4). Thus the asymmetric distribution and labeling by satellite 1 is consistent with random initiation at one of the five histone gene promotors.

Direct support for this model has been obtained recently by hybridizing histone gene single strands to lampbrush chromosome nascent transcripts (23). The important conclusions from these experiments are the following:

1. With a few exceptions, histone gene probes label all of a TU, or none of it. This is the result expected if transcription always begins within a histone gene cluster. Note that if transcription began within the satellite DNA, all of the probes should yield partially labeled TUs; the thin end of each TU should be unlabeled up to the point where it passes through a histone gene cluster, after which it should be labeled.
2. Probes containing only gene spacer sequences hybridize as well as those containing coding regions, as predicted.
3. Certain probes should, and do, label very few TUs. For instance, antisense transcripts of the H2A and H4 gene regions should never occur since these are upstream from the only promotor that can produce them, the H2B promotor ("antisense" is used here to denote the strand opposite to that of the mRNA). In fact, a few sites of hybridization have been detected with these probes. The labeling is always partial, and always to the thick end of a long TU. This evidently represents transcription that initiated normally at an H2B promotor, continued through the satellite 1 sequences, and into an adjacent histone gene cluster which did not itself contain an active initiation site. Only at this point did antisense transcripts of the H2A and H4 gene regions become attached to the nascent RNA. This finding requires that the original model proposed by Diaz et al. (4) be amended slightly to indicate the existence of histone gene clusters without active initiation sites.

This model may reflect the universal mechanism of transcription from lampbrush chromosomes. Transcription initiates at a structural gene promotor, proceeds through the gene into the downstream spacer, and terminates only when it meets an adjacent TU. The model provides a convenient explanation for the previous observations of the transcription of satellite and other highly repetitive sequences (24, 25) and for the extremely high sequence complexity of oocyte nuclear RNA (26).

What do these results tell us about the expression of newt histone genes themselves? There is little direct information, but analogy with other animals suggests useful hypotheses. For instance, newt oocytes contain histone mRNA (1, 27), as do the oocytes of sea urchins and *Xenopus* (11); the mRNA stored in newt oocytes is presumably used during early development. Is this maternal message the product of the major family of newt histone genes? By analogy with sea urchins, in which a large, homogeneous gene family is also responsible for the maternal mRNA (11), one is tempted to think so. As direct support for this notion is the observation that oocyte H1 and H4 mRNAs are completely homologous to the major gene family, as detected in S1 nuclease mapping experiments (Gall, unpublished). These genes are also known to be active during oogenesis, as detected in the *in situ* hybridization experiments (3, 4).

This brings up another question, namely, what is the relationship between the long lampbrush chromosome transcripts and the discrete histone mRNAs stored in the oocyte? One possibility is that normal-sized mRNAs are transcribed from the same clusters that initiate the long readthrough transcripts, the difference being that some polymerases recognize the normal termination signals and others do not. Short transcripts could not have been detected by *in situ* hybridization, and, at present, there is no other information to evaluate this idea. Another possibility is that histone mRNAs are processed from the long readthrough transcripts. Old et al. (28) advanced this idea to explain the pattern of labeling they saw after hybridization of sea urchin histone genes to lampbrush chromosomes of *Triturus cristatus*, the European newt. There is now doubt that Old et al. observed histone gene transcripts in these experiments. The sites they identified as being the *Triturus* histone gene loci do not correspond to the sites that hybridize to *Notophthalmus* histone genes (3), and the repeated sequences found in sea urchin histone gene spacers have since been found to hybridize to non-histone gene related sequences in the *Triturus* genome (9). There is no evidence that processing occurs during *Notophthalmus* histone gene transcription; if processing did occur, one would expect histone gene probes to label only the thin ends of transcription units, and such a labeling pattern is not observed. Of course, the possibility that histone mRNAs are processed from the readthrough transcripts after the nascent RNAs are released from the chromosome cannot be excluded with the available data. The final possibility worth considering is that lampbrush chromosomes produce no mRNA at all, the mRNA content of the egg having come from the pre-lampbrush stages. There is some evidence that the mass and sequence complexity of the stored mRNA reaches a plateau well before the lampbrush chromosome stage (29, 30), and little direct evidence that lampbrush chromosomes produce any functional mRNA. Recent data (31), however, suggest that lampbrush chromosomes produce message sufficient to replace that which turns over during oogenesis, the extremely large message pool of the egg requiring the high rate of transcription observed in lampbrush chromosomes.

Interesting answers to these developmental questions should be forthcoming when more biochemical experiments are attempted. In particular, experiments to define the patterns of histone mRNA and protein synthesis, as have been performed in sea urchins and *Xenopus* should be helpful. The stumbling block here is that the most commonly studied urodeles do not produce eggs and embryos in quantities sufficient for the application of large-scale biochemical techniques. Study of urodeles other than newts or use of more sensitive methods may well circumvent these problems.

ACKNOWLEDGMENTS

We thank Dr. Joseph Gall for reading and commenting on the manuscript, and Dr. Gall and Dr. Manuel Diaz for providing unpublished information. Experimental work reported here was supported by NIH grant GM 12427. The author is supported by NIH postdoctoral fellowship GM 07947.

REFERENCES

1. E. C. Stephenson, H. P. Erba, and J. G. Gall, *Nucl. Acids Res.* **9,** 2281 (1981).
2. E. C. Stephenson, H. P. Erba, and J. G. Gall, *Cell* **24,** 639 (1981).
3. J. G. Gall, E. C. Stephenson, H. P. Erba, M. O. Diaz, and G. Barsacchi-Pilone, *Chromosoma* **84,** 159 (1981).
4. M. O. Diaz, G. Barsacchi-Pilone, K. A. Mahon, and J. G. Gall, *Cell* **24,** 649 (1981).
5. J. Sommerville, in S. Rosenthal, H. Bielka, C. Coutrelle, and C. Zimmer, Eds., *Gene Function, FEBS Meeting*, Vol. 51, Pergamon, Oxford, 1979, p. 265.
6. V. A. Hilder, R. N. Livesay, P. C. Turner, and M. T. Vlad, *Nucl. Acids Res.* **9,** 5737 (1981).
7. E. Olmo, *Carylogia* **26,** 43 (1973).
8. W. von Dongen, L. de Laaf, A. Moorman, and O. Destree, *Nucl. Acids Res.* **9,** 2297 (1981).
9. H. G. Callan, R. W. Old, and K. W. Gross, *Eur. J. Cell Biol.* **22,** 21 (1980).
10. L. H. Kedes, *Cell* **8,** 321, (1976).
11. L. H. Kedes, *Ann. Rev. Biochem.* **48,** 837 (1979).
12. E. O. Long and I. B. Dawid, *Ann. Rev. Biochem.* **49,** 727 (1980).
13. G. P. Smith, *Science* **191,** 528 (1976).
14. T. Ohta, *Evolution and Variation of Multigene Families*, Springer-Verlag, Berlin, 1980.
15. L. Hereford, K. Fahrner, J. Woolford, M. Rosbash, and D. B. Kaback, *Cell* **18,** 1261, (1979).
16. J. W. Wallis, L. Hereford, and M. Grunstein, *Cell* **22,** 799 (1980).
17. P. Boseley, T. Moss, M. Machler, R. Portmann, and M. Birnstiel, *Cell* **17,** 19 (1979).
18. E. O. Long and I. B. Dawid, *Nucl. Acids Res.* **7,** 205 (1979).
19. N. V. Fedoroff and D. D. Brown, *Cell* **13,** 701 (1978).
20. C. Tschudi and V. Pirrotta, *Nucl. Acids Res.* **8,** 441 (1980).
21. E. Davidson, *Gene Activity in Early Development*, 2nd ed., Academic, New York, 1976.
22. H. C. Macgregor, *Heredity* **44,** 3 (1980).
23. M. O. Diaz and J. G. Gall, in preparation.
24. J. M. Varley, H. C. Macgregor, and H. P. Erba, *Nature (London)* **283,** 686 (1979).
25. J. M. Varley, H. C. Macgregor, I. Nardi, C. Andrews, and H. P. Erba, *Chromosoma* **80,** 289 (1980).
26. J. Sommerville and D. Malcolm, *Chromosoma* **55,** 183 (1976).

27. J. V. Ruderman and M. L. Pardue, *Dev. Biol.* **60,** 48 (1977).
28. R. W. Old, H. G. Callan, and K. W. Gross, *J. Cell Sci.* **27,** 57 (1977).
29. M. Rosbash and P. Ford, *J. Mol. Biol.* **85,** 87 (1974).
30. L. Golden, U. Schafer, and M. Rosbash, *Cell* **22,** 835 (1980).
31. D. M. Anderson, J. D. Richter, M. E. Chamberlin, D. H. Price, R. J. Britten, L. D. Smith, and E. H. Davidson, *J. Mol. Biol.* **155,** 281 (1982).

PART 3

HIGHER EUCARYOTES

CHAPTER 10

CHICKEN HISTONE GENES AND THEIR VARIANTS

R. P. HARVEY AND J. R. E. WELLS
Department of Biochemistry
University of Adelaide
Adelaide, South Australia

CONTENTS

1. INTRODUCTION

2. ORGANIZATION AND SEQUENCE
   2.1. Evolution
   2.2. Sequence
   2.3. An H2B Gene 5′-Specific Sequence

3. LINKER HISTONE VARIANTS—THE CHICKEN H5 GENE
   3.1. Relationship of H5 and H1° Histones
   3.2. Isolation and Characterization of Chicken H5 Gene Sequences
   3.3. Chromosomal Organization of the Chicken H5 Gene

4. CORE HISTONE VARIANTS
   4.1. An Extremely Variant Histone H2A

5. FUTURE STUDIES

REFERENCES

1. INTRODUCTION

This review attempts to put our own work in context with other areas of research on histones and their genes. It is meant to complement the chapter by J. D. Engel which emphasizes our current understanding of the organization of the chicken histone genes. This chapter focuses on the fine structure of chicken histone genes and preliminary analysis of variant histone genes.

The role of abundant core histones in nucleosome structure is well known, whereas that of the histone variants is not clearly defined. In different species, variants have been shown to be tissue specific or developmentally regulated (1). Recently a further class of variant subtypes called basal histones have been described (2). These are synthesized throughout the cell cycle rather than in S-phase alone. Clearly then, there are different classes of histones regulated in different ways. The fact that expression of histone variants is regulated strongly implies that they have functional significance, perhaps in modifying chromatin structure. It is therefore relevant to investigate the structure of different classes of histone genes, to identify the elements that control their temporal expression, and finally to assess the influence of variant histones on chromatin structure. A key question to be answered is whether the presence of variant histones in chromatin influences tissue-specific gene expression during development.

Our interest in the chicken histone gene system stemmed from the fact that chicken erythrocytes contain the tissue-specific histone, H5. When we started investigations on H5 genes, vertebrate histone genes, in general, were uncharacterized. The tandemly reiterated structure of the sea urchin histone genes was well known (3), and the maintenance of a similar arrangement for the histone genes of other invertebrates led to the belief that all histone genes would be in the same form. The incorporation of H5 genes into this format posed an interesting structural and regulatory problem in the chicken system. Since then, the complexities of the vertebrate histone gene system have become apparent. We discuss here what we have learned from the sequence analysis of histone genes where it specifically relates to organization and regulation, and then some structural and expressional aspects of one linker histone variant gene (H5) and one core histone variant cDNA (H2A.F).

2. ORGANIZATION AND SEQUENCE

In Chapter 11 on chicken histone genes, J. D. Engel describes the organizational details of the histone genes contained within several chicken genomal clones. Some of the genes within these clones have been sequenced by us and by Engel's collaborators.

2.1. Evolution

The first major conclusion to be drawn from the analysis of the chicken histone genomal clones was that the genes are clustered, but arranged in an apparently random fashion (4, 5). The ordered structure of the invertebrate histone genes does not occur in chickens, and this was also reported to be the case in the mouse (6) and human (7) systems. An explanation is therefore required to account for the dispersal of histone genes through evolution. The highly conserved nature of the sea urchin and *Drosophila* histone genes (3) was first thought to be intimately related to their expression. For instance, in these gene systems, because all five histone genes occur in a tandemly reiterated cluster, relative gene stoichiometry is maintained, and this could be interpreted as a partial requirement for coordinate expression. However, the dispersal of the conserved histone gene cluster through evolution would argue that this tandem arrangement is not obligatory and that coordinate expression (or a stoichiometric gene number) can be achieved in other ways. It also suggests that individual histone genes are controlled independently. DNA sequencing data shows that in the 5′ leader regions of sea urchin histone genes there are sequences that also occur in the promoter regions of most other genes transcribed by polymerase II (8) and that the 3′ region of each gene contains a characteristic terminator sequence (9). These tandemly arranged genes also function independently when introduced into heterologous transcription systems such as *Xenopus* oocyte nuclei. There is evidence, then, that histone genes are independently controlled, and so we cannot attribute the evolutionary dispersal of histone gene organization in higher eukaryotes to a recognizable change in their mode of expression.

The mechanism by which reiterated genes remain invariant (concerted evolution) poses a problem, given the inevitability of mutation. However, there is now evidence that recombinational events, such as unequal crossing over and gene conversion, play a significant role in the homogenization of reiterated sequences (10). In the case of invertebrate histone genes that are highly reiterated (300–1000 times), correctional events are probably also responsible for the conserved nature of each repeat unit. Because reiteration frequency is high, recombination events between clusters will be relatively frequent, and DNA sequence variation will be suppressed. Thus sequence homogeneity is maintained between reiterated genes, but the conservation, as such, may be of little consequence to the control of expression.

In the case of vertebrate histone genes, where reiteration frequency is relatively low, gene correction events will be infrequent and consequently ineffective in homogenizing the system. It may be useful to consider the globin gene system where there are only a few α-like and β-like globin genes in each locus. Gene conversion events have been detected between copies of the human γ genes ($^{G}\gamma$, $^{A}\gamma$) (11). However, such events must be rare, because throughout the evolution of each locus most gene copies have diverged to the point of functional difference.

In chickens, histone genes are reiterated about 10 times per haploid genome (12). Histone gene number is also low in the other species that have a disordered histone gene arrangement (mouse, 10–20 times; human 30–40 times). As a general thesis, we suggest that the evolutionary transition of histone gene organization from an ordered to a disordered disposition is a function of the declining frequency of recombinational events as the gene copy number becomes low, and does not reflect a change in the mode of expression of histone genes.

2.2. Sequence

We have determined the DNA sequence of several genes from the genomal clones λCH-01 and λCH-02 (see J. D. Engel, Chapter 11). Engel's collaborators have also sequenced some genes from their independently isolated genomal clones (from the same λ Charon 4A recombinant library), and this data has been made available to us. We have made comparisons of the coding region sequences of these genes and of the sequences immediately flanking these to search for sequences that may be important in gene expression or regulation. The general structure of these histone genes is similar to that of most other genes transcribed by polymerase II, except that the coding regions are not interrupted by intervening sequences. This seems to be a common feature of most histone genes (Chapter 11) and is shared by some other gene families, notably a number of human inteferon genes (13). In the 5′-flanking regions of the genes, sequences corresponding to general polymerase II promoter elements can be found. The modulator "CAT" sequence, the "TATA" box, and the "CAP" initiation site are, in general, well conserved in position and sequence. These observations strengthen the foregoing assertion that histone genes are controlled at the single gene level. In the 3′-untranslated regions, we have recognized the presence of a conserved dyad-symmetry element (14) which was observed in the equivalent region of all sea urchin histone genes (9). It is now apparent that this sequence is present in histone genes of this class from virtually all species in which genes have been examined (9, 14). Deletion mapping and gene manipulations have shown that this sequence is involved in the termination or maturation of histone mRNAs (9). Another homology block in the immediate 3′-flanking sequences of sea urchin histone genes (15) is also conserved in vertebrate histone genes, albeit weakly. Some 3′-flanking sequences, in addition to the dyad symmetry element in the untranslated region, are also necessary for efficient termination (9). In the chicken this second homology block is more extensive, particularly in two H2B genes (16). Although these termination/maturation elements are widely conserved, it is obvious that they are only a focus for the termination machinery. For example, although the sea urchin H3 gene contains the conserved sequences, its transcript is terminated inefficiently in *Xenopus* oocytes (17). Termination of H2A and H2B genes does, however, occur relatively normally in this system.

Detailed comparisons between iso-coding chicken genes have been made—in particular, two copies of unlinked H2B genes, three copies of unlinked H3 genes, and two copies of H2A genes which are part of the same genomal clones but separated by other genes. It is of interest to consider the sequence comparisons in the context of the evolution of histone gene organization.

So far, all chicken histone gene copies of a given type (H2A or H2B, etc.) that we have analyzed from overlapping genomal clones code for identical proteins. No evidence has been obtained for histone variant genes (the H1 genes have not yet been analyzed in detail). This observation suggests that so far we have only detected a specific class of histone genes and that variant genes must lie elsewhere in the genome. (This seems to be true for H5 and H2A.F, discussed later.) It also suggests that variant genes are distinct in function and not merely inconsequential fixed mutations within the predominant gene cluster. Comparisons of DNA sequences outside the protein coding regions reveal that, in general, the untranslated regions are totally divergent in sequence and often of different length. For example, in the two H2B genes compared (16), transcripts are likely to be about 29 bases different in length, and, apart from specific regions to be discussed, no homology can be seen in the 5′ and 3′ sequences. This indicates that there are very limited specific sequence requirements in these regions (the "CAP site" and 3′ dyad-symmetry element are conserved) and that these genes are quite distantly related. To some degree, remnants of homology in the untranslated regions can be used as a measure of the age of a gene duplication (18). That two chicken histone genes are distant relations is not surprising, considering the ancient nature of histones in eukaryotes.

If considered in the context of vertebrate histone gene evolution, this divergence demonstrates that in these chicken genes, recombinational interaction is indeed rare between regions of gene copies where there is no direct selection—a possible reason for the absence of the highly ordered histone gene arrangement seen in the invertebrates (discussed previously). It also demonstrates that conservation of sequences around the histone coding regions (as in the invertebrate systems) is not a requirement for histone gene expression. Most of our DNA sequences have been determined from the histone gene coding regions and adjacent flanking DNA. We have searched the sequences in the immediate vicinity of the polymerase II promoter elements for sequence homologies that may indicate DNA sequences important to regulation.

2.3. An H2B Gene 5′-Specific Sequence

In the two chicken H2B genes compared, we have located a 13 bp sequence, 5′ CTCATTTGCATAC 3′, that we now find is also con-

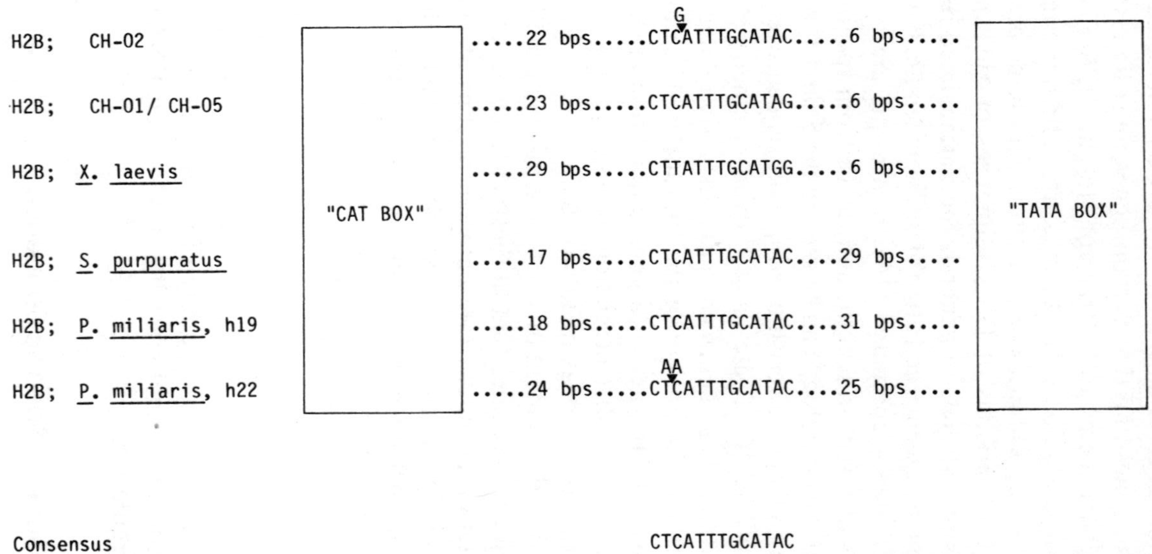

FIGURE 10.1. The H2B gene-specific sequence. The 13-bp conserved flanking sequence from the H2B genes of chickens, *Xenopus*, and sea urchins are tabulated (16). The position of this sequence relative to the "CAT" box and "TATA" box of each respective gene is also indicated.

served in all other H2B genes that have been sequenced (Figure 10.1) (16). It is placed between the "TATA" box and "CAT" box in H2B genes, but we have not found any related sequences in the 5′-flanking regions of other histone genes. To our knowledge, this is one of the first highly conserved gene-specific 5′-flanking sequences that has been recognized. Its location within the promoter region indicates that it may be related to specific transcription of H2B genes. Its gene-specific nature suggests that its function is indeed in regulation rather than general promotion or modulation of transcription. The presence of the H2B-specific sequence will be a focus to examine regulatory factors that may bind to promoter regions and could lead to a greater understanding of how H2B genes are controlled. No similarly placed regions of homology have been found in comparisons of other histone genes. This suggests two possibilities. Either sequences important for regulation will not always be found by intergene or cross-species comparisons (DNA stereochemistry may be more important than specific DNA sequence), or that regulatory regions exist further upstream than it has so far been possible to examine. Given the lack of constancy in chicken histone genes organization, the placement of regulatory sequences (those in addition to general promoter elements) is of considerable interest.

As mentioned, chicken histone genes show no constancy in their organization. One H2B gene (λCH-02), for instance, is relatively isolated, being several kilobase pairs from its nearest 5′ neighbor (5). By analogy with the expression of manipulated sea urchin histone (and other) genes in transcription systems, it is likely that sequences important to expression would be spread throughout this extensive 5′ region. On the other hand, another H2B gene that we have sequenced (λCH-01) is extremely closely linked to an H2A gene, and transcription of the gene pair is in a divergent fashion (16). Only 77 bps separate the respective TATA boxes of these two genes, so they are closely linked indeed. Such an arrangement profoundly limits the region of DNA in which regulatory sequences occur, unless these elements overlap with each other or lie within the coding regions of the adjacent gene. In this respect, it may be significant that within the H2A/H2B intergene region only the H2B gene has a true CAT box sequence, and this is located almost centrally between the TATA boxes of the two genes. No CAT sequence can be found for the H2A gene. It is an interesting possibility that both genes share this single CAT sequence as a modulator of transcription.

The point to be drawn from these observations is that the immediate environment of chicken histone gene copies varies to the extent that each individual gene may be a case for analysis. DNA sequence analysis of chicken histone genes has highlighted some distinct features such as an H2B-specific 5′ sequence, but the lack of such sequences in other genes suggests that answers to gene regulation may not be found in the DNA sequences alone. Tests of gene function are required.

3. LINKER HISTONE VARIANTS—THE CHICKEN H5 GENE

3.1. Relationship of H5 and H1° Histones

Although the H5 protein from chicken red blood cells has been well characterized (20–22), it is only very recently that gene sequences for this tissue-specific histone have been isolated (23–25). In this section we examine the possible roles of this tissue-specific histone, its relationship to the H1 histone group, and finally the H5 gene itself and its organization in the genome with respect to core and H1 genes.

The appearance of tissue-specific histones suggests a tissue-specific function. One characteristic feature of erythroid cells from birds, amphibia, and fish (26) is the presence of the H5 protein which accumulates during maturation of the cells. It is tempting to suggest (although not proven) that H5 plays a role in the condensation of chromatin in the mature cells of the series (27). The condensation may lead to suppression of transcription (28) and of replication (29).

Limited amino acid sequence data shows that chicken H5 has considerable homology (30) with a subset of H1 proteins from a number of sources, (including mammalian cells) and designated H1°. It is likely that H5 and H1° are functionally equivalent and should be classified as a histone category distinct from the usual H1 species (30). In general, histone synthesis is tightly coupled to DNA synthesis, whereas neither H5 (31) nor H1° (32) expression is restricted to S phase. In fact, a relative increase in the level of these histones is correlated with a lack of DNA synthesis (33, 34). There is evidently a much wider tissue distribution of H1° in mammalian tissue than that of the red-cell-specific H5, and in the unexpected case in which H5 is absent from red cells of the bullfrog (*rana catesbeiana*), minor components H1B, R1, and R2, are found, and of these H1B appears to be very similar to H1° (35). The H1° protein has been found in a wide variety of tissues including bovine liver and kidney (30), mouse neuroblastoma cells (36), and human Hela cells (35). An extensive study of H1° in rodents has been published recently (37) in which it is clear that H1° is associated with chromatin in a wide range of fully differentiated cell types. The appearance of H1° relative to H1 is tissue dependent and developmental, and hormone-sensitive responses are noted (37). The molecular mechanisms associated with expression of this unique subset of H1 proteins can be examined at the gene level with the availability of the appropriate gene sequences. This is now possible for the chicken H5 system (23, 25).

3.2. Isolation and Characterization of Chicken H5 Gene Sequences

The availability of the protein sequence of chicken H5 (22) allowed us to choose a region which, on back-translation, would predict a relatively unambiguous nucleotide sequence. An 11-base deoxynucleotide sequence complementary to the mRNA region was synthesized and used to prime cDNA synthesis on chicken reticulocyte RNA. The extended primer was used as a probe to screen a cDNA library constructed from reticulocyte RNA and H5-specific sequences isolated (23, 38). In a separate study, Ruiz-Vazquez and Ruiz-Carrillo used specific antibody to identify cDNA clones containing H5 sequences (24). We have used H5 cDNA as a probe to isolate the genomal gene from a λ Charon 4A library. The main points arising from sequence analysis of H5 cDNA clones (38) and some 1.8 kb of flanking and coding regions of the H5 genomal gene (39) are as follows:

1. The genomal H5 gene within the recombinant λCH5-01 is not a pseudogene and codes for the expected protein sequence. There are no intervening sequences.
2. The gene contains a poorly conserved TATA box (CTTAAAT) in contrast with most chicken core histone gene sequences.

3. H5 mRNA lacks the 3′ AAUAAA sequence, although it is polyadenylated. The conserved hyphenated dyad symmetry element associated with termination of transcription in other histone gene systems (9) is not present in H5 transcripts. Chicken H1 genes do contain this element (our unpublished data).

In summary, the H5 gene is similar to other histone genes in having no intervening sequences. The lack of highly conserved 5′ and 3′ noncoding elements associated with transcription of other histone genes distinguishes H5 from these.

3.3. Chromosomal Organization of the Chicken H5 Gene

Representatives of the four core histone genes and the H1 gene are present in the clone λCH-01 (5; and Chapter 11). Sequences from each of these genes have been subcloned and used to locate specific genes in the disordered array of genes in overlapping clones (Chapter 11). With these probes some 50 independent isolates of recombinants containing histone genes were isolated from a Charon 4A library, and although these have not been fully characterized, they can be grouped into 12 sets on the basis of core or H1 gene content. None of these isolates hybridized with H5 gene sequences. In a reciprocal experiment, an extensive region of DNA surrounding the H5 genomal gene (41 kb in all) failed to hybridize with core and H1 gene probes, supporting the notion that H5 is not closely linked (39). Southern blot analysis strongly suggests that the organization of the H5 gene in the recombinant λCH5-01 (39) is representative of the sequence in chromosomal DNA and that it is a single-copy gene.

4. CORE HISTONE VARIANTS

The major core histones are highly conserved proteins that constitute the structural scaffolding of the eukaryotic nucleosome particle. However, the widespread occurrence of histone variants leads us to believe that modified or variant nucleosome particles have specific functions in cellular events. The occurrence of core histone variants has been reviewed by Isenberg (1). Apart from sperm-specific subtypes, it is difficult to make correlations between the occurrence of core histone variants and specific changes in cellular activity. However, the nature of the amino acid substitutions in some variants and the specificity of their regulation and relative tissue abundance suggest that variant nucleosomes have some influence on chromatin architecture in differentiated cells. Recently some experimental evidence has been obtained that supports this premise. In sea urchins, distinct types of the core histones H2A and H2B are synthesized at different developmental stages (reviewed in Reference 15). The mRNAs for these subtypes are synthesized from maternal, embryonic, and somatically expressed genes. In addition, sea urchin sperm has its own distinct subtypes (40, 41). Measurements of some physical parameters of nucleosomal core particles derived from sea urchin sperm (sperm subtypes), blastula (100% embryonic subtypes), and pluteus (75% "late" subtypes), have defined structural heterogeneity in variant nucleosomes (42, 43). These parameters include stability to, and site specificity of, DN'ase 1 digestion, thermal melting characteristics, and expansion of core particles in response to low ionic strength. The differences observed in no way prove a biological distinction between variant nucleosomes, but they do show that variant histones are not inconsequential to nucleosome structure.

As a complementary interest in the regulation and developmental effects of linker histone variants (namely, H5), we have begun to search for core histone variant genes in the chicken. Our initial approach to this has been to look for variant mRNAs synthesized in the chick embryo. We have constructed a cDNA recombinant bank and screened it with core histone probes derived from the genomal clone, λCH-01. Clones with large inserts, but which hybridized poorly to these probes, were selected as possible variant cDNAs. H2A and H2B clones were isolated using this criteria.

4.1. An Extremely Variant Histone H2A

We have characterized one cDNA which hybridized weakly to the H2A probe (44). The complete sequence of this cDNA reveals that it encodes an extremely variant H2A-like protein which we have called H2A.F. It is certainly an H2A because it contains the nonapeptide that has been conserved in all H2A molecules so far sequenced (2).

When compared to the amino acid sequence of H2A.I, the most abundant H2A in chicken erythrocytes, H2A.F shows 40% divergence. To contrast this, the H2A.I sequences of calf and chickens have only diverged to 4% (Figure 10.2).

Using the H2A.F cDNA as probe, we have investigated the H2A.F genes in genomal DNA by southern analysis, and the occurrence of H2A.F mRNA in different tissue samples by RNA-nitrocellulose hybridizations (44). Several points can be made from these data:

1. From southern analysis, DNA regions that hybridize to the H2A.F probe are extensive and include partially homologous sequences (washed off at high stringency). However, all H2A.F-like sequences are linked in chicken DNA.
2. H2A.F genes are not extensively conserved in diverse species—no hybridization with the chicken sequence can be detected in human, mouse, or sea urchin DNAs.
3. H2A.F mRNA is polyadenylated.
4. There is considerable variation in the level of H2A.F mRNA in different tissues.

Our analysis of this system is in an early stage. However, we have isolated a cDNA that codes for an extremely variant H2A molecule—one almost certain to have some influence on the structure of the nucleosomes that contain it. Interestingly, the expression of H2A.F mRNA may be limited in its tissue distribution. These factors make H2A.F a particularly interesting subject in an investigation of the role of core histone variants. The organizational and regulatory aspects of this system are also of great interest with respect to the chicken histone multigene family.

5. FUTURE STUDIES

There is more to be learned about the organization of major and variant histone genes in the chicken genome. However, data now available suggest new directions that may provide insights in two areas. The first concerns the control of gene transcription. Identification of an H2B gene-specific DNA sequence in the general promoter region

1.	**CHICKEN H2A.I**	**H_2N**			**SER**	**GLY**	**ARG**	**GLY**	**LYS**	**GLN**	**GLY**	**GLY**	**LYS**	**ALA**	**ARG**	**ALA**	**LYS**	**ALA**	**LYS**	**SER**	**ARG**	**SER**	**SER**	**ARG**	**ALA**	**GLY**	**LEU**	**GLN**	**PHE**	**PRO**
2.	CHICKEN H2A.F		ALA	GLY	GLY	LYS	ALA			ASP	SER				LYS				VAL				GLN							
3.	CALF																		THR	ARG	SER									
4.	TROUT									THR																				
5.	RAT																													
6.	MOUSE																													

1.	**VAL**	**GLY**	**ARG**	**VAL**	**HIS**	**ARG**	**LEU**	**LEU**	**ARG**	**LYS**	**GLY**	**ASN**	**TYR**	**ALA**	**GLU**	**-**	**ARG**	**VAL**	**GLY**	**ALA**	**GLY**	**ALA**	**PRO**	**VAL**	**TYR**	**LEU**	**ALA**
2.				ILE			HIS		LYS	THR	ARG	THR	THR	SER	HIS	GLY			THR		THR		ALA			SER	
3.																											
4.																											
5.																											
6.																								MET			

1.	**ALA**	**VAL**	**LEU**	**GLU**	**TYR**	**LEU**	**THR**	**ALA**	**GLU**	**ILE**	**LEU**	**GLU**	**LEU**	**ALA**	**GLY**	**ASN**	**ALA**	**ALA**	**ARG**	**ASP**	**ASN**	**LYS**	**LYS**	**THR**	**ARG**	**ILE**	**ILE**
2.		ILE								VAL								SER	LYS		LEU		VAL	LYS			THR
3.																											
4.																										-	
5.																											
6.																											

1.	**PRO**	**ARG**	**HIS**	**LEU**	**GLN**	**LEU**	**ALA**	**ILE**	**ARG**	**ASN**	**ASP**	**GLU**	**GLU**	**LEU**	**ASN**	**LYS**	**LEU**	**LEU**	**GLY**	**LYS**	**VAL**	**THR**	**ILE**	**ALA**	**GLN**	**GLY**	**GLY**
2.										GLY					ASP	SER		ILE	-		ALA				GLY		
3.																											
4.								VAL												GLY							
5.																				ARG							
6.																											

1.	**VAL**	**LEU**	**PRO**	**ASN**	**ILE**	**GLN**	**ALA**	**VAL**	**LEU**	**LEU**	**PRO**	**LYS**	**LYS**	**THR**	**ASP**	**SER**	**HIS**	**-**	**LYS**	**ALA**	**LYS**	**-**	**ALA**	**LYS-COOH**
2.		ILE		HIS		HIS	LYS	SER		ILE	GLY			GLY	GLN	GLN	LYS		THR		-	-	-	-
3.															GLU			HIS					GLY	
4.																-------	-------					GLY		
5.																								
6.																								

FIGURE 10.2. Comparison of H2A sequences. Chicken H2A.I is shown in full. All other H2As are compared to the chicken H2A.I, and amino acids are only indicated where they differ.

(Section 2.3) is suggestive of a regulatory role, and genetic manipulation of this region in conjunction with *in vitro* transcription studies is now possible. It will obviously be of interest to see if specific factors interact with such a region.

A separate facet of histone gene expression is the role of variant gene products in remodeling chromatin structure. We have isolated chicken genes for linker (H5) and core (H2A.F) variants. These genes can be introduced into cells in high-expression vectors such as SV40 (45) or bovine papilloma virus (46). It may be possible to study the effects of the variant gene products on the chromatin of the vector and the host cell chromosome in both structural and biological terms.

The histone gene system, although one of the most intensely studied multigene families, still offers great scope for further study in its own right, and also because histones may affect the expression of other eukaryotic genes.

REFERENCES

1. I. Isenberg, *Ann. Rev. Biochem.* **48,** 159–191 (1979).
2. R. S. Wu and W. M. Bonner, *Cell* **27,** 321–330 (1981).
3. L. H. Kedes, *Ann. Rev. Biochem.* **48,** 837–870 (1979).
4. J. D. Engel and J. B. Dodgson, *Proc. Natl. Acad. Sci. USA* **78,** 2856–2860 (1981).
5. R. P. Harvey, P. A. Krieg, A. J. Robins, L. S. Coles, and J. R. E. Wells, *Nature* **294,** 49–52 (1981).
6. D. B. Sittman, I. Chiu, C. Pan, R. H. Cohn, L. H. Kedes, and W. F. Marzluff, *Proc. Natl. Acad. Sci. USA* **78,** 4078–4082 (1981).
7. N. Heintz, M. Zernik, and R. G. Roeder, *Cell* **24,** 661–668 (1981).
8. M. Busslinger, R. Portmann, J. C. Irminger, and M. L. Birnstiel, *Nucl. Acids Res.* **8,** 957–977 (1980).
9. C. Birchmeier, R. Grosschedl, and M. L. Birnstiel, *Cell* **28,** 739–745 (1982).
10. D. Baltimore, *Cell* **24,** 592–594 (1981).
11. J. L. Slighton, A. E. Blechl, and O. Smithies, *Cell* **21,** 627–638 (1980).
12. R. J. Crawford, P. Krieg, R. P. Harvey, D. A. Hewish, and J. R. E. Wells, *Nature* **279,** 132–136 (1979).
13. S. Ohno and T. Taniguchi, *Proc. Natl. Acad. Sci. USA* **78,** 5305–5309 (1981).

14. R. D'Andrea, R. P. Harvey, and J. R. E. Wells, *Nucl. Acids Res.* **9,** 3119–3128 (1981).

15. C. C. Hentschel and M. L. Birnstiel, *Cell* **25,** 301–313 (1981).

16. R. P. Harvey, A. J. Robins, and J. R. E. Wells, *Nucl. Acids Res.* **10,** 7851–7863 (1982).

17. C. Hentschel, E. Probst, and M. L. Birnstiel, *Nature* **288,** 100–102 (1980).

18. A. Efstratiadis, J. W. Psakony, T. Maniatis, R. L. Lawn, C. O. O'Connell, R. A. Spritz, J. K. DeRiel, B. G. Forget, S. M. Weissman, J. L. Slightom, A. E. Blenchl, O. Smithies, F. E. Baralle, C. C. Shoulders, and N. J. Proudfoot, *Cell* **21,** 653–668 (1980).

19. R. Grosschedl, B. Wasylyk, P. Chambon, and M. L. Birnstiel, *Nature* **294,** 178–180 (1981).

20. J. M. Neelin and G. C. Butler, *Can. J. Biochem. Physiol.* **39,** 485–491 (1961).

21. E. W. Johns and J. H. Diggle, *Eur. J. Biochem.* **11,** 495–498 (1969).

22. G. Briand, D. Kmiecik, P. Sautiere, D. Wouters, O. Borie-Loy, G. Biserle, A. Mazen, and M. Champagne, *FEBS. Lett.* **112,** 147–151 (1980).

23. P. A. Krieg, A. J. Robins, M. J. Gait, R. C. Titmas, and J. R. E. Wells, *Nucl. Acids Res.* **10,** 1495–1502 (1982).

24. R. Ruiz-Vazquez and A. Ruiz-Carrillo, *Nucl. Acids Res.* **10,** 2093–2108 (1982).

25. J. R. E. Wells, L. S. Coles, R. D'Andrea, R. P. Harvey, P. A. Krieg, A. J. Robins, and J. Whiting, in P. Nagley, A. W. Linnane, W. J. Peacock and J. A. Pateman, Eds., *Manipulation and Expression of Genes in Eukaryotes,* Academic, Sydney (in press).

26. L. J. Edwards and L. S. Hnilica, *Experimentia (Basel)* **24,** 228–229 (1968).

27. I. Dardick and G. Setterfield, *Exp. Cell Res.* **100,** 159–171 (1976).

28. K. Brasch, G. H. M. Adams, and J. M. Neelin, *J. Cell Sci.* **15,** 659–677 (1974).

29. W. H. Marsh and P. J. Fitzgerald, *Fed. Amer. Soc. Exp. Biol.* **32,** 2119–2124 (1973).

30. J. R. Pehrson and R. D. Cole, *Biochemistry* **20,** 2298–2301 (1981).

31. R. Appels and J. R. E. Wells, *J. Mol. Biol.* **70,** 425–434 (1972).

32. J. S. Zlatanova, *FEBS Lett.* **112,** 199–202 (1980).

33. S. Panyim and R. Chalkley, *Biochem. Biophys. Res. Commun.* **37,** 1042–1049 (1969).

34. B. J. Smith, J. M. Walker and E. W. Johns, *FEBS Lett.* **112,** 42–44 (1980).

35. T. Shimoda, Y. Okihama, C. Murata and R. Shukuya, *J. Biol. Chem.* **256,** 577–582 (1981).

36. J. R. Pehrson and R. D. Cole, *Nature* **285,** 43–44 (1980).

37. R. Gjerset, C. Gorka, S. Hasthorpe, J. J. Lawrence, and H. Eisen, *Proc. Natl. Acad. Sci. USA* **79,** 2333–2337 (1982).

38. P. A. Krieg, A. J. Robins, A. Colman, and J. R. E. Wells, *Nucl. Acids Res.* **10,** 6777–6785 (1982).

39. P. A. Krieg, A. J. Robins, R. D'Andrea, and J. R. E. Wells, *Nucl. Acids Res.* **11,** 619–627 (1983).

40. M. Strickland, W. N. Strickland, W. F. Brandt, and C. von Holt, *Eur. J. Biochem.* **77,** 263–275 (1977).

41. W. N. Strickland, M. Strickland, W. F. Brandt, and C. von Holt, *Eur. J. Biochem.* **77,** 277–286 (1977).

42. R. T. Simpson and L. W. Bergman, *J. Biol. Chem.* **255,** 702–709 (1980).

43. R. T. Simpson, *Proc. Natl. Acad. Sci. USA* **78,** 6803–6807 (1981).

44. R. P. Harvey, J. A. Whiting, L. S. Coles, P. A. Krieg, and J. R. E. Wells, *Proc. Natl. Acad. Sci. USA* **80,** 2819–2823 (1983).

45. M-J. Gething and J. Sambrook, *Nature* **293,** 620–625 (1981).

46. D. DiMaio, R. Treisman, and T. Maniatis, *Proc. Natl. Acad. Sci. USA* **79,** 4030–4034 (1982).

47. D. DiMaio, R. Treisman, and T. Maniatis, *Proc. Natl. Acad. Sci. USA* **79,** 4030–4034 (1982).

CHAPTER 11

ORGANIZATION AND EXPRESSION OF CHICKEN HISTONE GENES

JAMES DOUGLAS ENGEL
Department of Biochemistry,
Molecular Biology and Cell Biology
Northwestern University
Evanston, Illinois

CONTENTS

1. INTRODUCTION
2. THE VARIETY OF HISTONE GENE ORGANIZATION AMONG DIFFERENT ORGANISMS
3. ORGANIZATION OF CHICK HISTONE GENE CLUSTERS
4. HISTONE GENES EXPRESSED IN EMBRYONIC TISSUE AND THOSE EXPRESSED IN ALL TISSUE
5. THE MAJORITY OF THE ISOLATED CHICK HISTONE RECOMBINANTS ENCODE EMBRYONIC GENES
6. SUMMARY AND PERSPECTIVE

REFERENCES

1. INTRODUCTION

The aim of this chapter is to summarize the recent data pertinent to the organization and the expression of the histone genes of the chicken, and to relate these data to histone gene expression in other organisms. We do not attempt to review either the area of specific histone gene control elements and what such sequence studies have told us about the regulation of expression of these particular genes or the most recent data regarding the tissue-specific histone H5 gene expression; both of these subjects are within the purview of chapter 10.

The histone genes of the chicken appear to be a typical example of vertebrate histone gene organization. The term "higher" eucaryote is frequently used in describing the organization of the histone genes in chicken, mouse, and human genomes since this organization differs greatly from the classical examples of sea urchin early genes (1, 2) and the histone genes of *Drosophila* (3). However, it is not clear whether or not the organization of *Xenopus* histone genes lies strictly within one or the other example (4), whereas it is equally clear that the lowest eucaryote, yeast, also has the dispersed genomic organization pattern of histone genes found in the vertebrate organisms. Within this context, it has also become quite clear that the late embryonic histone genes in the sea urchin are similarly not organized into the classical tandem repeat pattern (5). Thus, most accurately, sea urchins appear to have characteristics of both classes. Therefore the distinction, in many cases, between "higher" and "lower" eukaryotic organization is for convenience and simply describes the majority (or historically most accurate) example, and is not therefore the most accurate description.

Both for the sake of historical accuracy and ease of explanation, we first introduce the histone gene organization of a variety of organisms. This description gives a brief overview and is used simply as a mechanism to set the stage for what we and others have found about histone gene organization and expression in the chicken. More recent and more detailed descriptions of the findings alluded to in this introduction are found in other chapters in this book, as well as a number of excellent reviews which have appeared in the literature during the last several years (6, 7).

2. THE VARIETY OF HISTONE GENE ORGANIZATION AMONG DIFFERENT ORGANISMS

The first documented example of histone gene organization differing from that seen in the sea urchin and *drosophila* was in yeast (8), in which two copies of each of the four core histone genes are found at dispersed chromosomal locations. These clusters of two genes are not closely linked physically within the yeast genome. When examined in detail, the H2A/H2B histone gene pairs were found to be similar only within the coding sequence, but extremely divergent outside of these sequences. The same is also true for the H3/H4 gene pairs (9).

During the early stages of the investigation into vertebrate histone gene arrangement, the first clue that the structural analysis was going to yield different information was the finding that (in the small number of higher eukaryotic organisms examined) the copy number of histone genes was substantially lower. Histone genes are present in *Drosophila* at roughly 100 copies per haploid genome and in sea urchins at 400–800 tandemly duplicated embryonic genes per haploid genome. Early studies in HeLa cells (10), in mice (11), and in chickens (12) suggested that there were perhaps only 10–20 copies of each histone gene subtype. *A priori*, there was no reason to expect that the genes in these vertebrates would also be tandemly duplicated.

When first discovered, the histone gene organization in vertebrates which we and others reported was unexpected. Essentially simultaneous publications on the organization of histone genes in chickens (13, 14), in mice (15), and in humans (16) demonstrated that these organisms indeed encoded histone genes that were physically contiguous to one another but which nonetheless did not follow the tandemly duplicated organization expected from the sea urchin and *Drosophila* examples. Similarly, and more recently, it has been reported that the "late" histone genes [those active before and during blastula formation (5)] expressed in sea urchins were also organized in this typical "higher" eukaryotic dispersed gene pattern. However, contrary to observations in individual chick embryos, which consistently show the same blot hybridization pattern to specific histone gene probes, different individuals within a colony of sea urchins have quite different hybridization patterns when probed with "late" H2b genes, thereby indicating substantial polymorphism in late gene organization (19). Thus our perception of the interrelationship of histone gene organization and expression during development became considerably more interesting, since it became equally and immediately clear that different organisms had solved the problem of genomic response to varying needs in histone biosynthesis throughout the life cycle of that particular organism by different mechanisms. On the one hand, *Drosophila* and sea urchins have duplicated one set of embryonic genes many times, whereas most vertebrates have accomplished the same result for histone polypeptide biosynthesis during early stages of rapid cell division using many fewer genes. Whether this difference resides in promoter and transcriptional activity, in RNA processing activity, or in mRNA half-lives in vertebrates is still a question to be resolved. However, the elucidation of the pathways leading to the developmentally specific activation of histone genes, as well as the evolutionary mechanism(s) by which histone genes became duplicated and stabilized in the genome, are central problems yet to be resolved in this field.

3. ORGANIZATION OF CHICK HISTONE GENE CLUSTERS

Original reports of the extremely diverse organization of chick histone gene clusters, both from our work and that of Wells, demonstrated that in the isolated bona fide histone encoding recombinants there was no common sequence organization (13, 14). A summary of observations on the organization of histone gene clusters (provided in published and unpublished studies by J. Dodgson, J. Wells, and ourselves) derived from the characterization of individual isolated histone recombinants is shown in Figure 11.1. The central theme of the original observation, that the histone genes of the chicken are clustered but not tandemly duplicated in the genome, is firmly borne out by the restriction maps shown. In only a few of these gene clusters has overlap with other recombinants been clearly demonstrated. The result of efforts to "walk" down the chromosome by Wells and colleagues is shown in Figure 11.1*B*. This figure demonstrates that within the organization of 36 kb of contiguous chromosomal DNA 10 histone genes are found. Although the linked genes are substantially dispersed (i.e., intergenic DNA segments are sometimes greater than

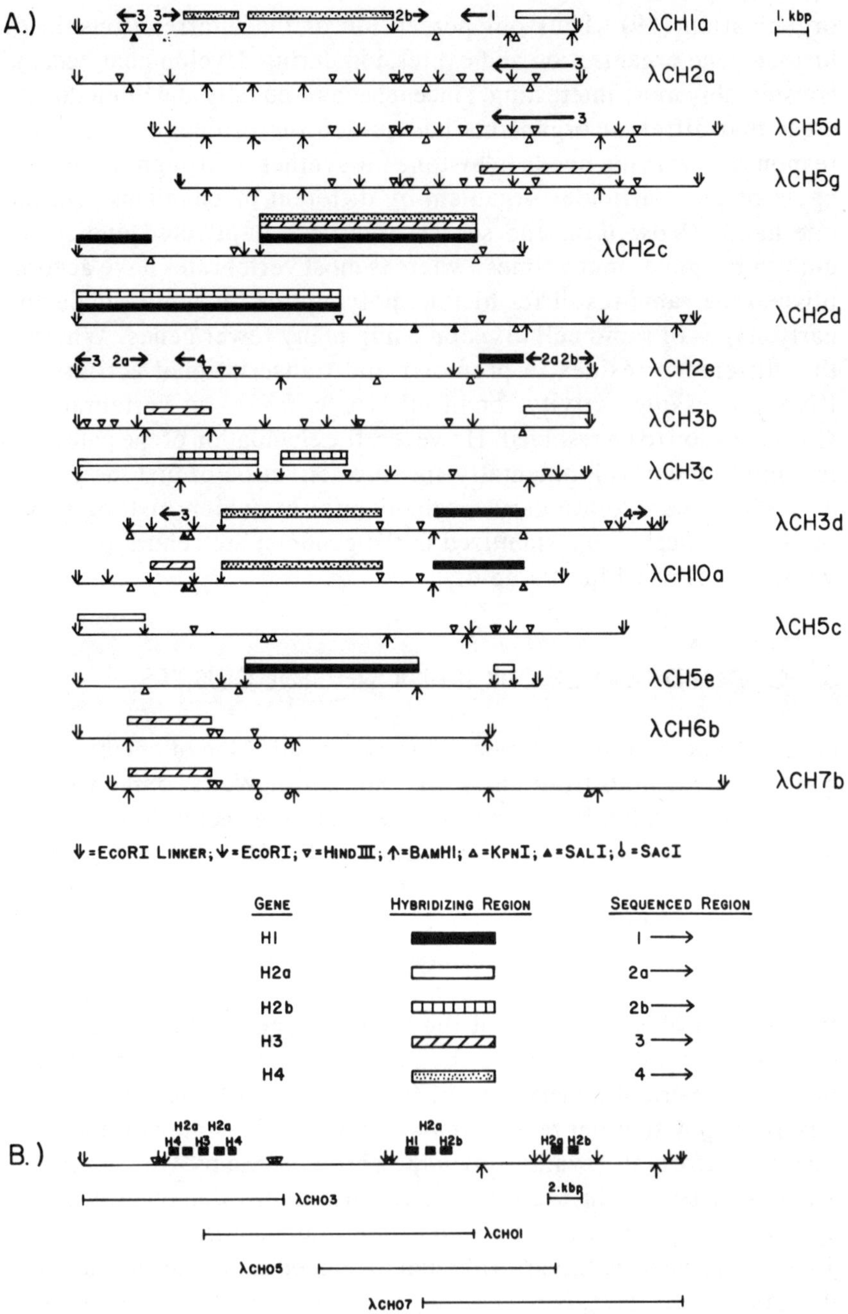
A.)
λCH1a
1. kbp
λCH2a
λCH5d
λCH5g
λCH2c
λCH2d
λCH2e
λCH3b
λCH3c
λCH3d
λCH10a
λCH5c
λCH5e
λCH6b
λCH7b
= EcoRI Linker; = EcoRI; = HindIII; = BamHI; = KpnI; = SalI; = SacI
Gene
Hybridizing Region
Sequenced Region
H1
H2a
H2b
H3
H4
1
2a
2b
3
4
B.)
H2a H2a
H4 H3 H4
H2a
H1 H2b
H2a H2b
2. kbp
λCHO3
λCHO1
λCHO5
λCHO7

10 kbp) within the chicken genome, and although we have not isolated large segments of overlapping chromosomal DNA containing these genes in a random screen (Figure 11.1*A*), as might also be expected from the clustered characteristics exhibited in the clones characterized in the "chromosomal walk" away from λCH02, other histone genes are present in segments of the chromosome directly adjacent to this single recombinant (J. R. E. Wells, personal communication).

As can also be discerned from the figure, hybridization with either heterologous sea urchin or *Drosophila* coding sequence probes (13) or with homologous chick histone gene probes (17–19) demonstrates that the same pattern of gene organization is observed within each individual restriction fragment. Therefore the early histone genes from sea urchin (from which our original chick histone recombinants were generated), as well as the *Drosophila* histone gene probes, hybridize at low stringency to the same sequences within each of these chicken chromosomal recombinants. Furthermore, in every case that has been examined in detail, the DNA sequence of each gene detected in such heterologous or homologous gene hybridizations is consistent with the expectation that all of the genes studied are normal RNA polymerase II transcribed genes (i.e., no histone pseudogenes have been detected; see Wells, Chapter 10). The genes in these recombinants appear to encode every kind (core and H1) of histone polypeptide sequence interspersed in a random pattern within the genome. Preliminary observations also suggest that these genes

FIGURE 11.1. Chicken chromosomal recombinants encoding histone genes. (*A*) Restriction maps for 15 independently isolated histone encoding recombinants are shown. Clones were isolated from the HaeIII/AluI linker library (30) as described (13, 19). Recombinants λCH1a and λCH2e are almost certainly identical siblings of λCHO2 and λCHO1, respectively, isolated from the same recombinant library by Wells and colleagues (14, 27, 28). In each case the left arm of the λ Charon 4A vector (31) is on the left (not shown). The position of hybridization of homologous, sequenced chicken histone gene probes is shown above the restriction enzyme fragment to which the probe is complementary (19). Arrows above any particular locus show the identity, position, and orientation (5′→3′) of completely sequenced chicken histone genes. (*B*) A short "walk" in the vicinity of λCHO1 (≡CH2e in *A*) demonstrates linkage of other histone genes in directly contiguous segments of the chromosome. Orientation of all genes and the restriction map of recombinant λCHO7 are not yet fully determined (data from J. R. E. Wells; personal communication).

(with the exception of the H3.3 gene in λCh5d and overlapping recombinants) are only expressed early in development (Section 5), and appear to all be located on a single chicken macrochromosome (J. D. Engel, unpublished observations). While the histone H5 gene is found in the same fractions of sucrose gradient-separated metaphase chicken chromosomes as the embryonic genes, it should be explicitly stated that this tissue-specific histone H5 gene expressed only in the terminally differentiated avian erythroid cell line, does not appear to be interspersed with the genes shown here (B. J. Sugarman, J. B. Dodgson, and J. R. E. Wells, personal communications). That is to say (and is addressed directly in Chapter 10), the histone H5 genes of chickens appear not to be closely linked to those genes shown in Figure 11.1.

Finally, in one of our earlier attempts to isolate the tissue-specific H5 genes, we identified one locus encoding a gene whose RNA was expressed in the poly(A)$^{+}$ fraction of cellular RNA in adult red cells. Further identification of that locus demonstrated that the gene in question (Figure 11.1; λCH2a, λCH5d) did not contain a histone H5 gene, but rather encoded a structurally novel histone H3 gene.

4. HISTONE GENES EXPRESSED IN EMBRYONIC TISSUE AND THOSE EXPRESSED IN ALL TISSUE

The question of whether we have detected the majority of histone genes in the chicken is interesting because of the recent finding that there may be another specific set of histone genes that are expressed at different developmental stages and that, under very stringent blotting conditions, are probably undetected in conventional chromosomal blotting or library screening analysis. The question therefore arises: How accurate is the estimate that there exist 10 copies of each histone gene in the chicken genome?

This question was first addressed by Crawford et al. (12) in an examination of the solution hybridization characteristics of purified embryonic histone-specific cDNA versus chromosomal chicken DNA. They demonstrated that the histone cDNA probe (tracer) reacted approximately an order of magnitude faster than reassociation of chicken single copy DNA alone, thus indicating approximately 10 copies of each gene per haploid genome. This question was addressed once again in a less quantitative way by Engel and Dodgson (13), who demonstrated that a partially sequenced subclone containing a single chicken histone H3 gene hybridized to a distinct and specific number of chromosomal restriction fragments on southern blots, using a variety of restriction enzymes. The copy number derived from this less precise estimate was again roughly 9 or 10 copies of the chicken histone H3 gene used as probe (17). This approximation, combined with the precise knowledge of the number of genes that have been isolated in overlapping and nonoverlapping chromosomal recombinants (Figure 11.1), is consistent with the previous findings from C_0t analysis, and therefore bears out the expectation that there exist approximately 10 copies of each of the five (core plus histone H1) genes per haploid genome when probed with homologous (12, 13) or heterologous (13) embryonic histone gene copies.

Since all of the studies alluded to used embryonic histone mRNA preparations and probes to identify the genes in question, we wondered whether or not the same set of histone genes used by embryos at the stages chosen (usually 4.5–5 days after fertilization) would be representative of all of the histone genes expressed in the chicken. To address this question, the completely sequenced histone H3 gene in λCH3d (Figure 11.1; 17), was used in northern blot analysis to determine the temporal expression of homologous histone H3 mRNA. As pointed out in Chapter 10, this histone H3 gene does appear to be a typical representative of the histone H3 genes expressed at this early stage of development since another histone H3 gene isolated from the same set of recombinants shown in Figure 11.1 contains the identical coding sequence to the H3 sequence in pCH3dR1 (J. D. Engel, unpublished observations). When RNA was isolated from stage-specific embryos at 4.5 days, 11 days, and 15 days after fertilization as well as from adult hens, electrophoresed on agarose gels (20), and then probed with the H3 subclone pCH3dR1 (17), it is clear that the probe detects information encoded by genes expressed only in embryonic tissue (Figure 11.2). In this representative northern blot, one can see that only a single histone H3 RNA sequence is detected by the H3 coding sequence in pCH3dR1, and furthermore that this band, and therefore the abundance of that mRNA species, decreases with increasing time of development. By quantitative anal-

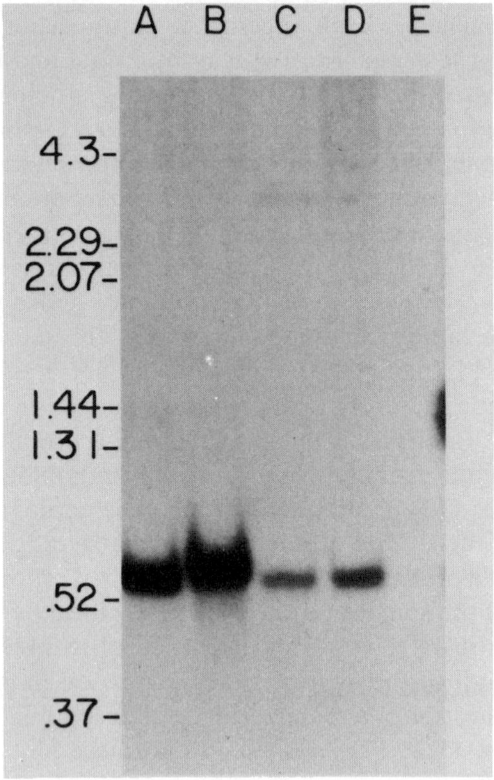

FIGURE 11.2. Northern analysis of RNA complementary to the H3 gene in pCH3dR1. Ten micrograms of RNA was isolated from different tissues, denatured with glyoxal in dimethylsulfoxide (20), electrophoresed on a 10 mM phosphate-buffered 1.2% agarose gel, and transferred directly to nitrocellulose. Filters were baked, prehybridized, hybridized, and washed as described (32). RNA was isolated from: (*A*) 4.5-day postfertilization whole embryos; (*B*) 4.5-day embryonic red cell cytoplasm; (*C*) 11-day embryonic brain; (*D*) 15-day embryonic liver; and (*E*) anemic adult hen red cell cytoplasm. Hybridization was to the sequenced H3 gene in λCH3d described by subclone pCH3dR1 (17). The washing stringency imposed was equal to approximately 3°C below the calculated homologous T_m (17). On the left the marker fragment sizes of pBR322 DNA digests are shown in kilobase pairs (kbp).

ysis of various exposures of the blot shown in Figure 11.2, one can conclude that the amount of homologous histone H3 gene RNA present in adult red cells is less than 0.1% of that present at the earliest embryonic times examined. From this kind of analysis, the most immediate question was the following: Where are the histone H3 genes expressed later in development to replace those embryonic genes that are coordinately diminished in RNA copy number with increasing maturation of the organism?

We have only recently gained a few clues as to where the genes that replace these embryonic genes might be in the chromosome and what their physical characteristics are. As stated in the previous section, we have isolated one gene that may obey the characteristics expected of an adult (or, by analogy to frog 5s genes, somatic) histone gene, which could replace the embryonic histone gene set at later times of avian development (Figure 11.3).

Several pieces of suggestive, but not conclusive, data argue that this is indeed an expressed gene and that it is expressed at all stages and in all tissue types during avian development. First, the gene was initially detected by hybridization to an embryonic sea urchin histone recombinant (13), and subsequently by hybridization to cDNA prepared from poly $(A)^+$ fraction of adult red cell mRNA (J. D. Engel and J. B. Dodgson, unpublished observations). Thus poly $(A)^+$ sequences within *adult* mRNA preparations contain complementarity to this locus. Second, a sense strand oligo (A) stretch, lying 3′ to the gene, would (if transcribed) account for the oligo d(T) cellulose selection of this particular mRNA. Third, the coding sequence of the gene predicts the amino acid sequence for a polypeptide, with only an amino acid inversion, which is identical to that of the chicken H3.3 histone, a ubiquitous minor variant of the most abundant H3 sequence (21, 22). Of more than passing interest, it had previously been shown that the histone H3.3 mRNA from HeLa cells is also preferentially selected by oligo d(T) cellulose chromatography (23). Thus several pieces of data and indirect observations strongly suggest that this locus is transcribed to finally yield the somatic low abundance H3.3 polypeptide.

This presumptive first example of a somatic histone H3 gene differs in several obvious features from those embryonic genes previously analyzed. First, and most striking, it contains intervening DNA se-

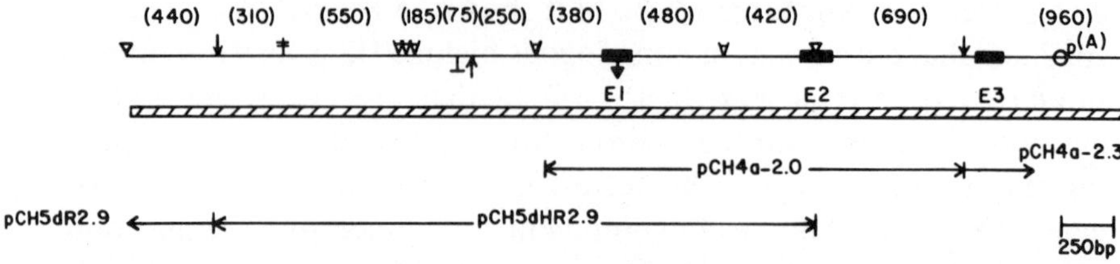

FIGURE 11.3. The split chicken histone H3 gene. A diagrammatic representation is shown of the chicken chromosome in the vicinity of the sequenced histone H3 gene with introns. The three exons (E1, E2, E3) are shown relative to the position of the sense strand oligo (A) tract [p(A)] and a number of restriction enzymes used for sequence analysis of subclones pCH5dR2.9, pCH5dHR2.9, pCH4a-2.0, and pCH4a-2.3, which were derived from λCH4a (≡λCH2a) and λCH5d (Figure 11.1). Numbers above the line (in parentheses) correspond (±5 base pairs) to the distance between adjacent restriction enzyme sites (J. D. Engel, unpublished observations). Legend: ↓ = EcoRI; ↑ = BamHI; ▽ = HindIII; ∀ = AvaI; ⊥ = Ta9I; ‡ = HinfI; ↓ = PvuII.

quences that interrupt the coding segments of the gene at positions corresponding roughly to those that would be expected to divide the gene into functional domains (24). These introns contain splice junctions that fall in accordance with Chambon's rule (25) and are thus typical of splice junctions seen in the majority of polymerase II transcribed eucaryotic genes containing introns (26). The second feature of this gene which is rather striking is that the DNA sequence varies by 19% from the histone H3 gene encoding embryonic histone H3 RNA (in pCH3dR1; 17), although the DNA sequence predicts a polypeptide that differs by only four amino acid residues from the major chick erthrocyte histone H3 polypeptide (22). The histone amino acid sequence predicted by this split gene is nonetheless very similar [and may, for technical reasons cited (17), be identical] to a low-abundance histone H3 polypeptide (H3.3) detected in all avian tissues examined (22). Whether or not this histone H3 gene is transcribed to yield a functional gene product *in vivo* in avian tissue remains to be firmly resolved. If this is an example of genes that replace the embryonic histone gene set, we should expect other somatic genes with many of the same general features to be found. However, as a cautionary experimental note, one should also expect that other somatic histone genes will probably not be isolated easily using conventional methodology (e.g., with heterologous histone gene probes or embryonic probes) since they may have diverged greatly from embryonic genes while still encoding roughly the same amino acid sequence. Thus isolation of other somatic histone genes will require strategies different from those used in the past.

5. THE MAJORITY OF THE ISOLATED CHICK HISTONE RECOMBINANTS ENCODE EMBRYONIC GENES

With the precise structural knowledge in hand of many of the histone genes represented in Figure 11.1 [i.e., at least one and in most cases several, of the core and H1 histone genes shown in Figure 11.1 have been entirely sequenced (17–19, 27, 28)], it was of interest to determine whether these various gene copies were expressed only in embryos, only in adults, or perhaps throughout the lifetime of a developing chick embryo to maturity. To distinguish between genes expressed in the embryo and those expressed in the adult, RNA samples were isolated from 4.5-day embryonic chicken red cells and from anemic adult hen red cell cytoplasm. Sequenced subclones representing each of the five histone genes from the isolated recombinants (Figure 11.4) were hybridized to filters to which these RNAs had been fixed, and these northerns were then washed at stringencies that should discriminate between RNAs expressed at the two different stages. As shown in Figure 11.5, each of the single gene-specific subclones detects only a single RNA species, and that species exists only in the lanes containing embryonic RNA; therefore, each of the sequenced recombinant DNA subclones contains sequence information complementary to RNA present only in embryonic tissue.

In several cases, the RNAs produced in the 4.5-day red cell were further examined by the much more stringent experimental criteria of 5′ end-protection analysis to further ensure that the gene being examined in northern blots was indeed the gene expressing that particular RNA species. The H1, H3, and H4 subclones used for RNA blot analysis (Figure 11.4) all protected singular and unique RNA species in 4.5-day embryos (19; B. J. Sugarman and H. B. Hughes, III, unpublished observations). Analogous blots using each of the λ histone recombinants hybridized versus the same RNA samples dem-

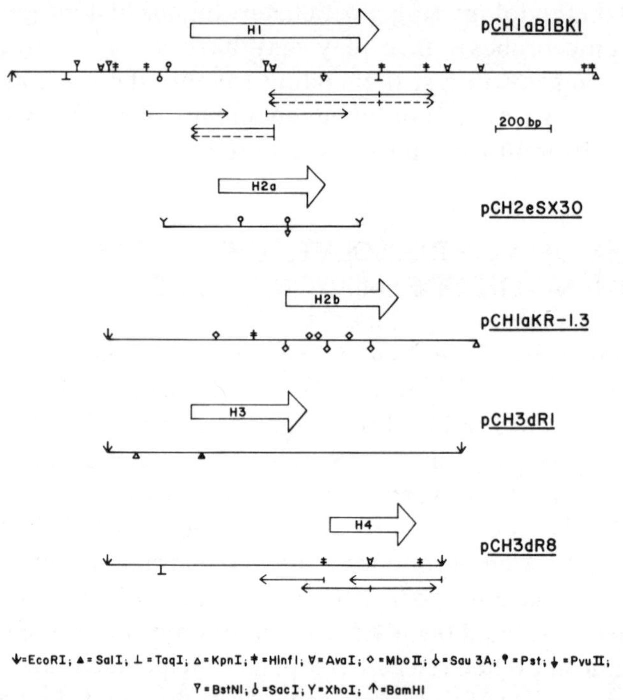

FIGURE 11.4. Individual subclones containing single histone genes. Chicken histone subclones were created as described and the coding sequence determined by the Maxam/Gilbert partial chemical degradation procedure (see Reference 19). The reports of each DNA sequence can be found in the following references: H1 and H4 (19); H2a (27); H2b (18, 28); and H3 (17). The extent of the reactions used to determine the H1 and H4 sequences are shown as arrows below each map. A broken line denotes a 3′ end-labeled species, whereas a solid line denotes 5′ end labeling.

onstrates that at least half of these recombinants (i.e., all of those we have tested, with the exception of λCH2a and λCH5d) hybridized exclusively and with high affinity to RNAs of precisely the same size as those detected in the controls (Figure 11.4; 19) and to those genes predicted by the hybridization of heterologous and homologous probes to each of the λ recombinant chromosomal restriction enzyme maps (Figure 11.1). Thus, with the exception of the single H3 gene represented in λCH5d (and overlapping recombinants thereof), all of the histone recombinants shown in Figure 11.2 hybridize exclusively to RNA samples that are embryonic in origin (19; B. J. Sugarman, unpublished observations).

6. SUMMARY AND PERSPECTIVE

In summary, the salient features of our current knowledge of histone gene organization and expression can be stated simply by the following points: first, there appears to be substantial agreement between the estimate of the copy number of embryonic histone genes in the chicken genome from embryonic cDNA-tracer C_0t analysis (12) and the actual number of histone gene recombinants isolated. Second, the majority of histone gene recombinants that we and others have isolated appear

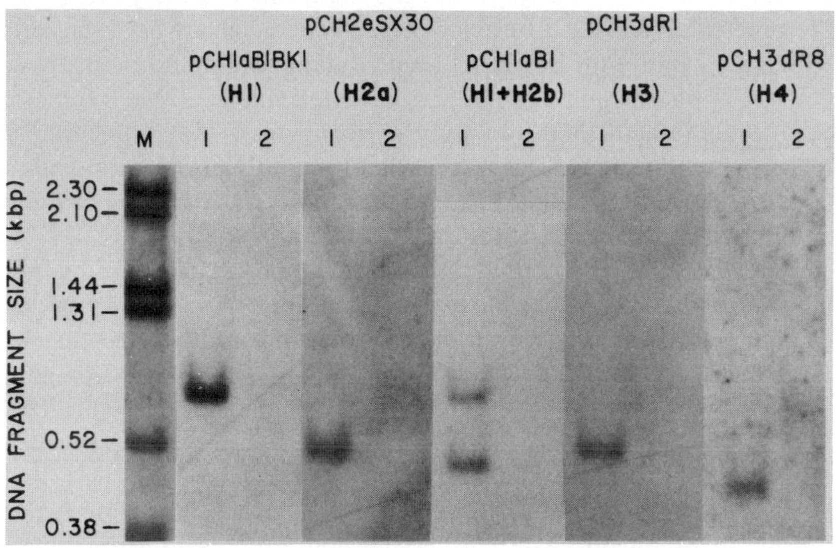

FIGURE 11.5. Northern blot analysis of the sequenced core and H1 genes in various genomic recombinants. RNA samples were the same as (*B*) and (*E*) in Figure 11.2; lanes 1 and 2, in each case, correspond to 10 μg of 4.5-day embryonic red cell cytoplasmic RNA and 10 μg of anemic adult hen red cell cytoplasmic RNA, respectively. Duplicate blotting lanes were hybridized versus the sequenced recombinant subclones shown in Figure 11.4 and exposed for autoradiography. On the left is shown the DNA size standard markers (pBR322 cleaved with EcoRI plus PvuII and pBR322 cleaved with TaqI; lane M). The H1 and H2b subclones shown in Figure 11.4 were both derived from plasmid pCH1aB1 (19).

to encode primarily embryonic histone genes (Figure 11.4) which are not tissue specific (Figure 11.2). Finally, the organization of the embryonic histone genes (core and histone H1 genes) appears as a dispersed and relatively random pattern in the chicken chromosome (Figure 11.1) in contrast to the classical tandem repeat pattern demonstrated in the *Drosophila* and the sea urchin early embryonic gene families.

One important structural question remains to be resolved before we can actually begin to examine the detailed temporal expression of these genes. Since it is clear that the majority of the isolated genes are embryonic in origin, where are the rest of the genes expressed later in development in the chicken? One alternative might be that the distinction within this chapter between "embryonic" and "adult" genes is an artifice of the stages of cells examined to initially categorize the genes. Perhaps the differences actually detected are between cells which are rapidly proliferating (and contain substantial numbers of cells in S phase) and cells which have terminated DNA synthesis (i.e., anemic adult hen immature erythrocytes). This immediately testable hypothesis would suggest that in rapidly proliferating adult tissues the "embryonic" histone gene mRNAs would be present abundantly, as they are in the embryonic brain or liver (Fig. 11.2). A second alternative is that, indeed, new and undiscovered genes are responsible for histone biosynthesis in "adult" tissues. A possible example of such a proposed structurally and temporally distinct histone gene may be the histone H3 gene in recombinant λCH5d (Figure 11.3). Since this gene is structurally novel, and since it hybridizes to both embryonic and adult RNA samples (J.D. Engel, unpublished observations), it is possible that the split H3 gene could be a first representative example of those that replace the embryonic genes during development. Alternatively, it could represent a low-abundance, constitutively expressed (i.e., cell-cycle independent) histone gene. In fact, recent evidence reported by Wu and Bonner (29) suggests that some types of variant histone biosynthesis (including H3.3 in Chinese hamster ovary cells) are not exclusively restricted to S phase; rather, they are synthesized at a basal level throughout the cell cycle. However, this biosynthesis of the H3.3 protein has not been conclusively demonstrated to be due to constitutive "basal" level RNA synthesis as opposed to the possibility of a greatly increased half-life of H3.3 mRNA.

Since, by northern analysis, the majority of the genes that have been isolated encode only RNAs expressed in embryonic tissue (Figure 11.4), it is clear that the methodology used thus far for histone gene isolation may have allowed us to select only a subset of the total number (embryonic and somatic) of histone gene recombinants. Thus the question remains open as to the total number of histone genes actually contained within the chicken genome. It is clear from the number of isolated embryonic recombinants that the number of embryonic histone genes will closely approximate that initially predicted by Wells (12). However, it is equally clear from the data presented here and elsewhere (17, 22) that more copies of somatic histone genes must exist. Specifically, even though the histone H3 polypeptide sequence predicted by the subclone pCH3dR1 is that of the major erythrocyte histone H3 protein (21), that subclone does not hybridize to adult RNA. This logically dictates that if a gene coding for the major adult chicken histone erythrocyte H3 polypeptide exists elsewhere in the chicken genome, despite amino acid sequence identity, the DNA sequence of this adult gene must be highly divergent from its embryonic counterpart.

ACKNOWLEDGMENTS

I thank Julian Wells, Richard Harvey, Barry Sugarman, Hugh Hughes, David Grandy, and Jerry Dodgson for access to information published herein as personal communications prior to publication and especially thank Barry Sugarman and Hugh Hughes for critical evaluation of the manuscript. This work was supported by grants from the NIH (HL 24415 and CA 15145) and the Leukemia Research Foundation. I also gratefully acknowledge the support of the American Cancer Society for a Faculty Research Award.

REFERENCES

1. L. H. Kedes and M. L. Birnstiel, *Nature New Biol.* **230,** 165 (1971).
2. L. H., Kedes, R. H., Cohn, A. C. Y. Chang, and S. Cohen, *Cell* **6,** 359 (1975).
3. R. P. Lifton, M. L. Goldberg, R. W. Karp, and D. S. Hogness, *Cold Spring Harbor Symp. Quant. Biol.* **42,** 1047 (1977).
4. M. Zernick, N. Heintz, I. Boime, and R. G. Roeder, *Cell* **22,** 807 (1980).

5. R. Mason, T. Mohun, G. Gormezano, G. Childs, and L. Kedes, *Nature* **301,** 120 (1982).
6. C. Hentschel and M. Birnsteil, *Cell* **25,** 301 (1981).
7. L. Kedes, *Cell* **8,** 32 (1976).
8. L. Hereford, K. Fahrner, J. Woolford, Jr., M. Rosbash, and D. B. Kaback, *Cell* **18,** 1261 (1979).
9. M. M. Smith and O. S. Andresson, *J. Mol. Biol.* **169,** 663 (1983).
10. M. C. Wilson, M. Melli, and M. L. Birnstiel, *Biochem. Biophys. Res. Comm.* **61,** 404 (1974).
11. E. Jacob, *Eur. J. Biochem.* **65,** 275 (1976).
12. R. J. Crawford, P. Krieg, R. P. Harvey, D. A. Hewish, and J. R. E. Wells, *Nature* **279,** 132 (1979).
13. J. D. Engel and J. B. Dodgson, *Proc. Natl. Acad. Sci. USA* **78,** 2856 (1981).
14. S. Bruschi, and J. R. E. Wells, *Nucl. Acids Res.* **9,** 1591 (1981).
15. D. B. Sittman, I.-M. Chiu, C.-J. Pan, R. H. Cohn, L. H. Kedes, and W. F. Marzluff, *Proc. Nat. Acad. Sci. USA* **78,** 4078 (1981).
16. N. Heintz, M. Zernik, and R. G. Roeder, *Cell* **24,** 661 (1981).
17. J. D. Engel, B. J. Sugarman, and J. B. Dodgson, *Nature* **297,** 434 (1982).
18. D. Grandy, J. D. Engel, and J. B. Dodgson, *J. Biol. Chem.* **257,** 8577 (1982).
19. B. J. Sugarman, J. B. Dodgson, and J. D. Engel, *J. Biol. Chem.* **258,** 9005 (1983).
20. P. Thomas, *Proc. Nat. Acad. Sci. USA* **77,** 5201 (1980).
21. W. F. Brandt and C. Von Holt, *Eur. J. Biochem.* **46,** 419 (1972)
22. M. K. Urban, S. G. Franklin, and A. Zweidler, *Biochemistry* **18,** 3952 (1979).
23. T. W. Borun, K. Ajiro, A. Zweidler, T. W. Dolby, and R. E. Stephens, *J. Biol. Chem.* **252,** 173 (1977).
24. J. D. McGhee and G. Felsenfeld, *Ann. Rev. Biochem.* **49,** 501 (1980).
25. R. Breathnach, C. Benoist, K. O'Hare, F. Gannon, and P. Chambon, *Proc. Natl. Acad. Sci. USA* **75,** 4853 (1978).
26. K. Mount, *Nucl. Acids Res.* **10,** 459 (1982).
27. R. D'Andrea, R. P. Harvey, and J. R. E. Wells, *Nucl. Acids Res.* **9,** 3119 (1981).
28. R. P. Harvey, A. J. Robins, and J. R. E. Wells, *Nucl. Acids Res.* **10,** 7851 (1982).
29. R. S. Wu and W. M. Bonner, *Cell* **27,** 321 (1981).
30. J. B. Dodgson, J. S. Strommer, and J. D. Engel, *Cell* **17,** 879 (1979).
31. F. R. Blattner, B. G. Williams, A. E. Blech, K. Denniston-Thompson, H. E. Faber, L. A. Furlong, D. G. Grunwald, D. O. Kiefer, D. D. Moore, J. W. Schumm, E. W. Sheldon, and O. Smithies, *Science* **196,** 161 (1977).
32. M. Dolan, J. B. Dodgson, and J. D. Engel, *J. Biol. Chem.* **258,** 3983 (1983).

CHAPTER 12

ORGANIZATION AND EXPRESSION OF MOUSE HISTONE GENES

WILLIAM F. MARZLUFF
REED A. GRAVES
Department of Chemistry
Florida State University
Tallahassee, Florida

CONTENTS

1. INTRODUCTION
2. ISOLATION OF THE MOUSE HISTONE GENES
3. STRUCTURE AND ORGANIZATION OF ISOLATED MOUSE HISTONE GENES
4. SEQUENCE OF THE MOUSE HISTONE GENES
5. STRUCTURE OF THE mRNAs
6. SIGNALS FOR TRANSCRIPTION INITIATION AND TERMINATION
7. CODON USAGE IN THE MOUSE HISTONE GENES
8. NUMBER OF DIFFERENT HISTONE mRNAs
9. SEQUENCE CONSERVATION OF HISTONE GENE SEQUENCES AMONG SPECIES
10. CONTROL OF EXPRESSION OF THE MOUSE HISTONE GENES
11. HISTONE mRNA LEVELS IN CELLS ARRESTED IN G1
12. NATURE OF SIGNAL REGULATING HISTONE mRNA LEVELS
13. SUMMARY AND CONCLUSION

REFERENCES

1. INTRODUCTION

The histone proteins are among the most conserved proteins in evolution. The striking (>95%) homology in amino acid sequence of these proteins in a wide variety of species allows one to predict a great deal of homology at the nucleic acid level. The histone mRNAs are unusual among eucaryotic mRNAs since they are not polyadenylated (except in yeast and *Xenopus* eggs) and do not contain intervening sequences (but for a possible exception see Engel, Chapter 11). Since the histone genes in both sea urchins and *Drosophila* make up a sizable percentage of the genome, these gene clusters were among the first genes isolated and cloned (1, 2). These genes have in turn been used to aid in the isolation of genes from higher eucaryotes. As discussed later, the structural organization of histone genes in higher eucaryotes is very different from that of sea urchin or Drosophila.

Despite the conservation of histone protein sequences in evolution, there are several nonallelic variants of the histone proteins found in mammalian cells which vary in several amino acids (see Zweidler, Chapter 14; 3). The amount of different histone variants present varies in chromatin of different tissues, and presumably the expression of these genes is regulated somewhat independently. Much attention has been focused on the regulation of histone protein synthesis and histone mRNA levels in mammalian cells. The availability of homologous mouse histone genes has allowed the direct measurements of the steady-state levels and rates of synthesis of histone mRNA in different cell lines. This chapter summarizes what is known about histone gene expression and organization in cultured mouse cells. The data indicate that a major class of histone genes is regulated both at the level of transcription and mRNA turnover. The intracellular regulatory signals are likely to be connected to the metabolism of deoxynucleotides. A minor class of histone mRNA, coding for a replacement variant histone, H3.3, is regulated independently from the other histone genes.

2. ISOLATION OF THE MOUSE HISTONE GENES

The mammalian histone genes have been isolated directly from gene libraries without the intermediate step of cDNA cloning (4, 5). In our case, we made a cDNA from a partially purified preparation of histone mRNA. Since the histone mRNA did not contain poly A, we used a random primer to prime the cDNA synthesis. The resulting cDNA was used to screen a mouse gene library. Each phage that hybridized to the cDNA was then individually hybridized to the sea urchin (*S. purpuratus*) histone H3 gene. This procedure resulted in selection of two clones containing H3-like genes. One is likely to code for an unusual histone variant or is a pseudogene, and the other phage contains two authentic H3 genes as well as an H2b and H2a gene (4). Using these "homologous" genes we have been able to isolate many other clones containing mouse histone genes, and eventually all the histone genes may be isolated by this approach. However, it has become apparent that the different histone protein variants, while very similar in amino acid sequence, have surprisingly different nucleic acid sequences. Thus a single histone gene probe is likely to isolate a particular class of histone genes (e.g., H3.1 genes) and not others (e.g., H3.3 genes or spermatocyte histone variants).

In addition, as the genes are sequenced, it is likely that a number of different potential histone protein sequences will be found. These minor variations in histone proteins will be most readily documented by DNA sequencing coupled with evidence that the genes are really expressed *in vivo*. We describe here an H2b protein differing in one amino acid from the major H2b protein and an amino acid substitution in H2a that has not previously been reported. Both these genes are expressed in cultured mouse cells.

3. STRUCTURE AND ORGANIZATION OF ISOLATED MOUSE HISTONE GENES

Upon isolation of the mouse and human genes (4, 5), it was obvious that there was no simple repeating structure and further that there is no obvious clustering of genes coding for particular histones [e.g., H3-H4 and H2b-H2a gene clusters as are found in yeast (6, 7)]. The genes that have been found on one phage, MM221, part of a larger gene cluster, containing mouse histone genes are shown in Figure 12.1. The four genes on this phage are two H3 genes, an H2b gene, and an H2a gene. The two H3 genes and the H2a gene are coded on

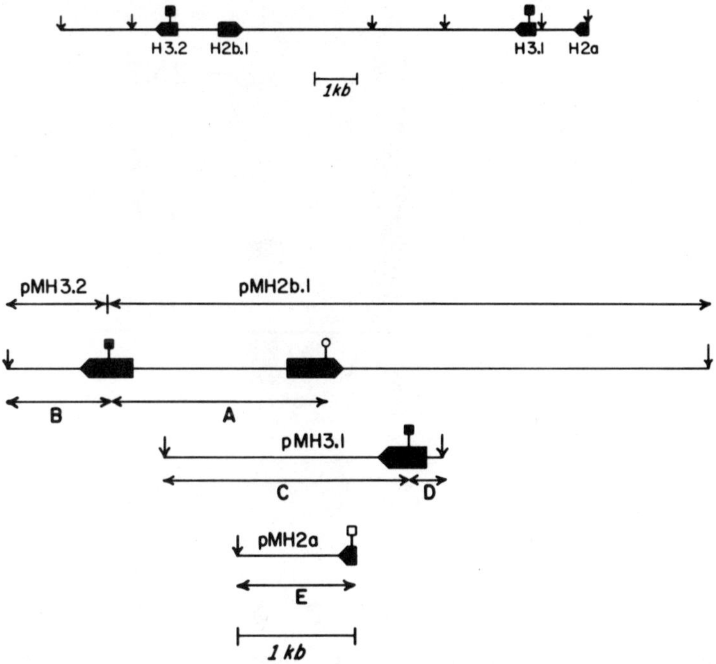

FIGURE 12.1. The location of the histone genes in MM221. The arrows indicate the direction of transcription. The plasmids constructed by subcloning the fragments from MM221 and the sites used for S1 nuclease mapping are shown. ↓ = Eco R1; ■ = Sal I; ♀ = Ava I; ♀ = Sau 961.

one strand of the DNA, and the H2b gene is coded on the other strand. Using one of the H3 genes, H3.2, as a hybridization probe, the number of H3 genes with homology to this gene can be estimated. The results are shown in Figure 12.2. There are 10 distinct major bands which hybridize to this H3 probe. One of these bands corresponds to the H3.2 and another to the H3.1 gene in MM221. The probe used in this experiment was a 900-base fragment containing 250 bases of coding region and 650 bases of nontranscribed 3′ flanking region. The increased intensity of hybridization to the 6.0 kb Eco R1 fragment containing the H3.2 gene is due to hybridization of the flanking region sequences which are not present elsewhere in the genome. The increased intensity of the hybridization to the 23 kb and 15 kb Eco R1 fragment is probably due to the presence of more

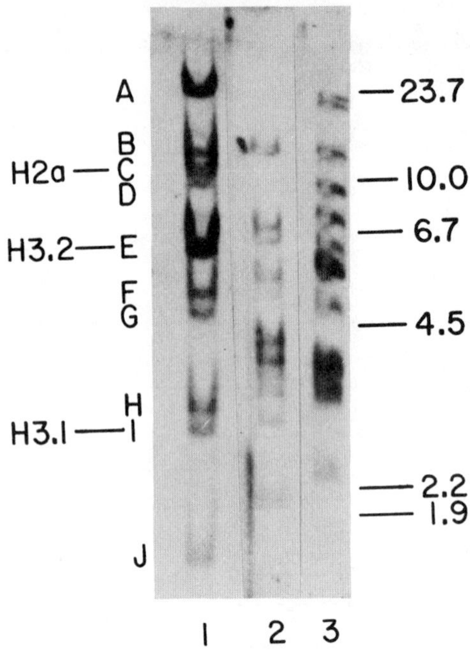

FIGURE 12.2. Organization of histone genes. Mouse, rat, and hamster DNAs were digested with Eco R1 and resolved on by agarose gel electrophoresis. The DNA was transferred to nitrocellulose and hybridized to $^{32}PO_4$-labeled plasmid pMH3.2. The size of the DNA markers are given in kilobases. The bands that correspond to the Eco R1 fragments from MM221 (H3.2, H3.1, and H2a) are labeled. Lane 1: mouse DNA; lane 2: hamster DNA; lane 3: rat DNA. The blot is courtesy of D. Bieber and V. Groppi (UCSF).

than one H3 gene in these fragments. This probe hybridizes to the same number of fragments in hamster, rat (Figure 12.2), and human (not shown) DNA, with similar intensity, indicating the high conservation of the coding region of this class of H3 genes among different species. The different sizes of the Eco R1 fragments containing H3 histone genes in mouse and hamster allows the assignment of the chromosomal location of these groups of histone genes. The majority of these genes map to chromosome 13, and the rest map to chromosome 3 in the mouse (D. Cox et al., unpublished observations). The total gene number detected by this probe is in good agreement with the estimate of 20 given by Jacob from solution hybridization (8). Using as a probe the 1.1-kb Eco R1 fragment containing the H2a genes and an H2b specific probe, we detected two other families of fragments, most of which also map to chromosome 13, suggesting that the majority of the replication-variant mouse histone genes reside on this chromosome.

4. SEQUENCE OF THE MOUSE HISTONE GENES

The genes on the clone MM221 have been completely sequenced. The termini of the mRNAs were identified by S1 nuclease mapping. Figure 12.3 shows the sequences of the two H3 genes in MM221. These genes code for two different H3 protein variants, H3.1 and H3.2, which differ from each other in a single amino acid 96. The H3.2 variant is predominant in rodents, whereas the H3.1 variant is predominant in other mammals. There are very few nucleotide changes in the coding region, only 10 third base changes and one first base change (from CGC to AGC in arginine 133, which did not change the amino acid) in addition to the three base changes around codon 96. However, when the regions immediately flanking the coding region, present in the untranslated regions of the mRNA, are examined, there is little detectable homology between the two genes. This region of nonhomology starts at the first nucleotide prior to the ATG coding for the initiating methionine and at the first nucleotide after the TAA termination codon. The changes that occur in the coding region are minimal except at codon 96, the one amino acid that has been changed, where all three nucleotides have changed. This striking homology suggests a strong selective pressure on the coding region nucleotides, independent of the selective pressure at the level of the amino acid sequence. The presumptive location of the start and stop of the mRNA has been determined by S1 nuclease mapping. These flanking regions are discussed further later.

The H2b gene has also been completely sequenced. The amino acid sequence is that predicted for the H2b.1 variant, with glycine at position 102(3). There is a single amino acid change from the reported H2b sequence of calf thymus, a valine for leucine at position 18 (Figure 12.4). As shown below we believe mouse histone H2b proteins are a mixture of at least these two variants. The untranslated regions are relatively short and do not share detectable homology

```
        180                              150
3.2                                    C ACAAATTTGA AGTTGAGACC TGTTATCCAA
3.1     CTCAGCCAAT AGGACTACTG CGCGGGACAC TTGAAAAGCA GACACGCCTA TCAGGATGCT

        120                              90
3.2     TTACCAAGTA CTTCCGCATA CATCATCATA GGCATTTGAA GATTTCAACC AATCAGGAGC
3.1     TTCTCGGTGG GAAGGAGGGG TACGAGCGCG GGTACGTGTG TTGCGCGTGT GCGACGCAAG

        60                               30       *
3.2     ATGTTCCTTC TATAAAGGAA CCCAGAACCT AACCTCTGCA TTTCCTATTT CTTTGTAGAA
3.1     CGTACTTAAA GGCCAAAGTG CGCTACTTAG GTATCTCACT TTTCCCTACG GTTACTTGCC
                                           *
H3-2                                              C   T
H3-1    ATG GCT CGT ACT AAG CAG ACC GCT CGC AAG TCT ACC GGC GGC AAG GCC CCG CGC
        MET Ala Arg Thr Lys Gln Thr Ala Arg Lys Ser Thr Gly Gly Lys Ala Pro Arg

H3-2
H3-1    AAG CAG CTG GCC ACC AAG GCC GCC CGC AAG AGC GCC CCG GCC ACC GGC GGC GTG
        Lys Gln Leu Ala Thr Lys Ala Ala Arg Lys Ser Ala Pro Ala Thr Gly Gly Val

H3-2          A                               C       G
H3-1    AAG AAG CCT CAC CGC TAC CGT CCC GGC ACT GTG GCA CTG CGC GAG ATC CGG CGC
        Lys Lys Pro His Arg Tyr Arg Pro Gly Thr Val Ala Leu Arg Glu Ile Arg Arg

H3-2
H3-1    TAC CAG AAG TCG ACC GAG CTG CTG ATC CGC AAG CTG CCG TTC CAG CGC TTG GTG
        Tyr Gln Lys Ser Thr Glu Leu Leu Ile Arg Lys Leu Pro Phe Gln Arg Leu Val

H3-2
H3-1    CGC GAG ATC GCG CAG GAC TTC AAG ACC GAC CTG CGC TTC CAG AGC TCG GCC GTC
        Arg Glu Ile Ala Gln Asp Phe Lys Thr Asp Leu Arg Phe Gln Ser Ser Ala Val

H3-2                            Ser
H3-2                          G A C               T
H3-1    ATG GCT CTG CAG GAG GCC TGT GAG GCC TAC CTC GTG GGT CTG TTT GAG GAC ACC
H3-1    Met Ala Leu Gln Glu Ala Cys Glu Ala Tyr Leu Val Gly Leu Phe Glu Asp Thr

H3-2
H3-1    AAC CTG TGC GCC ATC CAC GCC AAG CGT GTC ACC ATC ATG CCC AAG GAC ATC CAG
        Asn Leu Cys Ala Ile His Ala Lys Arg Val Thr Ile Met Pro Lys Asp Ile Gln

                                                                        20
H3-2              C   T           C     C             TAGGCACGCT TTCTACACTG
H3-1    CTG GCC CGT CGC ATC CGC GGG GAG AGG GCT TAA  GGGTTTCTGT TAATCCACAC
        Leu Ala Arg Arg Ile Arg Gly Glu Arg Ala  .

                                      50  *                            80
H3-2    GCACGTAAAC CAAAACGGCT CTTTTAAGAG CCACCTCCAT TATCCACCAA AGATGCTTGA
H3-1    AACCACTTTA AAGGCTCTTC TTAGAGCCAC CCATCTTCCA AAAAAAGAAC TGTGCGCTTT
                                     *
                                     110                              140
H3-2    AGTACAAGTT GTGAGAGTTT TCTAGGGTTT CCTATTATAG CCTTTCTTGA CAATGTGAGC
H3-1    TTCCAAACTT GTGGGTATTA ATCAGTTTCA TTTGTCAAAA GTGCTAGGTC TCC
```

FIGURE 12.3. Sequence of the H3 genes showing the sequence of the H3.1 and H3.2 genes and the base changes only in the coding region for H3.2. The sequence of the flanking regions is also given. The positions of the beginning and end of the mRNA are marked with an asterisk.

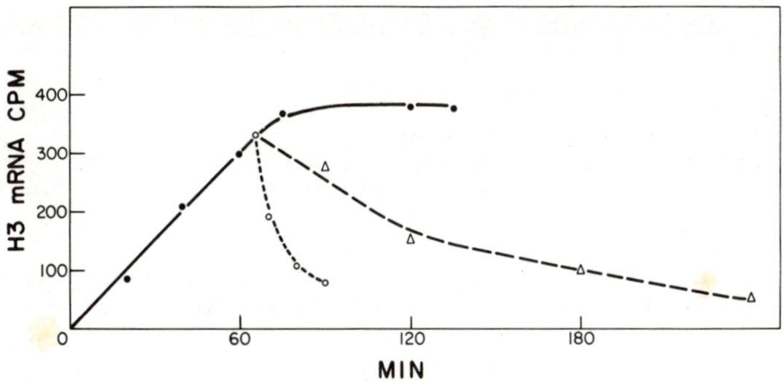

(a)

```
H2B GENE

                                     150
          TGATACTTAC GCAACTAACC ACTGAGCGAA TATGCTTCCT TGATGGACAG

120                                  90
TTAGTGCTTG ACGTTTGCAG ACTCTCTGAC AAGGACAGCC ACCGCTTTAT TTAAAGAGCA

60                    *              30
GGAAAGGAAC GGAACAGTTC AATATCTCTT TTCTTGGCCT ACCTTCATTC TCTGTTCACT

ATG CCT GAG CCC GCC AAG TCC GCT CCT GCC CCG AAG AAG GGC TCC AAG AAG GCC
MET Pro Glu Pro Ala Lys Ser Ala Pro Ala Pro Lys Lys Gly Ser Lys Lys Ala

CTG ACC AAG GCC CAG AAG AAG GAC GGC AAG AAG CGC AAG CGC AGC CGC AAG GAG
Leu Thr Lys Ala Gln Lys Lys Asp Gly Lys Lys Arg Lys Arg Ser Arg Lys Glu

AGC TAC TCG GTG TAC GTG TAC AAG GTG CTG AAG CAA GTG CAC CCC GAC ACC GGC
Ser Tyr Ser Val Tyr Val Tyr Lys Val Leu Lys Gln Val His Pro Asp Thr Gly

ATC TCC TCC AAG GCC ATG GGC ATC ATG AAC TCG TTC GTG AAC GAC ATC TTC GAG
Ile Ser Ser Lys Ala Met Gly Ile Met Asn Ser Phe Val Asn Asp Ile Phe Glu

CGC ATC GCG GGA GAG GCG TCC CGC CTA GCG CAT TAC AAC AAG CGC TCG ACC ATC
Arg Ile Ala Gly Glu Ala Ser Arg Leu Ala His Tyr Asn Lys Arg Ser Thr Ile

ACG TCC CGG GAG ATC CAG ACG GCC GTG CGC CTG CTG CTG CCC GGG GAG CTG GCC
Thr Ser Arg Glu Ile Gln Thr Ala Val Arg Leu Leu Leu Pro Gly Glu Leu Ala

AAG CAC GCC GTG TCG GAG GGC ACC AAG GCT GTC ACC AAG TAC ACC AGC TCC AAG
Lys His Ala Val Ser Glu Gly Thr Lys Ala Val Thr Lys Tyr Thr Ser Ser Lys

                              20                               *     50
TGA GTGCTCAAGA   CTCAGCTCTT   AACCCAAAGG   CTCTTTTCAG   AGCCACTCAA

                                 80                                   110
GACTTCAAAA   TTGGAGCTTT   AATGCTACCA   AGCGACTTAG   TGACTACCGG   GAAAATAACC

                              137
GACTTCATCG   CAGGATGTGT   ACAACAC
```

(b)

FIGURE 12.4. The sequence of the H2b gene and the portion of the H2a gene found in MM221. The amino acids that differ from the calf H2a and H2b gene are underlined. The beginning and end of the mRNA is marked with an asterisk.

(Figure continues on p. 290.)

```
H2A GENE
92                                      100
CTG AAC AAG CTG TTG GGC CGC GTG ACC ATC GCG CAG GGC GGC GTC CTG CCC AAC
Leu Asn Lys Leu Leu Gly Arg Val Thr Ile Ala Gln Gly Gly Val Leu Pro Asn

110                                         120
ATC CAG GCC GTG CTG CTG CCC AAG AAG ACC GAG AGC CAC CAT AAG CCC AAG GGA
Ile Gln Ala Val Leu Leu Pro Lys Lys Thr Glu Ser His His Lys Pro Lys Gly

128                             20                                 50
AAG TAA    GCCAGTGAGC TAAGTTTTTT TTTTTTTTTT TTTTTTTTTT TAAACAAAAC
Lys   .

                      *         80
CCAAGGCTCT TTTCAGAGCC ACCACTTCTT CATATAAGAG
```

(c)

FIGURE 12.4. *(continued)*.

with the corresponding regions of the H3 genes except in the region directly preceeding the 3′ end of the mRNA (see Section 6).

Only a portion of the H2a gene is represented in clone MM221 (Figure 12.4). The Eco R1 linkers were attached at amino acid 92, and only the 3′ end of the gene is present. The predicted protein sequence in the carboxyl region of the protein does not match any of the known H2a sequences. There is a proline at position 126 that is not found in any of the reported mammalian H2a sequences. It is likely that this gene, like the other genes in this clone, is expressed. There is a consensus sequence in the 3′ end untranslated region, and S1 nuclease mapping suggests the 3′ untranslated region is expressed.

5. STRUCTURE OF THE mRNAs

Since there are many gene copies of the H3 genes in the mouse genome, it is possible that only some of these genes are actually expressed. Since the coding regions are presumably highly conserved among the different gene copies, the only way to demonstrate that a gene is actually expressed is to show that the nontranslated regions flanking the genes at both the 5′ and 3′ ends are expressed. The untranslated regions on histone mRNAs are generally very short, and it is difficult to make a specific probe from them. S1 nuclease mapping provides a convenient, sensitive assay of whether a particular gene is expressed. It also allows one to map the initiation and termination sites of the mRNA precisely. In addition, slight changes in the coding sequence of different histone mRNAs can be detected. The results of S1 mapping of the genes in MM221 follow. All of these genes are expressed in cultured mouse cells.

The two H3 genes were labeled at the Sal I site at amino acid 58. To map the 3′ end the gene was labeled with $\alpha^{32}PO_4$ dCTP and the Klenow fragment of DNA polymerase I (9). The H2b gene was mapped by labeling the 5′ end of the Ava I site at amino acid 92 and the 3′ end of the other Ava I site at amino acid 103. Each of these fragments was hybridized to cytoplasmic RNA from mouse myeloma cells. Identical results were obtained with both 9S polysomal poly A^- RNA and total cytoplasmic RNA.

The results of the S1 mapping (10) confirm the great similarity in the coding regions of the histone genes. RNA protected the H3.1 and H3.2 gene probes at the 5′ end for 180 bases (Figure 12.5). This corresponds to the AUG codon at the start of the gene. In addition, there was a longer fragment protected in each of the genes which was 5–10% as intense as the major protected fragment. These longer protected DNA fragments presumably extended to the 5′ end of the mRNA. The H3.2 mRNA has a 21-base untranslated region at the 5′ end, and the H3.1 mRNA has a 28-base 5′ untranslated region. There is no detectable homology with the other H3 genes in the 5′ untranslated region. Similar results were found with the 3′ labeled probe for each of these genes. Once again, the major protected DNA fragment extended to the TAA termination codon (Figure 12.5). Minor protected fragments extending about 50 bases past the termination codon presumably represented the 3′ untranslated regions of these genes. The H3.1 untranslated region is 50 nucleotides long and that of H3.2 is 54 nucleotides long (see Section 6). In addition, the H3.1 probe protected a fragment only about 110 bases long. This fragment extends to the codon at amino acid 96 which differs between H3.1 and H3.2. This band is intense, presumably because the H3.2 protein makes up 80% of the total H3 protein in these cells (Zweidler, Chapter 14) and may represent the H3.2 mRNAs that differ in more than one nucleotide at this position.

Similar results were obtained with the H2b gene. The major protected fragment at the 5′ end maps to the ATG initiation codon, and

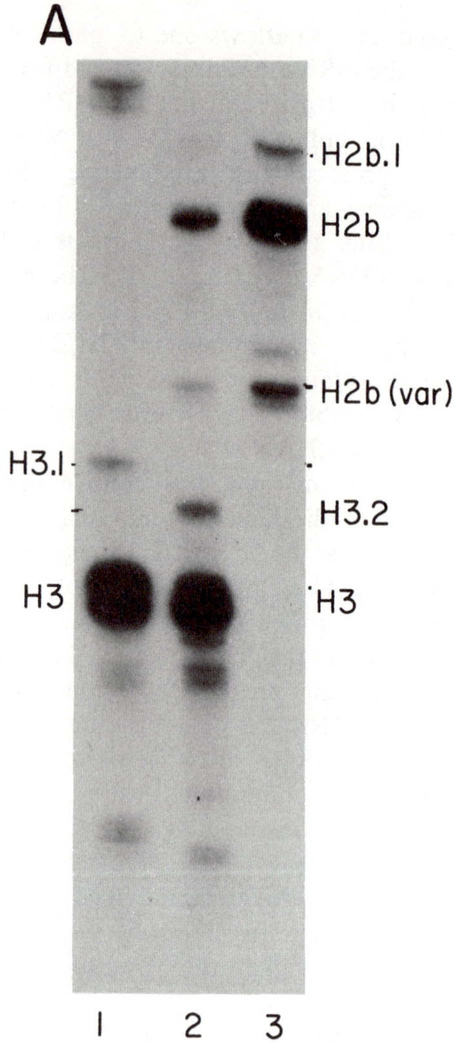

FIGURE 12.5. S1 mapping of the genes in MM221. The H3 genes were labeled at both the 5′ and 3′ end of the Sal I sites. The H2b genes were labeled at the 5′ end at the Ava I site amino acid 92 and at the 3′ end of the Ava I site at amino acid 103. The H2a was labeled at the 3′ end of the Sau 961 site at amino acid (see Figure 12.1). The protected fragments were analyzed by polyacrylamide gel electrophoresis in 7 M urea. The protected bands are H3: protected to the AUG codon or UAA codon of H3 genes; H3.1 and H3.2: DNA protected to the start and end of the mRNA derived from these genes; H3.1[96] DNA which protects to the amino acid

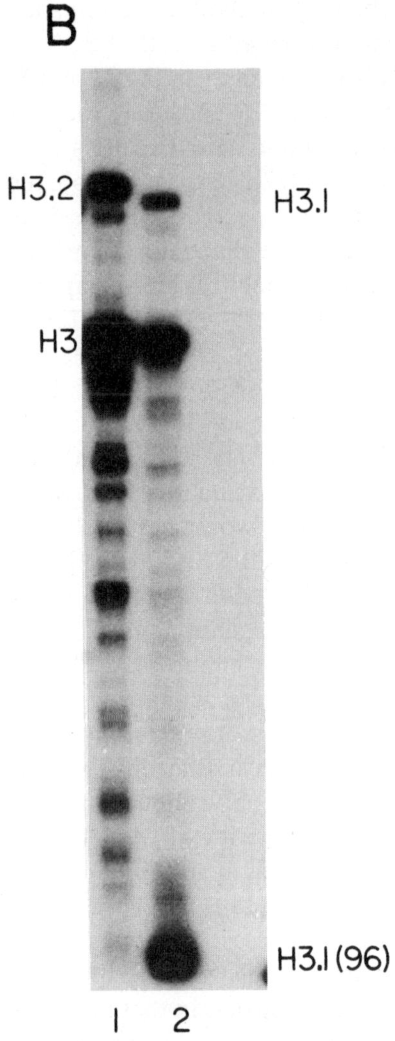

change at cysteine 96; H2b: DNA protected to the AUG or UAA codon of the family of H2b genes; H2b.1-DNA; protected to the start and end of the H2b gene in MM221; H2b (var): DNA protected to the amino acid change at amino acid 18; H2a: DNA protected to the UAA codon in the H2a genes; H2a.1: DNA protected to the end of the H2a mRNA. (*A*) 5′ end: lane 1 = H3.1; lane 2 = H3.2; lane 3 = H2b.1. (*B*) 3′ end: lane 1 = H3.2; lane 2 = H3.1; lane 3 = H2a; lane 4 = H2b; lane 5 = H2b + tRNA.

the major protected fragment at the 3′ end maps to the TAA termination codon. There are also longer fragments protected at both the 5′ (36 bases) and 3′ (47 bases) end presumably extending to the 5′ and 3′ end of the mRNAs derived from this gene. In addition, there is another protected fragment that maps to amino acid 18, the site of the single amino acid change in this histone H2b gene. It is probable that this marks a histone H2b gene with the expected valine residue in this position and at least one other nucleotide change in the nearby region. The relatively low intensity of the 3′ end fragment mapping to the TAA codon of the H2b gene is due to inefficient formation of this short hybrid.

It should be pointed out that under our conditions a single base mismatch in the RNA–DNA hybrid is not sufficient for S1 nuclease cleavage (Lobo and Marzluff, unpublished results). Thus a majority of the H2b mRNAs could have the expected valine residue at position 18, but only a single base change. These hybrids would not be cleaved at the single mismatch. However, an mRNA with an additional base change in the codon for amino acid 18 or in codon 17 or 19 would probably be cleaved at this site. Thus the H3.1 genes show a strong cleavage at amino acid 96, where there is an amino acid change probably accompanied by at least two single-base changes from the bulk of the histone H3 mRNA (Figure 12.3).

S1 mapping of the H2a gene fragment suggests that this gene is also expressed in myeloma cells. The H2a gene was cut with Sau 961 at amino acid 98 and labeled at the 3′ end with $^{32}PO_4$dCTP and hybridized to cytoplasmic RNA. There was a fragment protected that extended to the TAA termination codon. In addition, there was a minor fragment that extended an additional 68 bases. This presumably marks the end of the mRNA. This indicates that this gene is expressed. The 3′ untranslated region of this mRNA is unusual in that there are 27 consecutive uridine residues from position +13 to +40 in the mRNA. The H2a and H2b genes code for proteins that are slightly different in amino acid sequences from the bulk of the H2a and H2b proteins.

Similar results on the size of the histone H4 mRNA have been obtained by Seiler and Birnstiel (11). They isolated a 5kb Eco RI fragment containing a mouse histone H4 gene. This fragment did not contain any other histone genes. This gene is expressed as judged by S1 nuclease mapping and shares several characteristics with the genes in MM221.

6. SIGNALS FOR TRANSCRIPTION INITIATION AND TERMINATION

At the 3′ end of each of the genes in MM221 there is a 15-base pair sequence identical to that found at the end of each of the other histone genes. These are shown in Figure 12.6. There is a pyrimdine-rich sequence followed by a purine-rich segment which potentially can form a hairpin. The end of the mRNAs maps to the end of this hairpin. This common sequence may be a transcription termination site, or transcription may extend beyond the end of the mRNA. The 3′ end is then formed by an endonucleolytic cleavage, as is likely for other RNA polymerase II products (12, 13). This is the only homologous region among these four genes, indicating that this is the region necessary for termination and/or endonucleolytic cleavage.

```
                22         32         42         52         62         72
H3-2 CTACACTGGC ACGTAAACCA AAACGGCTCT TTTAAGAGCC ACCTCCATTA TCCACCAAAG ATGCT
                18         28         38         48         58         68
H3-1 GTTAATCCAC ACAACCACTT TAAAGGCTCT TCTTAGAGCC ACCCATCTTC CAAAAAAAGA ACTGT
                14         24         34         44         54         64
H2-B TCAAGACTCA GCTCTTAACC CAAAGGCTCT TTTCAGAGCC ACTCAAGACT TCAAAATTGG AGCTT
                41         51         61         71         81         91
H2-A TTTTTTTTTT TAAACAAAAC CCAAGGCTCT TTTCAGAGCC ACCACTTCTT CATATAAGAG
                26         36         46         56         66         76
H4   TCCCCCCCCC CCCCCATCCC TAACGGCCCT TTTTAGGGCC AACCACAGTC TCTTCAGGAG AGCTG
```

```
      T - T                  T - T
     /     \                /     \
    T       T              T       C
     \     /                \     /
      T-A                    T-A
      C-G                    C-G
      C-G                    T-A
      C-G                    C-G
      G-C                    G-C
  TAACG-CAACC            CAAAG-CACCC

  H4 GENE                MM221 GENES
```

FIGURE 12.6. The 3′ sequences at the end of the mouse histone genes are aligned so that the termination sequences at the end of each gene are in register. The distance from the termination codon is indicated. The dyad symmetry is underlined. Below the sequences is the postulated structure of the 3′ end of the mRNA. The sequence of the mouse H4 gene is taken from Seiler et al. (11). The genes in MM221 end at the A or C directly after the hairpin structure.

The histone H4 gene (11) shows a similar structure but a slightly different sequence at this position. Similar structures are present at the 3′ end of histone mRNAs in other species (14).

There is no such clear-cut region of homology at the 5′ end of the mRNA. There is a short 5′ untranslated region of less than 40 nucleotides in the three mRNAs in MM221 and in the mouse H4 gene. The only striking feature of these untranslated regions is their high pyrimidine content. The presumed capping site is not conserved to the extent that the 3′ regions are. There is also no clear-cut TATAA sequence 5′ to the cap site of all these genes, characteristic of most other genes transcribed by RNA polymerase II. However, there is an AT-rich sequence common to each 5′ end 30 bases upstream from the cap site (marked in Figure 12.7). Again the relative variability in flanking sequences is striking, compared to the striking conservation of coding region sequences. It should be pointed out that the presumed initiation site has been deduced by S1 nuclease mapping. We do not yet have direct evidence that these sites represent transcription initiation sites. Again, a similar 5′ untranslated region is found in the histone H4 gene, but it shows no obvious conserved sequences with the other histone genes.

7. CODON USAGE IN THE MOUSE HISTONE GENES

The histone coding regions, both in MM221 and the histone H4 gene of Seiler and Birnstiel (11), are striking in one regard. There has been no discrimination in the use of codons containing CG doublets or in the appearance of CG doublets formed from the third base of

```
               70         60         50         40         30         20
H3-2 ATTTCAACCA ATCAGGAGCA TGTTCCTTCT ATAAAGGAAC CCAGAACCTC AACCTCTGCA TTTCC
               75         65         55         45         35         25
H3-1 GTGTGTTGCG CGTGTGCGAC GCAAGCGTAC TTAAAGGCCA AAGTGCGCTA CTTAGGTATC TCACT
               48         58         68         58         48         38
H2B  ACTCTCTGAC AAGGACAGCC ACCGCTTTAT TTAAAGAGCA GGAAAGGAAC GGAACAGTTC AATAT
               78         68         58         48         38         28
H4   AGTTTTCAAT CTGGTCCGAT CCTCTCATAT ATTAGTGGCA CTCCACCTCC AATGCCTCAC CAGCT
```

FIGURE 12.7. The 5′ sequences at the end of the mouse histone genes are aligned so that the conserved TATAA sequences are in register. The bases found in the mRNA are underlined. The sequence of the mouse H4 gene is taken from Seiler et al. (11). The distance from the AUG codon is indicated.

TABLE 12.1. Dinucleotide Frequency in Histone Coding and Flanking Regions[a]

	Number	Percent	Expected Percent
Coding Region (64.7% GC)			
AA	85	5.2	4.4
AC	84	5.1	7.1
AG	130	8.0	6.5
AT	43	2.6	3.0
CA	131	8.0	7.1
CC	169	10.4	11.5
CG	157	9.6	10.4
CT	96	5.9	4.9
GA	95	5.8	6.5
GC	209	12.8	10.4
GG	122	7.5	9.5
GT	75	4.6	4.5
TA	31	1.9	3.0
TC	89	5.5	4.9
TG	94	5.8	4.5
TT	22	1.3	2.1
Histone Flanking (45.6% GC)			
AA	141	8.4	7.4
AC	108	6.4	6.6
AG	127	7.5	5.8
AT	82	4.9	7.4
CA	122	7.2	6.6
CC	103	6.1	5.8
CG	56	3.3	5.2
CT	127	7.5	6.6
GA	102	6.0	5.8
GC	95	5.6	5.2
GG	87	5.2	4.6
GT	78	4.6	5.9
TA	93	5.5	7.4
TC	102	6.0	6.6
TG	92	5.5	5.9
TT	173	10.2	7.5

[a] The dinucleotide frequencies for the coding and flanking regions (starting at the AUG and the termination codon) are given along with the expected values based on the base composition. The genes analyzed were the gene in MM221 and the H4 gene sequenced by Seiler et al. (11).

one codon with the first base of the rest. The doublet frequency for the coding regions of H3.1, H3.2, and H2b.1, H4, and H2a are shown in Table 12.1. There is not the great discrimination against CG doublets seen in other genes. In particular, the frequency of arginine codons (CGX versus AGR) is what you would expect on the basis of probability. The use of G as the third base in codons NCX has not been discriminated against. There is less than the expected number of CG doublets in the flanking regions. The only Sal I sites we have found in all the histone genes clusters are the sites in the H3 genes. In addition, the CG frequency in the flanking regions we have sequenced is much lower than in the coding regions, and is significantly lower than expected on the basis of probability even with the low AT content of these regions (Table 12.1). Thus there is some mechanism by which the cell has not discriminated against CG doublets in the mouse histone genes, which may extend throughout the whole cluster, but seems to be limited to the regions close to the

TABLE 12.2. Codon Usage in Mouse Histone Genes[a]

TTT-Phe	2	TCT-Ser	2	TAT-Tyr	0	TGT-Cys	1
TTC-Phe	10	TCC-Ser	9	TAC-Tyr	15	TGC-Cys	2
TTA-Leu	0	TCA-Ser	0	TAA- ·	3	TGA- ·	1
TTG-Leu	4	TCG-Ser	8	TAG- ·	1	TGG-Trp	0
CTT-Leu	1	CCT-Pro	4	CAT-His	4	CGT-Arg	10
CTC-Leu	4	CCC-Pro	11	CAC-His	8	CGC-Arg	42
CTA-Leu	2	CCA-Pro	0	CAA-Gln	1	CGA-Arg	0
CTG-Leu	34	CCG-Pro	7	CAG-Gln	22	CGG-Arg	5
ATT-Ile	0	ACT-Thr	4	AAT-Asn	0	AGT-Ser	0
ATC-Ile	28	ACC-Thr	31	AAC-Asn	9	AGC-Ser	9
ATA-Ile	0	ACA-Thr	0	AAA-Lys	2	AGA-Arg	1
ATG-MET	11	ACG-Thr	2	AAG-Lys	61	AGG-Arg	1
GTT-Val	0	GCT-Ala	13	GAT-Asp	1	GGT-Gly	8
GTC-Val	11	GCC-Ala	35	GAC-Asp	13	GGC-Gly	26
GTA-Val	0	GCA-Ala	2	GAA-Glu	0	GGA-Gly	6
GTG-Val	21	GCG-Ala	9	GAG-Glu	25	GGG-Gly	3

[a] The codon frequencies for the genes in MM221 and the mouse H4 gene are given.

coding regions. The use of other codons is certainly nonrandom, with every amino acid showing one or two highly preferred codons. Consisently, C or G has been used as the third base of the codon, rather than A or T (Table 12.2).

8. NUMBER OF DIFFERENT HISTONE mRNAs

S1 nuclease analysis of both the H3 and H2b genes yields the following striking result: There is a major intense band that maps to the AUG initiation codon (5′ labeling) or to the UAA termination codon (3′ end). In addition, there is a minor band that extends to the 5′ or 3′ end different distances for different genes. For H3.1, H3.2, and H2b.1 the minor bands are 5–10% of the intensity of the major band (Figure 12.5), and the 5′ and 3′ bands are of equal intensity. Our interpretation of this result is that there are a number of genes with coding regions sufficiently identical to give protection with S1 nuclease. (One base change in an otherwise homologous region is not sufficient to render the hybrid sensitive to nuclease S1.) The flanking regions are very different, however, and the minor band detected for each gene corresponds to the mRNA product of that particular gene, which comprises 5–10% of the total H3 (or H2b) mRNA of this class. In the case of H3 a class of mRNA is defined by the Sal I site at amino acid 58, and in the case of H2b by Ava I sites at amino acid 93 and 102. The H2a gene gives a similar result at the 3′ end with respect to the Sau 961 site at amino acid 98. This result suggests the possibility of 10–20 H3 and H2b genes of similar coding but divergent flanking sequences. It is possible that all of the genes detected with the H3-2 probe by hybridization to the mouse genome are expressed, each giving rise to an mRNA with different flanking regions. There is one mouse H3 gene (MM614) we have recently isolated that gives a different result on S1 mapping. (Figure 12.8) The band at the end of the mRNA is as intense as the band at the AUG codon and the same relative intensities are found at the 3′ end. Presumably this gene accounts for the major histone H3 mRNA (about 50% of the total) in mouse myeloma cells. It is an H3.2 gene since it does not show the band at 120 nucleotides characteristic of H3.1 genes.

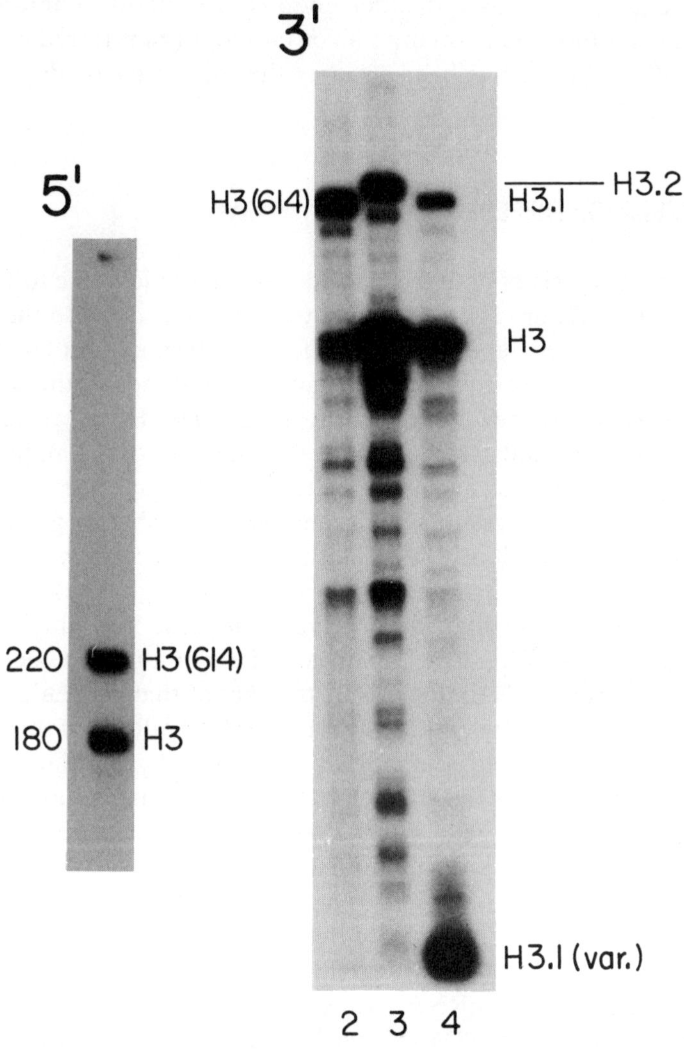

FIGURE 12.8. S1 Mapping of the H3 Gene from MM614. The H3 gene from MM614 was cut with Sal I, labeled with either polynucleotide kinase (5′) or the Klenow fragment of DNA polymerase (3′). The DNA was hybridized to 10 g of myeloma cytoplasmic RNA. For 3′ mapping the H3 bands map to the AUG codon (5′) and TAA codon (3′). Lane 1 = MM614 (5′); lane 2 = MM614 (3′); lane 3 = H3.2 (3′); lane 4 = H3.1 (3′).

9. SEQUENCE CONSERVATION OF HISTONE GENE SEQUENCES AMONG SPECIES

The histone coding regions have also been highly conserved among different species. S1 nuclease mapping using the mouse genes and hamster or human RNA yields the same protected fragments up to the AUG initiation codon for both H3 and H2b genes (Figure 12.9). However, there are no protected bands visible beyond the initiation codon with any of the mouse gene probes, including MM614. This result strongly suggests that the nucleotide sequence in the coding region has been strongly conserved among species, but that the flanking regions vary greatly. These results again suggest that the similarity in coding regions one sees within a species (e.g., H3.1 versus H3.2) is not due to a gene conversion event. These results provide further support for the idea that there has been selective pressure to maintain the coding sequence. The flanking regions in the major H3 gene, MM614, are not present in the hamster or human RNA either.

10. CONTROL OF EXPRESSION OF THE MOUSE HISTONE GENES

It has been recognized for many years that the levels of histone mRNA are regulated. There is an extensive literature on their regulation in HeLa cells. Most of this work suggests that histone mRNA is present only during S phase (15–20). Groppi and Coffino (21) have recently challenged this viewpoint. They reported that in continuously cycling S49 mouse lymphoma cells histones are synthesized in both G1 and S phase cells. Their results with other cells (see Chapter 13) suggests that there is only a loose coupling between DNA synthesis and histone synthesis. In this section we summarize recent results that suggest that the level of histone mRNA may be controlled by a metabolite involved in deoxynucleotide metabolism. The levels of histone mRNA are regulated both by regulating the transcription of the histone genes and altering the half-life of the histone mRNA.

The synthesis of all the histone proteins is not regulated coordinately. Zweidler (see Chapter 14), has classified the histone proteins into replication and replacement variants. The replication variants include those proteins (or genes) whose expression is regulated coordinately

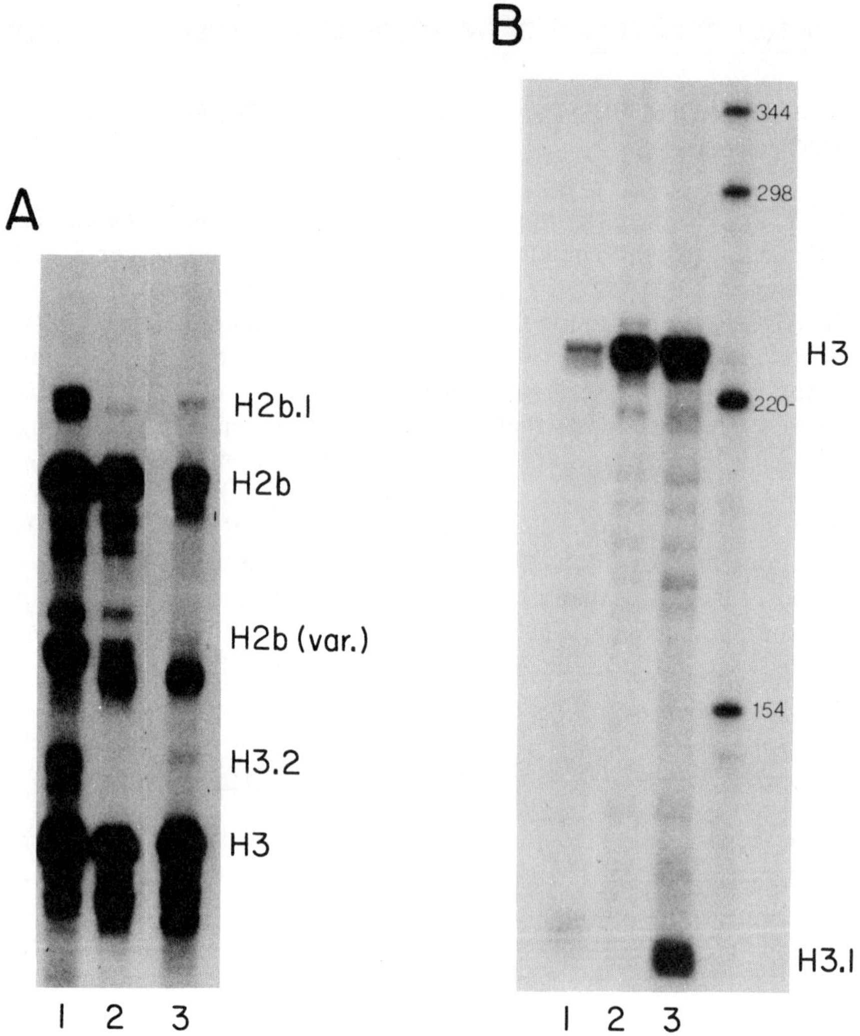

FIGURE 12.9. Conservation of the histone coding sequences in different species. (*A*) The Ava I-Sal I fragment (Figure 12.1*A*) was labeled with polynucleotide kinase and hybridized with mouse, hamster, and human RNA. The S1-protected fragments were resolved by polyacrylamide gel electrophoresis. The bands are identified as in Figure 12.5. Lane 1 = mouse RNA; lane 2 = hamster RNA; lane 3 = human RNA. (*B*) Hamster RNA was hybridized with H3.1, H3.2, and 614 DNA labeled at the 3′ end of the Sal I site with ^{32}P dCTP. The protected bands map to the termination codon and to amino acid 96 (Figure 12.5). Lane 1 = MM614; lane 2 = H3.2; lane 3 = H3.1.

with cell growth, and the replacement variants are expressed in nondividing cells (and possibly constitutively in all cells). The H3 genes that we have isolated code for replication variants, and their regulation is consistent with this role.

It was reported by Gallwitz and co-workers (17, 18) and Borun and co-workers (19) that treatment of HeLa cells with inhibitors of DNA synthesis results in a rapid reduction in the amounts of translatable histone mRNA in the cytoplasm. Using the isolated histone genes from MM221, we have measured the transcription rates and stability of histone mRNA under different conditions. These studies conclusively show that the replicative histone mRNA levels are regulated at both the level of transcription and mRNA stability.

When mouse myeloma cells are treated with various DNA synthesis inhibitors, the levels of cytoplasmic histone mRNA fall rapidly. The levels of histone mRNA were measured by S1 nuclease mapping (Figure 12.10). After a 5-min lag the levels of histone mRNA drop precipitously. The apparent half-life of the message is 12 min. However, the mRNA is not completely destroyed but rather a new steady-state level is reached which is 5–10% of the original level. This is seen most clearly from the decay plot in Figure 12.10*A*. All of the replicative histone mRNAs assayed show the same behavior. Thus this class of mRNAs becomes unstable when DNA synthesis is blocked.

Alteration in histone mRNA levels could be due to a change in the rate of mRNA synthesis or to a change in mRNA half-life or some combination of the two. To determine how histone mRNA levels are controlled, we measured the rates of transcription and the half-life of the histone mRNA. The rates of transcription were measured both by pulse labeling and by measuring the rate of transcription of the histone genes in isolated nuclei. After treating cells for 1 hour with either hydroxyurea or cytosine arabinoside, the rate of histone gene transcription was reduced to about 20% of the rate in exponentially growing cells (22). This is due to a real reduction in the rate of transcription and not to alterations in histone RNA turnover. Identical results were found with both 5-min and 10-min pulses in cultured cells. The same result was found when RNA synthesis was measured in isolated nuclei. There is no degradation of histone RNA synthesized in isolated nuclei. A "chase" of labeled RNA in isolated nuclei was performed by adding unlabeled GTP after 10 min incubation followed

A

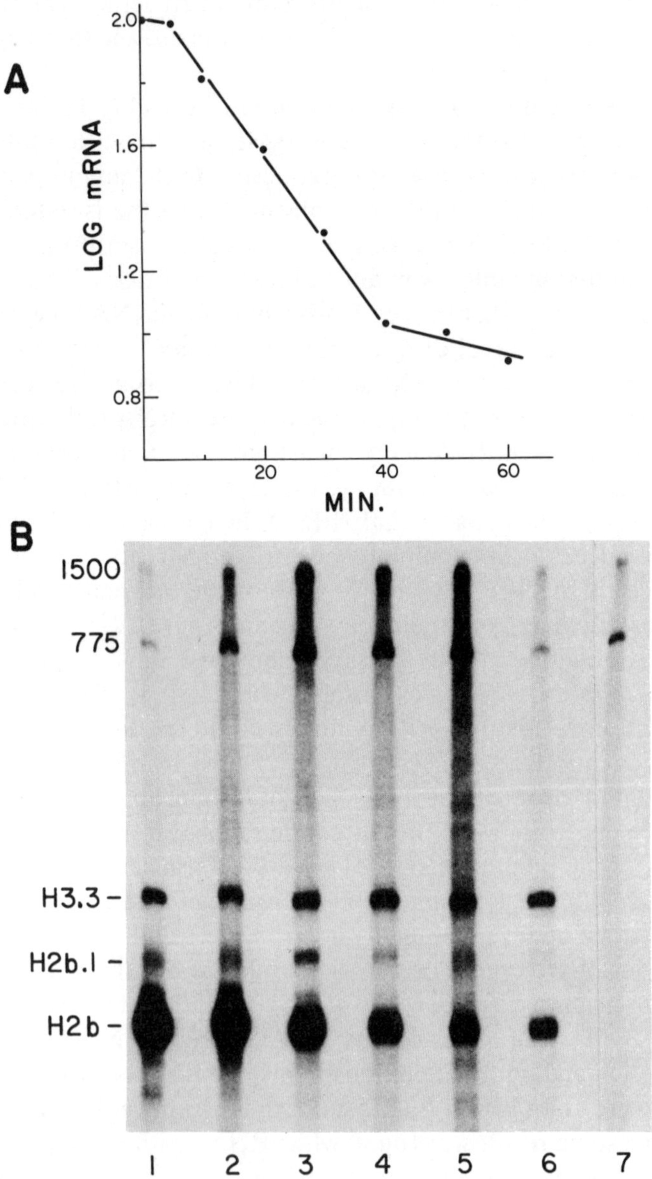

by further incubation for 20 min. There was no degradation of histone mRNA during the incubation with nuclei isolated from control cells or cells treated with cytosine arabinoside. Thus the changes observed are due to changes in histone gene transcription (Table 12.3).

There was no alteration in rate of the synthesis of total RNA in hydroxyurea or cytosine arabinoside treated cells or in the rate of RNA synthesis in nuclei isolated from these cells (Table 12.3). Histone H3 RNA makes up about 0.01–0.02% of the total RNA synthesis in exponentially growing cells as measured both by pulse labeling and synthesis in isolated nuclei. These cells produce an immunoglobulin light chain mRNA. The rate of synthesis of this RNA was unchanged by treatment with DNA synthesis inhibitors. The same is true of the synthesis of ribosomal RNA (22). The 80% reduction in the rate of histone gene transcription is not by itself sufficient to account for the drop in histone mRNA levels. The second component of the reduction in histone mRNA levels is a change in the half-life of histone mRNAs.

The half-life of histone mRNA in growing cells was estimated by two methods: the "approach to steady state" and a "label-chase" experiment. Myeloma cells were labeled continuously with ^{3}H-uridine and the amount of labeled histone RNA measured at different times. When the amount of labeled histone RNA becomes constant, a steady state has been reached balancing the rate of synthesis and degradation. This will occur rapidly for RNAs with a short half-life and will take a long time for RNAs with a long half-life. Histone mRNAs reach a steady state within 2 hours. From this rate of accumulation the half-life of histone mRNAs in exponentially growing myeloma cells is estimated at about 1 hour. Similar results were obtained with a label-chase experiment (Figure 12.11). Cells were labeled for 90 min with ^{3}H-uridine to reach a steady state and then chased with unlabeled

FIGURE 12.10. Effect of DNA synthesis inhibitors on histone mRNA levels. (*A*) The levels of H2b RNA were quantitated by densitometry. The decay of H2b RNA is plotted on a log scale and shows an apparent half-life of 12 min. (*B*) The levels of the H2b and H3.3 RNA was measured at various times after treatment of mouse myeloma cells with hydroxyurea. 5 g of cytoplasmic RNA was hybridized to a mixture of end-labeled H2b and H3.3 DNA. The S1 resistant material was analyzed as in Figure 12.5.

TABLE 12.3. Transcription of Mouse Histone Genes[a]

	Pulse Label				Isolated Nuclei			
DNA	H3	H2a	K	U1	H3	H2a	K	U1
Control cells	1137	976	2244	391	1265	1007	680	371
(Input, %)	(0.016)	(0.014)	(0.032)	(0.006)	(0.020)	(0.016)	(0.011)	(0.006)
+ ARA C	200	170	1897	211	266	149	624	211
(Control, %)	(18)	(17)	(81)	(54)	(21)	(15)	(92)	(67)
ARA-C + CYCLO	633	363	2439	272	823	509	1326	277
(Control, %)	(56)	(37)	(110)	(70)	(65)	(51)	(200)	(75)

[a] Cells were treated with cytosine arabinoside (40 μg/ml) or cytosine arabinoside + cycloheximide (10^{-4}M) for 1 hour. The cells were divided and an aliquot pulsed labeled for 10 min with ^{3}H-uridine (50 μc/ml). Nuclei were prepared from the rest of the cells and incubated with α-$^{32}PO_4$-GTP for 30 min to label RNA synthesized *in vitro*. The RNA was hybridized to DNA dots containing H3, H2a, the light chain cDNA expressed by these cells (30), and a U1 RNA gene (31). The dots were counted to determine the amount of each RNA synthesized. The input counts were 7×10^6 in the pulse label and 6.4×10^6 in the $^{32}PO_4$-labeled RNA made in isolated nuclei.

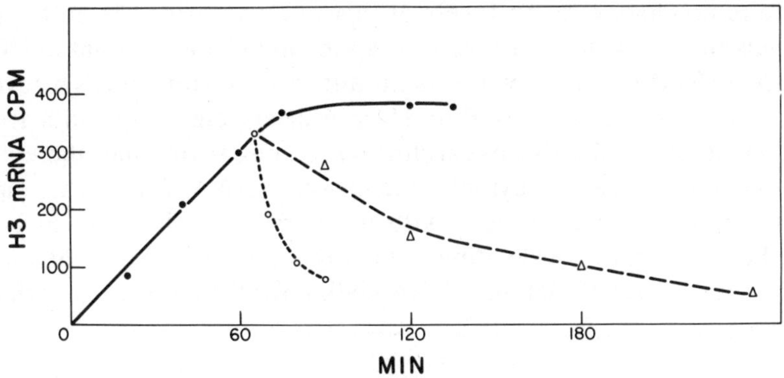

FIGURE 12.11. Half-life of histone mRNA. Mouse myeloma cells were incubated with ^{3}H-uridine for varying amounts of time. The cytoplasmic RNA was isolated and hybridized to DNA dots containing H3.2 DNA (● - - ●). After labeling the cells were chased with 5 mM cytidine and uridine in the absence (△ - - △) and presence of hydroxyurea (○ - - ○).

uridine and cytosine. The apparent half-life of the mRNA again was about an hour. This is probably an upper limit since the triphosphate pools in mammalian cells make chase experiments difficult. Glucosamine has been used to reduce the uridine pool (23) in HeLa cells, but glucosamine is toxic to myeloma cells and inhibits their growth (our unpublished results). Taken together, these experiments indicate the normal half-life of histone mRNA in these cells is about 1 hour, in agreement with the results of Heintz and Roeder for HeLa cells (23).

Inhibitors of DNA synthesis lower the half-life of histone mRNA. The disappearance of the bulk histone mRNA after hydroxyurea treatment gives a half-life of 12 min. Prelabeled histone mRNA also disappears very rapidly when cells are treated with hydroxyurea, with a half-life of less than 15 min (Figure 12.10). Thus there is a change of both the rate of transcription and the half-life of the mRNA by a factor of 4–6. It is possible that the same signal is responsible for both changes (see below).

The change in histone mRNA metabolism due to inhibition of DNA synthesis are specific for the replication histone variants. The levels of the "replacement" variant histone H3.3 mRNA did not change as a result of treatment with hydroxyurea (22). Similarly, there is no change in the levels of this mRNA during the cell cycle in cells that regulate their levels of replication histone variant mRNA.

The effects of DNA synthesis inhibitors on histone mRNA levels are rapidly reversible. Within 15 min after release of cells from treatment with 5-fluorodeoxyuridine by feeding thymidine, there was a detectable increase in cytoplasmic histone mRNA levels. This rapid recovery is again consistent with a reversal of inhibition of both synthesis and turnover leading to the rapid reestablishment of a new steady state level of histone mRNA coding for the replication variants (Figure 12.12).

11. HISTONE mRNA LEVELS IN CELLS ARRESTED IN G1

The levels of histone mRNA are not tightly coupled to the synthesis of DNA in all cells. As discussed by Coffino (see Chapter 13) some continuously cycling cells have large amounts of histone mRNA in G1 cells, whereas others regulate their histone mRNA levels more tightly. For example, the levels of histone mRNA change by a factor of 50 in mouse fibroblasts arrested in G1 by serum starvation and then released by feeding with 20% serum (Figure 12.13*A*). Thus the levels of histone mRNA can be regulated dramatically during the cell cycle. The rates of transcription of the histone genes also vary. However, the rate of transcription of the histone genes in the starved fibroblasts is 20% of the rate in S phase cells, the same amount of change that was observed in cells treated with DNA synthesis inhibitors (22, 24). The regulation of the histone mRNA levels during the cell cycle of serum-arrested fibroblasts is due to alterations in both the rate of synthesis of histone mRNA and the half-life of the mRNA, a

FIGURE 12.12. Reversibility of histone mRNA levels. Mouse myeloma cells were treated with 2 M fluorodeoxyuridine (FdU) for 45 min. The cells were then treated with 10 M thymidine and RNA prepared 15, 30, and 60 min after refeeding thymidine. The levels of histone mRNA were measured by S1 nuclease mapping using as a probe the Ava I-Sal I fragment (A in Figure 12.1) to measure the H2b and H3 genes. The bands are labeled as in Figure 12.5. lane 1, no treatment; lane 2, FdU − 45 min; lane 3, FdU + thymidine, 15 min; lane 4, FdU + thymidine, 15 min; lane 5, FdU + thymidine, 30 min; lane 6, FdU + thymidine, 60 min.

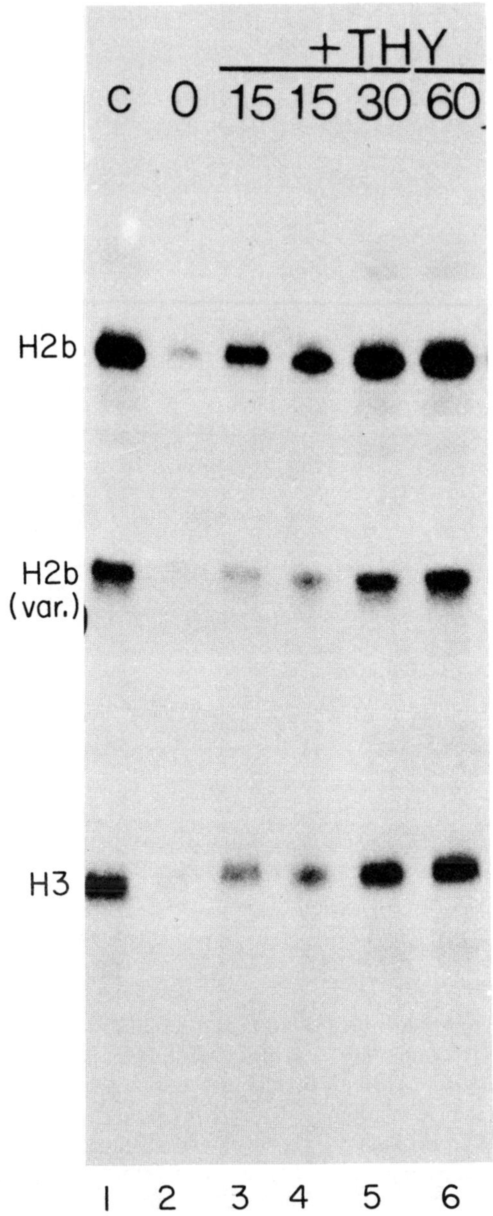
+THY
c
0
15
15
30
60
H2b
H2b
(var.)
H3
1
2
3
4
5
6

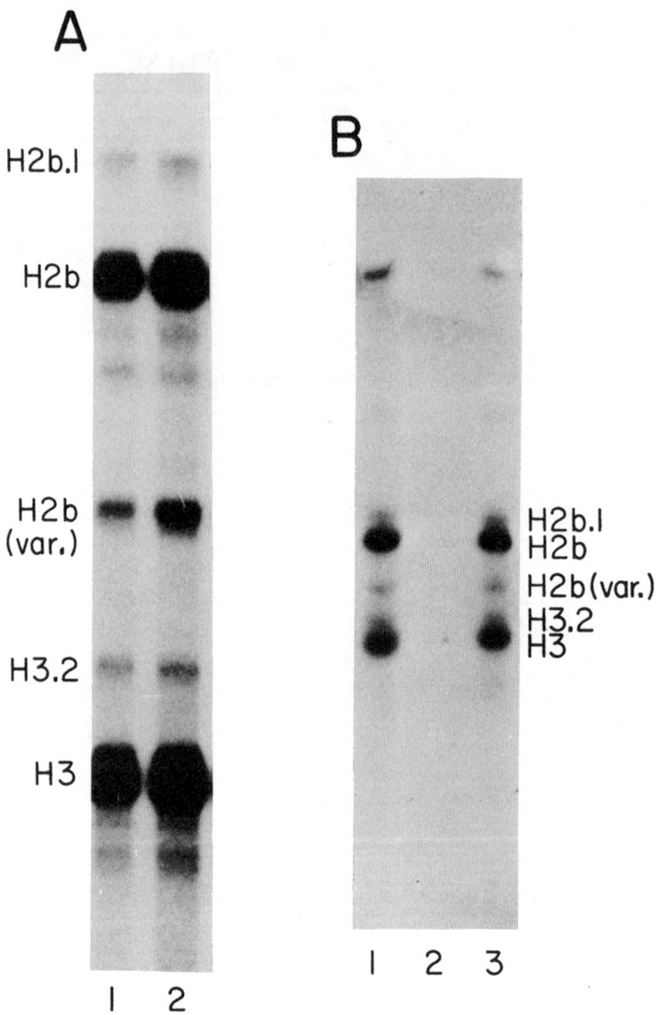

FIGURE 12.13. Levels of histone mRNA in two cell lines arrested in G1. The histone mRNA levels in G1 lymphoma cells [the "deathless" mutant of S49 cells (25)] arrested with dibutyryl cAMP, 3T3 cells arrested by serum starvation, and exponentially growing mouse myeloma cells were measured by S1 nuclease mapping as described in Figure 12.12. Lane 1, G1 lymphoma cells; lane 2, exponentially growing lymphoma cells; lane 3, exponentially growing myeloma cells; lane 4, arrested 3T3 cells; lane 5, S phase 3T3 cells (16 hours after serum stimulation). The protected DNA fragments in lanes 3–5 were analyzed by polyacrylamide gel electrophoresis in 99% fromamide. The protected fragments are labeled as in Figure 12.5.

situation similar to that in the cells treated with inhibitors of DNA synthesis (A. DeLisle, L. F. Johnson, R. A. Graves, and W. F. Marzluff, unpublished results).

In contrast to the serum-starved fibroblasts, another cell type that may be efficiently arrested in G1 shows a different regulation of histone mRNA levels. A derivative of S49 lymphoma cells, "deathless" cells, is arrested in G1 by treatment with cyclic AMP without any cell death (25). These arrested cells (>95% G1) contain 30–50% of the histone mRNA of exponentially growing cells (Figure 12.13*B*) in contrast to the fibroblasts which contain less than 5% of the S phase histone mRNA levels. The high levels of histone mRNA in G1 are a characteristic of S49 cells (21). When these G1 cells (which are not synthesizing DNA) are treated with an inhibitor of DNA synthesis, hydroxyurea, both the steady-state level of histone mRNA and the rate of histone mRNA transcription are reduced. The relative rate of histone mRNA (percent of total RNA polymerase II activity) synthesis in cAMP blocked cells is 60–70% of the control (unblocked) culture (22). This rate is reduced by hydroxyurea treatment to the same level in both the blocked and unblocked cells, "a basal" level of transcription. Thus the effect of hydroxyurea is not due to its effect on DNA synthesis but more likely to its direct effect on deoxyribonucleotide synthesis.

12. NATURE OF SIGNAL REGULATING HISTONE mRNA LEVELS

Several agents that interfere with deoxynucleotide metabolism exert similar effects on histone mRNA metabolism. These include hydroxyurea, which blocks ribonucleotide reductase and hence all deoxynucleotide synthesis; fluorodeoxyuridine, which blocks thymidylate synthetase (26) and hence only thymidine synthesis and not that of other deoxynucleotides; cytosine arabinoside, which may interfere with deoxycytidine synthesis; mycophenolic acid (27), which blocks conversion of inosine monophosphate to xanthine monophosphate; and aphidicolin, which blocks deoxynucleotide metabolism as well as DNA polymerase (28). All of these agents reduce deoxynucleotide levels rapidly in cultured myeloma cells (our unpublished results). All of these agents exhibit the same rapid effect on histone

RNA transcription and half-life, with the exception of mycophenolic acid, which has not been extensively tested yet.

All these results suggest that (1) there is a signal that acts both in the cytoplasm and nucleus and (2) that the metabolism of deoxynucleotides may be involved in altering the levels of the signal molecule. It is unlikely that the signal molecule is one of the deoxynucleotides involved in DNA synthesis, both on a theoretical basis (cells usually utilize a compound external to metabolic pathways, e.g., ppGpp or cyclic AMP, as regulators) and because of the range of inhibitors that give similar effects on histone mRNA synthesis. Hydroxyurea blocks all deoxynucleotide synthesis. Fluorodeoxyuridine affects thymidine synthesis specifically. Mycophenolic acid inhibits guanosine biosynthesis. Both cytosine arabinoside and aphidicolin affect deoxycytidine metabolism most strongly. The signal molecule is likely to be a metabolite whose levels are affected by the levels of other deoxynucleotides, perhaps an unusual deoxynucleotide. Thus the central signal is likely to be a small molecule regulated in some way by the levels of deoxynucleotides or the rate of deoxynucleotide synthesis. The finding that in G1 cells which contain histone mRNA, the histone mRNA is degraded after treating cells with the inhibitors of deoxynucleotide metabolism indicates that DNA synthesis is not required for histone mRNA stability.

The second component of the regulatory system is likely to be a protein that has a short half-life. It was observed several years ago that inhibitors of protein synthesis prevented the decrease in histone mRNA levels observed when DNA synthesis was inhibited (15, 17). There are two possibilities to explain this observation. (1) Autoregulation: histones themselves regulate the stability of the excess histone mRNA, as suggested by Butler and Mueller (15). (2) There is an unstable regulatory protein. Neither of these possibilities has been excluded, and there are arguments in favor of each one. The most striking feature is the rapidity of the effect. The inhibitors of protein synthesis block the degradation of histone mRNA within 5 min after their addition. This is consistent with the possibility that inhibitors of DNA synthesis block DNA synthesis leading to an increased accumulation of histones. The increased accumulation of histones could then lead to a destabilization of histone mRNA, according to an autoregulation model. In this model there is no necessity for a very unstable regulatory protein but a requirement that histones specifically recognize their mRNAs and possibly also their genes. It is hard to reconcile this model with the finding that the same effect on histone mRNA levels is found in cells not synthesizing DNA. Inhibitors of DNA synthesis have the same effect on histone mRNA levels in the G1-arrested lymphoma cells as they do in growing cells (Figure 12.12). In addition, some cells normally synthesize some histones when they are not synthesizing DNA (21), and the mRNA levels are affected similarly in these cells by treatment with DNA synthesis inhibitors.

Nevertheless, it is possible that there is a pool of histone protein that is compartmentalized in a complex (e.g., with a "nucleoplasmin"-like protein similar to that in *Xenopus* oocytes) (27). Histone proteins synthesized in excess of this pool could then regulate histone mRNA. If the size of this pool varied in different cells then the results could be explained in terms of an autoregulation phenomena.

Since there are at least five different classes of mRNA that are regulated coordinately, the simplest hypothesis is that common structures are recognized both in regulating rate of transcription and degradation. In the mRNA structures there are two different regions that are conserved in the genes sequenced thus far: (1) the hairpin at the 3′ end close to the coding regions; and (2) the short pyrimidine rich 5′ untranslated region. Either of these could be potential targets for the nuclease. It is possible that the structural signal is present in the coding region since all the H3 genes with similar coding regions are regulated coordinately. However, this would require similar structures in the coding region of all five histone genes. These structures may not be present in the H3.3 mRNA, which is insensitive to DNA synthesis inhibitors. Identification of the target sites for histone mRNA degradation will require manipulation of the DNA structure and reintroduction of the gene into mammalian cells.

Similarly, we predict conserved DNA structures, presumably at the 5′ end of the genes, common to replicative histone genes, which are involved in regulating transcription. As yet we have no direct evidence for the function of any DNA or RNA sequences in the regulation. Our current efforts are directed to this end.

13. SUMMARY AND CONCLUSION

1. The mouse histone genes are a family of genes repeated a moderate number of times in the mouse genome. The class of genes

described here appears to code for the mouse replication histone variants. There must be other histone gene clusters coding for the replacement histone variants and spermatocyte histone variants. These replication histone H3 genes show extensive sequence conservation in the coding region, and the same extensive conservation is seen in the coding regions of the H2a and H2b genes. On the other hand, the replacement H3 variants have diverged in nucleic acid sequences from the replication variants, and presumably the nucleotide sequences of the other histone replacement variants have also diverged.

2. There is a common sequence at the 3′ end of the mouse histone mRNA that may represent either a termination signal for RNA polymerase II or a recognition site for a nuclease that forms the mature end of the RNA by endonucleolytic cleavage.

3. The levels of the replication variant histone mRNAs are regulated both at the level of mRNA transcription and mRNA degradation. Both processes are probably regulated by the same signal. A common feature of this regulation is that compounds that interfere with deoxynucleotide metabolism alter histone mRNA levels. These effects require protein synthesis. The regulation could be mediated either by a metabolite involved in deoxynucleotide metabolism or autoregulation due to the histone proteins themselves.

ACKNOWLEDGMENTS

We acknowledge the help of collaborators in our group: Dr. Don Sittman, Ing Ming Chiu, Chi-Jiunn Pan, and Cindy Sprecher. We also thank Drs. David Cox, David Bieber, and Vince Groppi (UCSF) for the chromosome assignments; Dr. Ron Cohn for a gift of the human H3.3 clone; and Drs. Vince Groppi and Phil Coffino for the S49 cells and for helpful discussions. This work was supported by a grant from NIH.

REFERENCES

1. L. H. Kedes, *Ann. Rev. Biochem.* **48,** 837–870 (1979).
2. R. P. Lifton, M. S. Goldberg, R. W. Karp, D. S. Hogness, *Cold Spring Harbor Symp. Quant. Biol* **42,** 1047–1051 (1977).

3. S. G. Franklin and A. Zweidler, *Nature* **266,** 273–274 (1977).
4. D. B. Sittman, I. M. Chiu, C. J. Pan, R. H. Cohn, L. H. Kedes, and W. F. Marzluff, *Proc. Natl. Acad. Sci. USA* **78,** 4078–4082 (1981).
5. N. Heintz, M. Zernik, and R. G. Roeder, *Cell* **24,** 661–668 (1981).
6. L. M. Hereford, K. Fahrner, J. Woolford, and M. Robash, and D. Kalback, *Cell* **18,** 12261–12271 (1979).
7. M. M. Smith, Chapter 1.
8. E. Jacob, *Eur. J. Biochem.* **65,** 275–284 (1976).
9. A. Zweidler, Chapter 14.
10. R. Weaver and C. Weissman, *Nucl. Acids Res.* **7,** 1175–1193 (1979).
11. A. Seiler-Tuyns and M. L. Birnstiel, *J. Mol. Biol.* **151,** 607–625 (1981).
12. H. Weintraub, H. Larsen, and M. Groudine, *Cell* **24,** 333–344 (1981).
13. E. Hofer, R. Hofer-Warbinek, and J. Darnell, *Cell* **29,** 887–893 (1982).
14. M. L. Birnstiel and C. C. Hentschel, *Cell* **25,** 301–305 (1981).
15. W. B. Butler and G. C. Mueller, *Biochim. Biophys. Acta* **294,** 481–496 (1973).
16. T. W. Borun, M. D. Scharff, and E. Robbins, *Proc. Natl. Acad. Sci. USA* **58,** 1977–1982 (1967).
17. M. Breindl and D. Gallwitz, *Eur. J. Biochem.* **32,** 381–391 (1973).
18. H. Stahl and D. Gallwitz, *Eur. J. Biochem.* **72,** 385–392 (1977).
19. T. W. Borun, F. Gabrelli, K. Ajiro, A. Zweidler, and C. Baglioni, *Cell* **4,** 59–67 (1975).
20. R. Rickles, F. Marashi, F. Sierra, S. Clark, J. Wells, J. Stein, and G. S. Stein, *Proc. Natl. Acad. Sci. USA* **79,** 749–753 (1982).
21. V. E. Groppi and P. Coffino, *Cell* **21,** 195–204 (1980).
22. D. B. Sittman, R. Graves, and W. F. Marzluff, *Proc. Natl. Acad. Sci. USA* **80,** 1859–1863 (1983).
23. R. Levis and S. Penman, *Cell* **11,** 105–113 (1977).
24. I. Lamaire and P. Coffino, *Cell* **11,** 149–155 (1977).
25. N. Heintz, H. L. Sive, and R. G. Roeder, *Mol. Cell. Biol.* **3,** 539–550 (1983).
26. Y. Wataya, A. Matsuda, and D. V. Santi, *J. Biol. Chem.* **255,** 5538–5544 (1980).
27. R. C. Mulligan and P. Berg, *Proc. Natl. Acad. Sci. USA* **78,** 2070–2076 (1981).
28. J. A. Huberman, *Cell* **23,** 647–648 (1981).
29. R. A. Laskey, B. M. Honda, A. D. Mills, and J. T. Finch, *Nature* **275,** 416–420 (1978).
30. U. Schibler, K. B. Marcu, and R. P. Perry, *Cell* **15,** 4129–4137 (1978).
31. C. O. Card, G. F. Morris, D. T. Brown, and W. F. Marzluff, *Nucl. Acids Res.* **10,** 7677–7688 (1982).

CHAPTER 13

HISTONE EXPRESSION AND THE MAMMALIAN CELL CYCLE

PHILIP COFFINO
Department of Medicine
Department of Microbiology and Immunology
University of California
San Francisco, California

EVA STIMAC
VINCENT E. GROPPI
Department of Microbiology and Immunology
University of California
San Francisco, California

DAVID BIEBER
Department of Biochemistry and Biophysics
University of California
San Francisco, California

CONTENTS

1. INTRODUCTION

2. GENERAL CONSIDERATION OF HISTONE S-PHASE COUPLING

3. SYNCHRONIZED S49 CELLS
   - 3.1. Fraction 1: Small G_1 Cells
   - 3.2. Fraction 2: Larger G_1 Cells
   - 3.3. Fraction 3: Populations Enriched in S-Phase Cells

4. HISTONE SYNTHESIS IN G_1 CELLS

5. EVALUATION OF HISTONE MESSENGER RNA LEVELS

6. REGULATION OF HISTONE SYNTHESIS IN HeLa CELLS

7. DISCUSSION

REFERENCES

1. INTRODUCTION

The growth and division of nucleated cells is marked by biochemical and morphological events that occur in a distinct temporal pattern. This makes possible analysis of the cycle as a succession of discrete phases. One widely used scheme considers the cycle to be constituted by four phases: mitosis (M); gap one, preceding the period of nuclear DNA synthesis (G_1); synthesis of DNA (S); and gap 2 (G_2). The duration of G_1 varies among different cell types (1) and can be extended and contracted experimentally (2, 3). The commitment to initiate DNA synthesis probably takes place in the G_1 phase of the cycle. This decision is generally assumed to be the critical event in the cell cycle because, once committed to replication, a cell is usually required to complete the growth cycle and undergo mitosis (4). Therefore, much attention has been given to G_1-specific cellular events in the hope of uncovering regulatory signals.

G_1 phase itself can be temporally divided in a variety of ways. In yeast, analysis of the many cell-division-cycle (cdc) mutants has revealed that G_1 phase is composed of a temporal sequence of genetically separable events (5). The cellular landmarks used to distinguish and order these events are, sequentially, formation of the spindle pole body satellite, duplication of the spindle pole body, and separation of the spindle pole body (5). Distinct markers of temporal progression through the mammalian G_1 phase are less apparent, although parallels with yeast exist. Specifically, duplication of the centrosome, the mammalian equivalent of the spindle pole body, begins in Chinese hamster ovary cells in late G_1 and is independent of DNA synthesis (6). Progression through G_1 can be mapped in mammalian cells by the technique of premature chromosome condensation. When somatic hybrids are formed between mitotic and interphase cells, the degree of condensation induced in the interphase chromosomes decreases systematically as the cells bearing them progress through G_1 phase (7). Cycloheximide and actinomycin D, inhibitors of G_1, traverse block this progressive chromosome condensation (7). These experiments suggest that the G_1 compartment is not a uniform entity but instead represents a progression of specific morphogenetic events.

It is widely assumed by analogy with differentiating systems (8) that differential gene expression is associated with a cell's progression through the cycle and ultimately controls that progression. The expression of most gene products is not periodic (9, 10). Specific examples of periodic gene expression therefore may aid in elucidating the regulators of the growth cycle. The histones have been extensively examined from this perspective.

In certain yeast and mammalian cells, histone synthesis appears to be coincident with DNA synthesis (reviewed in Pederson, 11). This has been shown in several ways. The early experiments of Borun et al. showed that mitotically selected G_1 phase HeLa cells lacked a 7-9S RNA species found on the polysomes of S phase cells (12). This S phase-specific RNA was deduced to code for the histone proteins (13). Subsequent studies showed that this same size class of RNA was very labile when HeLa cells were treated with agents that inhibited DNA synthesis (14, 15). Moreover, recent work by Stein et al., using cloned histone genes as probes, confirmed the correlation found in earlier studies (16). Hereford et al., using cloned histone genes, have shown that G_1-arrested yeast cells have greatly diminished amounts of histone mRNA as compared to growing yeast (17). In the same study, unbudded yeast also had low amounts of histone mRNA as compared to budded cells in the same culture. Since the relationship between budding and cell cycle position was not examined, it was not possible to determine whether DNA synthesis was an absolute prerequisite for histone synthesis in these cells. In general, these and related studies have led to the prevailing view that histone gene expression is coordinate with DNA replication; that is, in the absence of DNA synthesis there is an insignificant amount of histone synthesis.

In contrast to this point of view, there are several good examples that show the continuation of histone synthesis in the absence of DNA synthesis. Oocytes of amphibians such as *Xenopus laevis* are quiescent with respect to DNA synthesis yet make and store 10,000 genome equivalents of histone protein (18). Histones are made in mature, nondividing sea urchin eggs, and expression is controlled at both the transcriptional and translational level (19). Furthermore, sea urchin eggs that have been activated parthenogenetically continue to synthesize histones at comparable rates before and after enucleation (20, 21). Some (22, 23) but not all (24) classes of G_1-arrested, temperature-sensitive cell cycle mutants in mammalian cells continue to synthesize histone proteins at similar rates when shifted from permissive temperatures to temperatures that restrict DNA synthesis. It has been shown repeatedly that inhibitors of DNA synthesis also inhibit but do not eliminate histone gene expression. Histones synthesized in the presence of inhibitors of DNA synthesis cause the rapid turnover of otherwise stable chromatin-bound histones (25, 26). These and other studies indicate that cells can synthesize histones in the absence of DNA synthesis. In this chapter we present our data on histone gene expression in HeLa and S49 cells (27) and consider the relationship between histone and DNA synthesis in cycling animal tissue culture cells.

2. GENERAL CONSIDERATION OF HISTONE S-PHASE COUPLING

There is wide agreement in the literature that cultured animal cells regulate their histone synthesis in accord with their cell cycle. The question of whether G_1 cells synthesize histones at all and, if so, to what degree is unsettled. The most uncompromising model, one that may be termed "strong coupling," declares that the onset of histone and DNA synthesis are simultaneous. "Weak coupling" models allow for the possibility that histone synthesis may begin at a substantial rate before the onset of DNA synthesis. Such a view is consistent with the possibility that onset of histone expression is deterministic, for example, occurs at some fixed time before DNA synthesis begins or, alternatively, is stochastic, for example, occurs with increasing probability as postmitotic cells age.

To rule out the strong coupling model, it suffices to display a G_1 cell population that makes histones at a high rate. Because of imperfections of cell synchrony methods, it is inevitable that G_1 populations will to some degree suffer contamination with S phase cells. It is therefore essential to demonstrate that the latter cannot account for the histones made by the G_1 population. Loose coupling therefore predicts the following: Less mature G_1 cells should synthesize less histones than more mature G_1 cells, and some G_1 cells should synthesize as much as S phase cells. Furthermore, the average G_1 and S phase cell should differ in their histone synthetic rate. The magnitude of this difference could well differ among cell lines, or within a given line when cells grow at different rates. For example, if histone synthesis begins one hour before S phase, the average G_1-S differential in rate of histone synthesis will be greater for cells whose mean duration of G_1 is 10 hours than for those whose mean duration is 2 hours.

Discrimination between strong and weak coupling models is a challenging task, demanding pure and defined cell populations, and scrupulous quantitative determination of biochemical parameters reflecting histone synthetic rates. To what degree is this a worthwhile endeavor? We believe significantly so, because temporal relationships among biochemical events may provide significant clues to mechanism, and thus guide further experimentation. We present here our evidence that supports a loose coupling model.

3. SYNCHRONIZED S49 CELLS

S49 is a mouse T lymphoma line (28). The cells grow in suspension culture and have the morphology of small lymphocytes. They have a 16-hour generation time, with mean durations for G_1, S, and G2 + M of approximately 2, 12, and 3 hours, respectively (29). To obtain populations synchronized with respect to cell cycle phase, we use centrifugal elutriation, a method that fractionates cells on the basis of size (30). The nature of the relationship between cell size and DNA content (29) makes it possible to use this method to generate acceptably pure G_1 populations and less pure but substantially enriched S phase populations. To evaluate synchrony, we use two methods, flow cytometry (FCM) and autoradiography of cells pulse-labeled with ^{3}H-thymidine (31). When a culture in asynchronous growth is subjected to centrifugal elutriation, successive fractions of cells of increasing size are obtained. By appropriate pooling of fractions, synchronous populations of distinguishable characteristics can be obtained.

3.1. Fraction 1: Small G_1 Cells

These comprise the class obtained by pooling the smallest cells in an exponential population, those that constitute some 3–6% of the cells. Their FCM profile shows an essentially pure G_1 population, but ^{3}H-thymidine autoradiography reveals 4–8% contamination with S phase cells. When these cells are returned to culture after fractionation, they exhibit a delay in synthesis of DNA. There is little change in the FCM profile for some 3 hours, then the cells begin to enter S phase with a population half-life of about 1.5 hours. Eventually almost all the cells enter S phase.

3.2. Fraction 2: Larger G_1 Cells

This population is generated by pooling the larger cells in succeeding fractions from the elutriator, some 10–20% of the initial unfractionated population. By FCM profile, these G_1 cells appear also to be very pure. When assessed by autoradiography, they may be contaminated by about 15% S phase cells. These larger G_1 cells exhibit no delay when returned to culture, but enter S phase promptly, with a half-life of about 1.5 hours. It is a reasonable presumption, although one not proved, that these fraction 2 cells, on average larger, closer to initiation of S phase, and somewhat more S phase contaminated, have had either a longer duration of postmitotic maturation than the average fraction 1 cell, or have arisen as daughters of larger cells.

3.3. Fraction 3: Populations Enriched in S Phase cells

As one continues to collect larger cells, populations progressively enriched in S phase are obtained. FCM and ^{3}H-thymidine autoradiographic analysis show that these populations contain up to 70–80% S phase cells.

Figure 13.1 illustrates the size distribution, and Figure 13.2 the FCM profiles of unfractionated asynchronous S49 cells (*A*) and synchronized elutriated populations (*B–D*), obtained as described. By FCM analysis, fraction 1 (Figure 13.2*B*) and fraction 2 G_1 cells (Figure 13.2*C*) appear identical. As noted, however, they are distinguishable when the degree of S phase contamination is measured by ^{3}H-thymidine autoradiography, and they differ in their growth characteristics. As can be shown, they differ too in their capacity for histone synthesis. Fraction 3 cells (Figure 13.2*D*) are predominantly in late cycle.

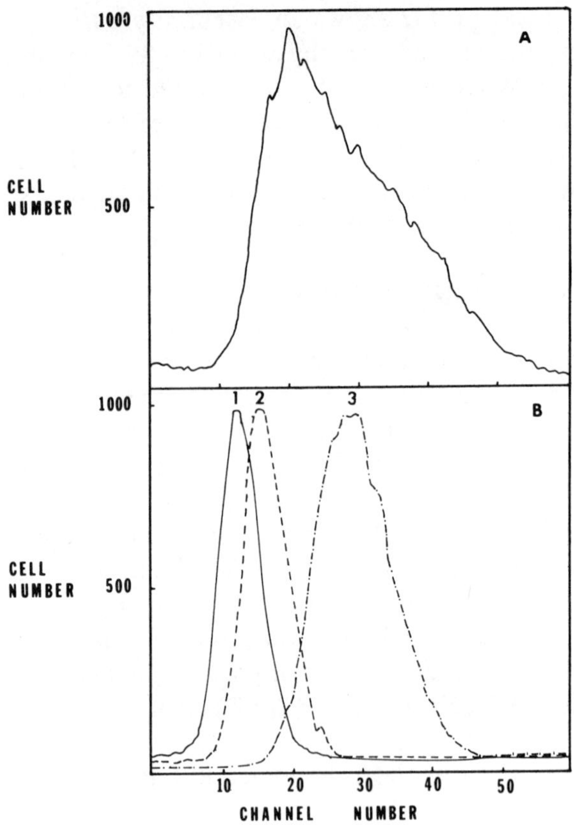

FIGURE 13.1. Coulter volume distributions of S49 cells. A population in exponential growth (*A*). Three fractions of S49 cells separated by centrifugal elutriation (*B*). Fraction 1 = smallest G_1 phase cells; Fraction 2 = next largest G_1 phase cells; Fraction 3 = S phase.

By thus fractionating cells on the basis of size, one obtains populations with distinct properties. Because cell size is not a precise index of cell properties, and because cells, even within a cloned tissue culture line, are heterogeneous with respect to their cell cycle parameters, this methodology will generate populations with less than perfect synchrony. For example, to obtain a well-defined fraction 2 G_1 population, one must steer an appropriate course between the Scylla of G_1 cells that are slow to enter S, highly represented in G_1

fraction 1, and the Charybdis of S phase contamination. Our investigations of histone synthesis in G_1 cells were initially limited to the study of fraction 2 populations. Our rationale for avoiding the fraction 1 cells at that time was the fear that the smallest cells represented an aberrant population of low abundance that were damaged or otherwise unrepresentative and not progressing toward S phase.

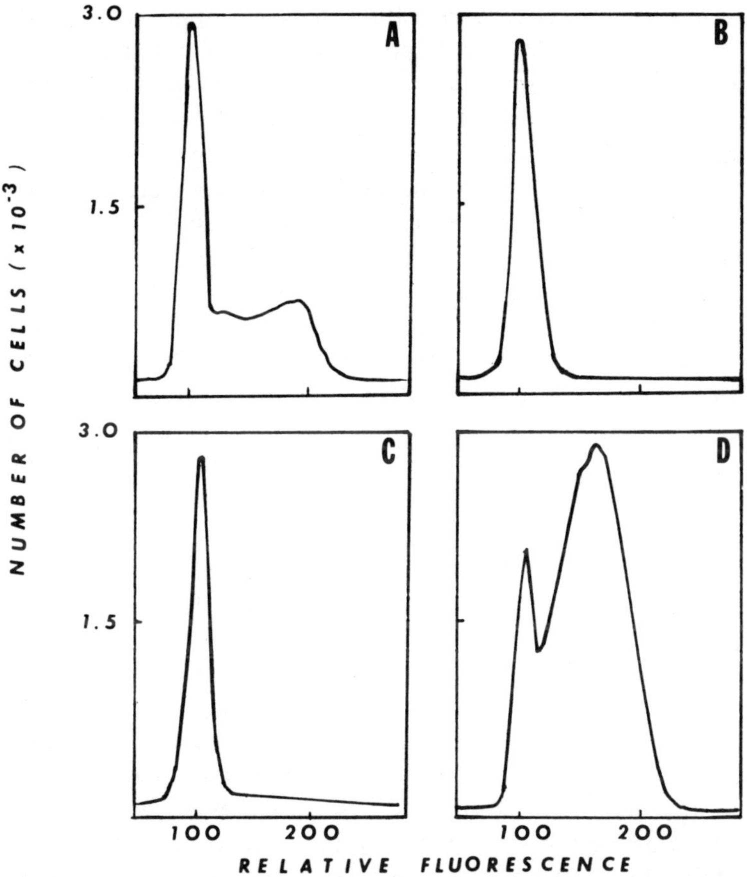

FIGURE 13.2. DNA content per cell histograms of S49 cells stained with chromomycin A_3. The populations are identical to those shown in Figure 13.1. Exponential (*A*), fraction 1(*B*), fraction 2(*C*), fraction 3(*D*). DNA content is proportional to fluorescence, expressed in arbitrary units and normalized so that G_1 phase cells have a modal fluorescence of 100. Autoradiography after a 10-min incubation with ^{3}H-thymidine (10 μCi/ml) showed the S phase content of the populations to be 59%, 5%, 13%, and 85%, respectively.

4. HISTONE SYNTHESIS IN G_1 CELLS

The relative rates of histone synthesis in synchronous and asynchronous populations were evaluated by pulse-labeling cells with ^{3}H-lysine or ^{35}S-methionine and analyzing extracts on nonequilibrium pH gradient electrophoresis (NEPHGE) gels. In most experiments analysis was performed on whole cell extracts, using for solubilization a lysis buffer containing detergents, high salt, and protamine (32). Under these conditions, recovery of cellular histones approaches 100% regardless of subcellular localization or state of assembly (32). G_1 fraction 2 and exponentially growing asynchronous cells were separately labeled for 15 min with ^{35}S-methionine or ^{14}C-lysine. Samples containing equal amounts of radioactivity were analyzed on NEPHGE gels (Figure 13.3). The positions of the histones were identified by similarly analyzing purified S49 histones. Mouse histone H1 is unlabeled and H2A is poorly labeled with ^{35}S-methionine. Although the exponential populations contained at least four times more S phase cells than did the G_1 populations, the intensity of labeling of the hisone polypeptides was indistinguishable. The equivalence in synthetic rates persisted, even when the elutriated G_1 cells were further fractionated by flow cytometry to reduce their degree of S phase contamination. These experiments imply that histone synthesis can occur at a substantial rate in G_1 phase S49 cells.

In experiments of the kind reported in Figure 13.3, histone synthetic rates were compared by analyzing equal amounts of radioactive protein and exposing autoradiograms for equal times. If the S phase contaminants in G_1 cell populations incorporate label at a significantly elevated rate, these alone could account for the histone synthesis. To examine the relationship between DNA content and the rate of incorporation of labeled precursors into protein and DNA, exponentially growing asynchronous S49 cells were pulse-labeled simultaneously with ^{35}S-methionine and ^{3}H-thymidine. Cells were stained with acriflavine and separated by flow cytometry on the basis of their DNA content into 12 fractions. ^{3}H-thymidine incorporation occurred predominantly in cells with S phase DNA content (27). ^{35}S-methionine incorporation, on the other hand, was nonperiodic and displayed a smooth twofold increase between G_1 and G_2 + M phase cells (27). This implies that the increase in DNA synthesis at the beginning of S phase is not associated with a concomitant increase in the rate of protein labeling.

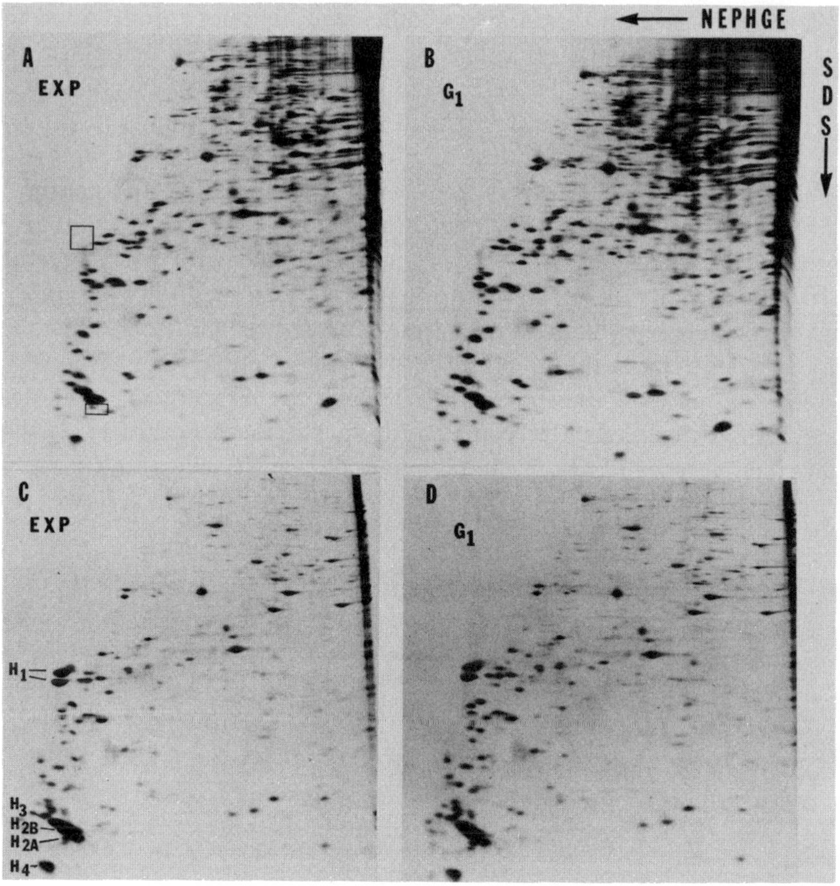

FIGURE 13.3. Synthesis of histones and other basic proteins during the S49 cell cycle. S49 cells were fractionated by centrifugal elutriation. The G_1 fraction 2 and unfractionated exponential culture were labeled for 15 min with either 100 μCi/ml ^{35}S-methionine or 40 μCi/ml ^{14}C-lysine. Whole cell lysates were prepared by solubilization in histone lysis buffer. Aliquots containing either 1.75×10^6 cpm of ^{35}S-methionine labeled exponential or G_1 elutriated cells were electrophoresed in two dimensions and visualized by autoradiography. (*A*) G_1 and (*B*) asynchronous populations labeled with ^{35}S-methionine. (*C*) G_1 and (*D*) asynchronous populations labeled with ^{14}C-lysine. The positions of ^{14}C-lysine labeled histone H1 and the nucleosomal histones are indicated in (*C*). Histone H1 was unlabeled and H2A poorly labeled with ^{35}S-methionine (*A*, *B*). The two boxes in (*A*) enclose the area of the autoradiogram normally occupied by these two histones. From Reference 27, with permission.

Studies with asynchronous cells or synchronized S phase cells have demonstrated that newly synthesized histones rapidly bind to DNA. The mechanism that drives this transport from cytoplasm to nucleus is unknown. To determine whether transport is associated with DNA synthesis, we performed subcellular fractionation experiments to ask whether the time of appearance in the nucleus of newly synthesized histones differs in G_1 and S phase cells. G_1 fraction 2 elutriated cells, S phase enriched elutriated cells, and exponentially growing asynchronous cells were pulse-labeled with ^{3}H-lysine for 10 min and chased for an additional 10 min in medium containing excess unlabeled lysine. Nuclei were prepared, solubilized in histone lysis buffer, and aliquots representing equal numbers of nuclei were analyzed on two-dimensional NEPGHE gels, as in Figure 13.3. Whole-cell lysates were similarly analyzed. As shown in Figure 13.3, the histone patterns of G_1, S, and exponential phase whole-cell lysates were not distinguishable. The pattern of nuclear histones, however,

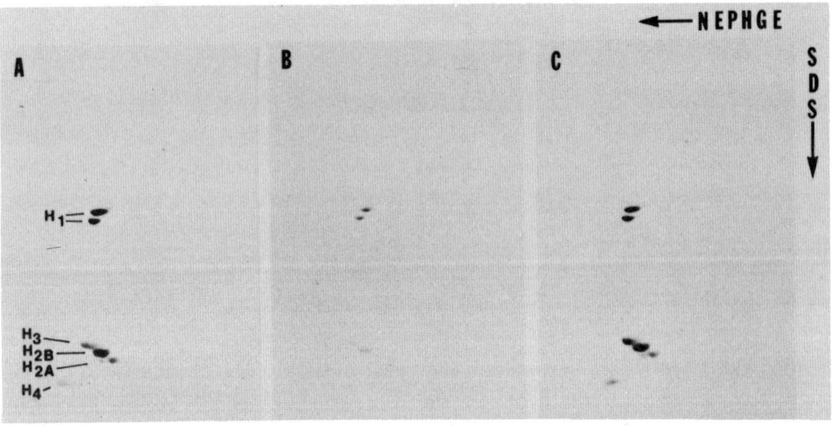

FIGURE 13.4. Newly synthesized histones in G_1 and S phase nuclei. G1- and S-elutriated populations, as well as nonelutriated exponentially growing S49 cells, were pulse-labeled with 100 μCi/ml ^{3}H-lysine for 10 min and chased for an additional 10 min by incubation in conditioned DME + 10% horse serum to which 0.8 mM lysine had been added. Nuclei were prepared from each fraction and solubilized in histone lysis buffer. Aliquots containing 7×10^5 nuclei from exponential (*A*), G_1-(*B*) or S-elutriated cells (*C*) were separated on two-dimensional gels. The distribution of labeled polypeptides was determined by fluorography. The positions of the histones are indicated in (*A*). From Reference 27, with permission.

differed strikingly (Figure 13.4). Nuclear histones in the G_1 population were much less labeled than those in the S phase enriched or exponential populations. Microdensitometry of the autoradiograms revealed that nuclei prepared from G_1 phase cells contained less than 13% of the labeled core histones present in nuclei of S phase or exponential cells. Similar comparative analysis of whole-cell lysates and nuclear fractions showed, under the labeling and fractionation conditions used, that in S phase cells 87% of labeled core histones appeared in the nucleus, but that in G_1 cells only 18% did so. These results indicate that histones made in G_1 cells do not associate rapidly with the nucleus. Because G_1 cells transport newly synthesized histones differently from S phase cells, these experiments serve further to rule out the possibility that histone synthesis in G_1 populations is entirely attributable to contamination with S phase cells.

5. EVALUATION OF HISTONE MESSENGER RNA LEVELS

To assess by independent means the results obtained in our studies of histone protein synthesis, and to extend our understanding of the regulatory mechanisms involved, the level of histone messenger RNA was measured in synchronized cell populations. This was done in two ways: by *in vitro* translation of extracted RNA and by quantitative RNA dot hybridization (33) using a cloned histone H3.2 DNA (34). Whole-cell RNA was extracted (35) from fraction 2 G_1 cell populations and from asynchronous populations. These were separately translated in a reticulocyte lysate translation system containing ^{35}S methionine. The radioactive polypeptides were analyzed on NEPGHE gels as in Figure 13.3. Once again, the patterns were similar (Figure 13.5). The two messenger RNA preparations direct the synthesis of a very similar array of polypeptides, including the histones.

RNA dot hybridization to a cloned histone H3.2 DNA was used to assess the relative amounts of hybridizable RNA in synchronized cells. In this procedure, RNA is extracted from the cells and applied to nitrocellulose paper under conditions that promote binding. The filters are allowed to hybridize with ^{32}P-labeled cloned probe, and binding is evaluated by autoradiography. By applying known amounts of RNA in a serial dilution series, the relative amounts of hybridizable

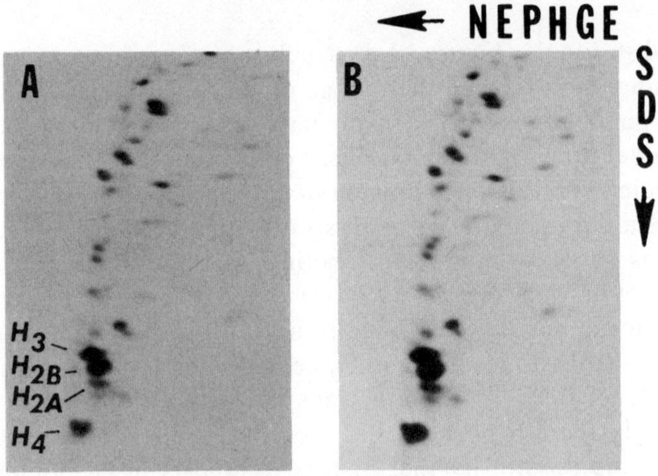

FIGURE 13.5. *In vitro* translation of total cell RNA from G_1 and S phase S49 cells. Total cell RNA was prepared from G_1 fraction 2 and S elutriated populations of S49 cells. Approximately 25 μg of RNA from G_1 (*A*) and S (*B*) phase cells were added to a rabbit reticulocyte *in vitro* translation assay. G_1 and S phase RNA had approximately the same translational efficiency. Aliquots of the *in vitro* translation products were analyzed on two-dimensional gels. The first dimension was a 4% NEPHGE gel, and the second dimension was a 12.5% SDS gel. The nucleosomal histones are indicated by H.

RNA in different cell preparations can thus be determined. Figure 13.6 shows the results of such an experiment, in which RNA extracts from G_1 fraction 2 and S phase cell populations were compared. The S phase content of these populations, assessed by ^{3}H-thymidine autoradiography, was 17 and 63%, respectively. Visual inspection, confirmed by microdensitometric scanning, shows both preparations to be equally rich in RNA specifically hybridizing to the probe. Also shown are the results of a similar analysis of RNA extracted from the identical cell populations that had been treated previously for 1 hour with hydroxyurea after elutriation. That treatment reduced the abundance of hybridizable sequences approximately tenfold in both populations. Hydroxyurea is believed to act by perturbing deoxyribonucleotide pools, thus preventing the progression of cells through S phase (36, 37). Treatment with this agent has been shown to reduce the histone synthetic rate and abundance of mRNA histone pools in several tissue culture lines (14, 15, 38). These results have been taken to imply the S phase specificity of histone synthesis. Interpreted in this way, our result with hydroxyurea implies that in both cell populations S phase cells were responsible for the presence of histone mRNA. Such an interpretation may, however, be untenable. Hydroxyurea's effects on cells are not well defined and may not be limited to ribonucleotide pool perturbation. Even if the effects are so limited, it is difficult to rule out the possibility that these pools exert regulatory effects on histone messenger levels in G_1 cells. In addition to its effects on histone mRNA pools, we have shown, as have others, that hydroxyurea reduces the rate of histone protein synthesis in both asynchronous and synchronous cell populations, when synthesis is evaluated by pulse-labeling experiments and NEPHGE gel analysis (results not shown).

Similar RNA dot hybridization analysis was applied to elutriated S49 cell populations collected in such a way as to encompass, among separate fractions, a range from the smallest to the largest class of cells. The result of such an experiment is shown in Figure 13.7, along with the percentage of S phase cells in each fraction, as determined by ^{3}H-thymidine autoradiography. This analysis revealed that the smallest class of cells had approximately one-tenth as much

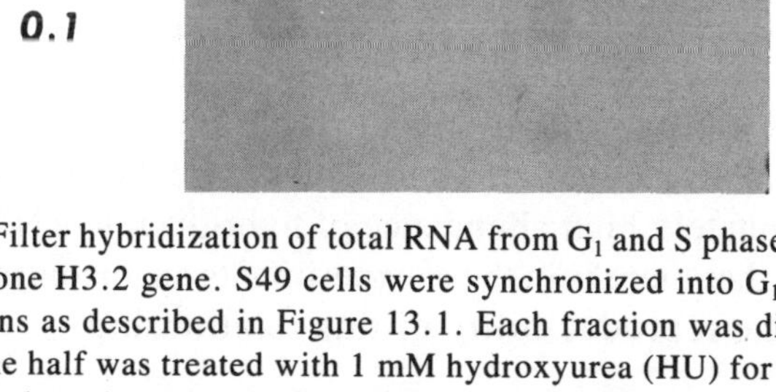

FIGURE 13.6. Filter hybridization of total RNA from G_1 and S phase S49 cells with the mouse histone H3.2 gene. S49 cells were synchronized into G_1 fraction 2 and S phase fractions as described in Figure 13.1. Each fraction was divided into two equal parts. One half was treated with 1 mM hydroxyurea (HU) for 30 min and the other half served as the control. Total RNA was separately prepared from each culture and dotted onto nitrocellulose filters in a threefold dilution series, using 1 µg RNA as the maximum amount in each lane. The filters were hybridized with cloned mouse histone H3.2 gene. Lane 1 = G_1 RNA (17% S); lane 2 = G_1 + HU RNA; lane 3 = S RNA (63% S); lane 4 = S + HU RNA.

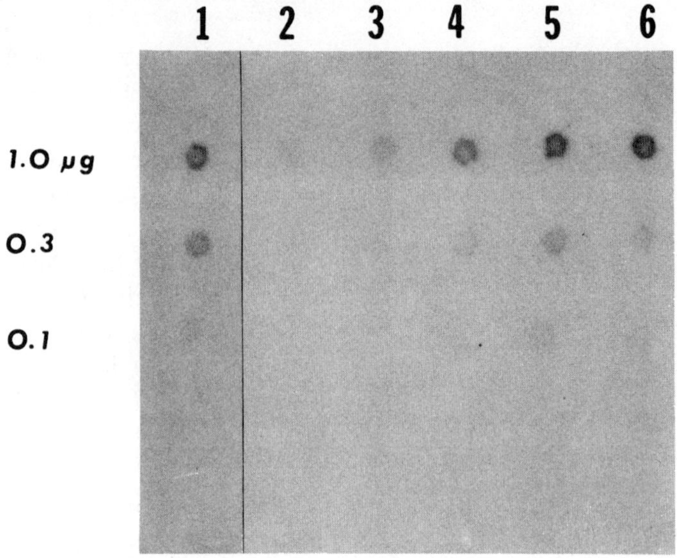

FIGURE 13.7. Filter hybridization of total RNA from elutriated S49 cells with the mouse histone H3.2 gene. S49 cells were separated into five size fractions by centrifugal elutriation. Total RNA was prepared from each fraction, as well as the starting exponential culture. The RNA was dotted onto nitrocellulose and hybridized as described in Figure 13.6. Lane 1 = asynchronous starting culture (61% S); lane 2 = fraction 1 (7% S); lane 3 = fraction 2 (27% S); lane 4 = fraction 3 (58% S); lane 5 = fraction 4 (87% S).

hybridizable RNA as the largest class. Furthermore, quantitative analysis indicated that, in this experiment, the histone RNA levels correlated well with the fractional S phase content. Clearly, this result contradicts that shown in the preceding figure. Among nine experiments in which synchronized cell populations were evaluated by both RNA dot hybridization and ^{3}H-thymidine autoradiography, most have demonstrated such a correlation, and the others have shown a disproportionately high abundance of hybridizable RNA in G_1 cells. It is apparent that there is in these experiments an uncontrolled variable that affects the reproducibility of this relationship.

6. REGULATION OF HISTONE SYNTHESIS IN HeLa CELLS

The mechanism of histone synthesis and its timing in the cell cycle has been extensively studied in HeLa cells (12–16). These cells have a 24-hour generation time, of which G_1 phase comprises about half. The same methodology used to analyze histone synthesis in S49 cells was applied to HeLa. Cells were fractionated by centrifugal elutriation. The smallest cells in the first fraction were 96–98% pure G_1, as determined by ^{3}H-thymidine autoradiography. After 12 hours of further culture, 80% of this population was making DNA by the same criterion. The second elutriation fraction contained slightly larger cells, of which 10% were already in S. These were about 2 hours ahead of the first fraction in their progression through the cycle, as determined by Coulter Channelyzer size analysis, ^{3}H-thymidine autoradiography, and FCM. To determine the rate of histone synthesis during G_1 and S, these elutriated populations were labeled for 20 min with ^{35}S-methionine or ^{3}H-lysine immediately after fractionation or after 12 hours of further culture. Extracts containing equal amounts of acid-insoluble radioactivity were analyzed by NEPHGE gel electrophoresis. Figure 13.8 shows autoradiograms from one such experiment. The spots corresponding to the core histones and H1 were cut out and their radioactivity measured by scintillation counting. The histones made by cells in G_1 contained 20% as much radioactivity as those made by S phase populations.

Figure 13.9 shows a more elaborate experiment in which RNA was prepared from both the first and second elutriated G_1 fractions

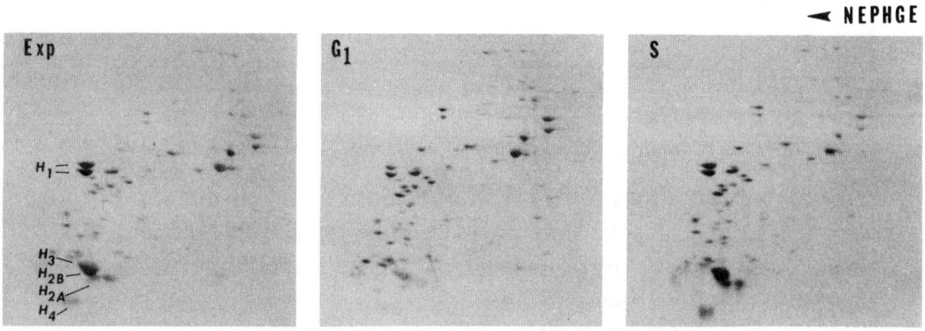

FIGURE 13.8. Histone synthesis in G_1 and S phase HeLa cells. Exponentially growing HeLa cells were fractionatd to obtain a G_1 elutriated population and chased for 12 hours into S phase. The G_1 population contained at most 2% S phase cells; the S phase population contained approximately 80% S phase cells. Both synchronous cultures as well as the asynchronous starting cultures were labeled for 20 min with 200 μCi/ml ^{3}H-lysine. Histone synthesis was analyzed as in Figure 13.3.

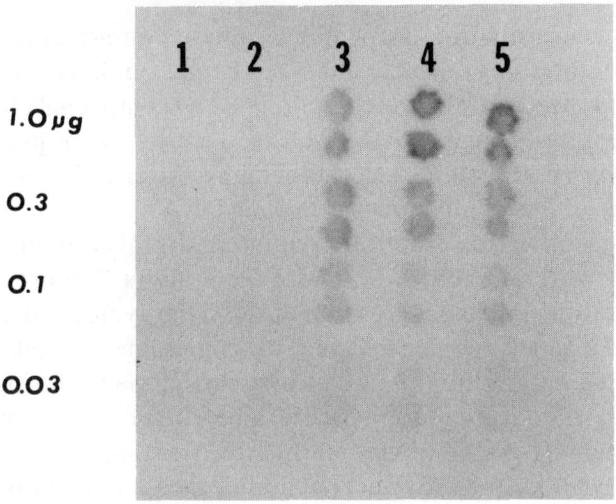

FIGURE 13.9. Filter hybridization of total RNA from elutriated HeLa cells with the mouse histone H3.2 gene. HeLa cells were synchronized as in Figure 13.8. Total RNA was prepared, dotted on nitrocellulose, and hybridized as described in Figure 13.6. Lane 1 = fraction 1 G_1, 4% S; lane 2 = fraction 2 G_1, 11% S; lane 3 = fraction 1 G_1 14 hours later, 58% S; lane 4 = fraction 2 G_1 14 hours later, 38% S; lane 5 = exponentially growing cells 14 hours later, 34% S.

and from exponentially growing cells immediately after elutriation and 14 hours later. Equal amounts of RNA from each preparation were analyzed by RNA dot hybridization, using the mouse histone H3.2 probe. The H3 histone mRNA content is approximately proportional to the fraction of cells in S phase; G_1 cells immediately after elutration contain less than one-tenth as much histone mRNA per total cell RNA as the same cells 14 hours later. Control experiments using a radioactive actin probe in place of the histone probe demonstrated that there were no significant cell-cycle-dependent changes in the content of actin mRNA (results not shown).

These results confirm the recent work of Stein and colleagues (16). As in S49 cells, G_1 cells make less histones and have smaller histone mRNA pools than S phase cells. The resolution of these experiments, however, does not exclude the initiation of histone synthesis in late G_1.

7. DISCUSSION

The preponderance of evidence suggests that, in S49 mouse lymphoma cells, the rate of core histone synthesis is low following mitosis, but attains the rates found in S phase before the initiation of DNA synthesis. We have less information regarding the regulation of H1 histones because most of our experiments have been performed using ^{35}S-methionine, which does not label the H1 histones in mouse cells, or with a histone H3 DNA probe. These results are not inconsistent with those obtained in certain animal cell lines, but are consistent with the demonstration that in the developing oocytes of frog and sea urchin large amounts of histones are synthesized and accumulated in the absence of DNA synthesis. The finding that histone synthetic rates are low following mitosis is based on the observation that when very small cells are purified from asynchronous populations, their histone synthetic rates, measured by pulse-labeling experiments and NEPHGE two-dimensional gel analysis, are low when normalized for overall amino acid incorporation. In addition, the histone messenger pools are low in these small G_1 cells, based on RNA dot hybridization using the mouse histone H3.2 probe.

S49 and HeLa cells in G_1, however, can make histones, and larger, presumably older, G_1 phase S49 cells do so. We base this conclusion on the following experimental evidence: the rate of histone synthesis in such G1 populations is greater than can be accounted for on the basis of the demonstrable contamination with S phase cells. For such an explanation to hold for S49 cells, the S phase contaminants would be required to synthesize histones at a rate up to four times that found in other cell populations. However, the protein synthetic rate in S49, as in other cells, is roughly proportional to cell size. Last, G_1 S49 cells synthesize histones in a qualitatively distinct fashion. Newly synthesized histones in S phase fractionate with the nucleus within minutes. They do so in G_1 cells only after the onset of DNA synthesis. It has recently been shown that in yeast the onset of histone gene transcription and histone mRNA accumulation occurs in late G_1, prior to DNA synthesis (39). It is reasonable to assume that the newly synthesized histone proteins in this system behave analogously to those described in G_1 phase S49 cells.

Results using RNA dot hybridization to analyze histone H3 message levels are more equivocal. In certain experiments there is an equivalence of messenger levels among cell populations widely divergent in their S phase content. In other experiments there is near proportionality between these two variables. This suggests a degree of independence of cell cycle position and capacity for histone synthesis; the basis of this variability remains to be defined experimentally.

We find, in confirmation of the work of others, that hydroxyurea reduces histone synthesis rate and H3 messenger pool size. As noted, this can be interpreted to provide evidence that histone synthesis is coordinated with DNA synthesis. We believe this interpretation to be premature until the biochemical effects of hydroxyurea are better understood.

Treatment of S49 and HeLa cells with inhibitors of protein synthesis induces prompt and marked increases in histone H3 mRNA pool size (41). We do not yet know whether this results from increased synthetic rate or stabilization of mRNA. We also do not know the identity of the protein or proteins that are presumably the target of this inhibition and whose depletion results in increases in mRNA pools. The effect is highly specific with respect to histone messenger. *In vitro* translation of messenger from cycloheximide-treated cells demonstrates that the core histones, but not other proteins demonstrable on NEPHGE gels, are enhanced. One might imagine the histones themselves to be the target proteins. Perhaps free histones have a negative autoregulatory effect on expression of histone genes (40). If so, the initiation of DNA synthesis, by providing a sink for free histones, could initiate transcription of their genes. Such a model would demand concurrent synthesis of DNA and histones. However, recent reports indicate the existence of highly acidic proteins that bind specifically to histones (42, 43). In further extension of our speculation, one may imagine in this autoregulatory model that it is these acidic proteins that control the level of free histones and thus their cell-cycle-dependent expression.

REFERENCES

1. A. P. Pardee, R. Dubrow, J. L. Hamlin, and R. F. Kletzien, *Ann. Rev. Biochem.* **47,** 715 (1978).

2. G. M. Stancel, D. M. Prescot, and R. M. Liskay, *Proc. Natl. Acad. Sci. USA* **78,** 6295 (1981).
3. R. F. Brooks, *Cell* **12,** 311 (1977).
4. J. A. Smith and L. Martin, *Proc. Natl. Acad. Sci. USA* **70,** 1263 (1973).
5. J. R. Pringle and L. H. Hartwell, in J. N. Strathern, E. W. Jones, and J. R. Broach, Eds., *The Molecular Biology of Yeast Sacchromyces*, Cold Spring Harbor Laboratory, New York, 1981, p. 97.
6. R. Kuriyama and G. G. Borisy, *J. Cell. Biol.* **91,** 814 (1981).
7. W. N. Hittelman and P. N. Rao, *J. Cell. Physiol.* **95,** 333 (1978).
8. E. H. Davidson, *Gene Activity in Early Development*, 2nd ed., Academic, New York, 1976.
9. R. Bravo and J. E. Cellis, *J. Cell. Biol.* **84,** 795 (1980).
10. S. G. Elliot and C. S. McLaughlin, *Proc. Natl. Acad. Sci. USA* **75,** 4384 (1978).
11. T. Pederson, in E. H. McConkey, Ed., *Protein Synthesis*, Vol. 2, Marcel Dekker, New York, 1976, p. 69.
12. T. W. Borun, M. D. Scharff, and E. Robbins, *Proc. Natl. Acad. Sci. USA* **58,** 1977 (1967).
13. E. Robbins and T. W. Borun, *Proc. Natl. Acad. Sci. USA* **57,** 409 (1967).
14. D. Gallwitz and G. C. Mueller, *J. Biol. Chem.* **244,** 5947 (1969).
15. D. Gallwitz, *Nature* **257,** 247 (1975).
16. R. Rickles, F. Marashi, R. Sierra, S. Clark, J. Wells, J. Stein, and G. Stein, *Proc. Natl. Acad. Sci. USA* **79,** 749 (1982).
17. L. M. Hereford, M. A. Osley, J. R. Ludwig II, and C. S. McLaughlin, *Cell* **24,** 367 (1981).
18. E. D. Adamson and H. R. Woodland, *J. Mol. Biol.* **88,** 263 (1974).
19. B. P. Brandhorst, *Dev. Biol.* **79,** 139 (1980).
20. D. Nishioka and D. Mazia, *Cell Biol. Int. Rep.* **1,** 23 (1977).
21. R. J. Arceci and P. R. Gross, *Proc. Natl. Acad. Sci. USA* **74,** 5016 (1977).
22. K. Gooderham and P. M. Naha, *Cell Biol. Int. Rep.* **5,** 1105 (1981).
23. R. Sheinin and P. N. Lewis, *Somatic Cell Genet.* **6,** 225 (1980).
24. A. M. Delegeane and A. S. Lee, *Science* **215,** 79 (1982).
25. L. R. Gurley and J. M. Hardin, *Arch. Biochem. Biophys.* **130,** 1 (1969).
26. G. Russev, L. Vassilev, and R. Tsanev, *Nature* **285,** 584 (1980).
27. V. E. Groppi and P. Coffino, *Cell* **21,** 195 (1980).
28. K. Horibata and A. W. Harris, *Exp. Cell. Res.* **60,** 61 (1970).
29. P. Coffino and J. W. Gray, *Cancer Res.* **38,** 4285 (1978).
30. M. L. Meistrich, R. E. Meyn, and B. Barlogie, *Exp. Cell. Res.* **105,** 169 (1977).

31. J. W. Gray and P. Coffino, in *Methods in Enzymology*, Vol. 48, Academic, New York, 1979, p. 233.

32. M. M. Sanders, V. E. Groppi, and E. T. Browning, *Anal. Biochem.* **103,** 157 (1980).

33. P. S. Thomas, *Proc. Natl. Acad. Sci. USA* **77,** 5201 (1980).

34. D. S. Sittman, I. Chin, C. Pan, R. H. Cohn, L. H. Kedes, and W. F. Marzluff, *Proc. Natl. Acad. Sci. USA* **78,** 4078 (1981).

35. J. M. Chirgwin, A. E. Przbyla, R. J. MacDonald, and W. J. Rutter, *Biochemistry* **18,** 52294 (1979).

36. Y. Engstrom, S. Eriksson, L. Thelander, and M. Akerman, *Biochemistry* **18,** 2941 (1979).

37. L. Skoog and G. Bjursell, *J. Biol. Chem.* **219,** 6434 (1974).

38. R. S. Wu and W. M. Bonner, *Cell* **27,** 321 (1981).

39. L. Hereford, S. Bromley, and M. A. Osley, *Cell* **30,** 305 (1982).

40. M. A. Osley and L. M. Hereford, *Cell* **24,** 377 (1981).

41. E. Stimac, U. E. Groppi, and P. Coffino, *Biochem. Biophys. Res. Commun.* **114,** 131 (1983).

42. W. C. Earnshaw, B. M. Honda, R. A. Laskey, and J. O. Thomas, *Cell* **21,** 373 (1980).

43. J. A. Kleinschmidt and W. W. Franke, *Cell* **29,** 799 (1982).

CHAPTER 14

CORE HISTONE VARIANTS OF THE MOUSE: Primary Structure and Differential Expression

ALFRED ZWEIDLER
The Institute for Cancer Research
The Fox Chase Cancer Center
7701 Burholme Avenue
Philadelphia, Pennsylvania

CONTENTS

1. HISTONE PROPERTIES, PRIMARY FUNCTION, AND EVOLUTION

2. RESOLUTION AND CHARACTERIZATION OF NONALLELIC FORMS OF CORE HISTONES
   - 2.1. Polyacrylamide Gel Electrophoresis in Presence of Nonionic Detergents (PAGEND)
   - 2.2. High Resolution Analysis of Core Histones
   - 2.3. Characterization of Individual Core Histone Components
     - 2.3.1. H2A Variants
     - 2.3.2. H2B Variants
     - 2.3.3. H3 Variants

3. DIFFERENTIAL EXPRESSION OF HISTONE VARIANT GENES
   - 3.1. Changes in Histone Variant Composition of Somatic Tissues during Development
   - 3.2. Histone Variant Metabolism in Differentiating Murine Erythroleukemia Cells
   - 3.3. Histone Variant Changes during Mouse Liver Regeneration
     - 3.3.1. Strictly Replication-Dependent Variants
     - 3.3.2. Partially Replication-Dependent Variants
     - 3.3.3. Replication Independent Variants
     - 3.3.4. Noncompeting Minor Histones
   - 3.4. Spermatocyte-Specific Changes in Histone Variants

4. GENETICS OF HISTONE VARIANT POLYMORPHISMS

5. SUMMARY, CONCLUSIONS, OUTLOOK

REFERENCES

1.—HISTONE PROPERTIES, PRIMARY FUNCTION, AND EVOLUTION

Histones are small, highly basic proteins that form strong and stable complexes with the DNA of all eucaryotes. They are the principal structural proteins of eukaryotic chromosomes. The primary function of histones is the packaging of DNA into compact structures without interfering with the necessary access to the genetic information and the exact distribution of replicated chromosomes during mitosis and meiosis. The chemical properties of histones and their molecular interactions have been reviewed by Isenberg (1) and McGhee and Felsenfield (2). There are two main classes of histones: (1) the *core histones*, consisting of four protein species called H2A, H2B, H3, and H4, which interact with each other in stoichiometric amounts to form an octameric complex around which the DNA is wound in two turns to form the nucleosome core, the basic structural unit chromatin, and (2) the *outer histones*, which bind to the DNA on the outside of the nucleosome core and between adjacent nucleosomes. The four core histones have probably evolved from a common ancestral DNA-binding protein such as the HTa protein of Thermoplasma acidophilum (3) or the HU protein of *E. coli* (4), at the time of evolution of the first eukaryotes. Although the amino acid sequences of the four core histones have diverged considerably, they can be shown to be related (5). More important, all four core histones have maintained a very similar overall architecture, with a highly basic and hydrophilic *N*-terminal region involved in DNA-histone interactions and a central hydrophobic region involved in specific interactions with other histones (Table 14.1; Reference 2).

H4 is one of the most highly conserved proteins known, differing in only two conservative changes between pea and calf (6). The amino acid sequence of *Tetrahymena* H4, however, has diverged considerably (7). H3 was also extremely conserved between pea and calf, with only four amino acid substitutions (8). However, H3 of yeast (9) shows 14 differences from calf H3. In vertebrates H3 has diverged into different nonallelic forms which are coexpressed in the same species (10). Two of these forms, H3.2 and H3.3, seem to have been completely conserved between fishes, birds, and mammals (10–13). A third form, H3.1, occurs in all mammals but not in other vertebrates (14). H2A and H2B evolve more rapidly than H3 and H4

TABLE 14.1. Primary Structure of Vertebrate Core Histones[a]

```
                                                                                                                     1         1         1         1
                   1         2         3         4         5         6         7         8         9         0         1         2         3
                   0         0         0         0         0         0         0         0         0         0         0         0         0

          p   a
            + +   + + + + +  +        +  +  ++    -+             -    -  -      +- ++ +   +      + ---  +   +                  ++ -   + + +
H2A      aSGRGKQGGKARAKAKTRSSRAGLQFPVGRVHRLLRKGNYAERVGAGAPVYLAAVLEYLTAEILELAGNAARDNKKTRIIPRHLQLAIRNDEELNKLLGKVTIAQGGVLPNIQAVLLPKKTESHHKAKGK-
NAV             TS V                                        M              G           T                                            E  V
AV             T         S                                  -     -    V   -  RIP      -        V           R                     D--- S AAK

              ap     a   a   a
              +     ++  ++   +  ++- +++++ ++-       +  +    -     +          -  -+   -  +     ++     +-     +     -  +    -  +   +    +
H2B      +PQPAKSAPAPKKGSKKAVTKAQKKDGKKRKRSRKESYSVYVYKVLKQVHPDTGISSKAMGIMNSFVNDIFERIAGEASRLAHYNKRSTITSREIQTAVRLLLPGELAKHAVSEGTKAVTKYTSSK-
NAV                     F          E R                                S              SQ
AV                --          TAG GD     A     AI                    -                S
```

```
                a     a    a     a     mp
      + +   ++     +  ++    +  ++         ++  + +      +- ++  +  -    ++    +  +-    - + - +             -  -           --          ++      +-     ++ + -+
H3    +ARTKQTARKSTGGKAPRKQLATKAARKSAPATGGVKKPHRYRPGTVALREIRRYQKSTELLIRKLPFQRLVREIAQDFKTDLRFQSSAVMALQEACEAYLVGLFEDTNLCAIHAKRVTIMPKDIQLARRIRGERA-
 NAV                                                                        A                IGG    S
  AV                                                                                                                                   T

       p  a  a   a   a  m
        + +  +   +   ++ ++  +-        +   ++  ++   ++        -- +   +    -    +-      -  +++              ++  +
H4    aSGRGKGGKGLGKGGAKRHRKVLRDNIQGITKPAIRRLARRGGVKRISGLIYEETRGVLKVFLENVIRDAVTYTEHAKRKTVTAMDVVYALKRQGRTLYGFGG-

                1       2       3       4       5       6       7       8       9       1
                0       0       0       0       0       0       0       0       0       0
                                                                                        0
```

[a] The amino acid sequences of the major component of the four core histones of calf thymus are given in the one-letter code, and the position of every tenth residue indicated by numbers and a dot above the residue. Charged amino acids (+, −), as well as the positions of post-translational modifications (a = acetylation, m = methylation, p = phosphorylation), are indicated above the sequence. Nonallelic (= NAV = coexpressed in inbred animals or all individuals) and allelic (= AV = not coexpressed) variation in the primary structure of vertebrate core histones are indicated below the sequence. The regions involved in histone–histone interactions (2) are underlined, with those regions that have a high probability to form helices (23) indicated by a wavy line. Amino acid sequence information for calf, trout H2A, H2B, carp H3, and so on, as reviewed by Isenberg (1), for chicken H2A, H2B, H3 (13, 24), for calf, mouse H2A, H2B, H3 variants (10, 20).

(1). In many organisms either H2A or H2B or both have diverged into different nonallelic forms (10, 13, 15–18). In contrast to the nonallelic forms of H3, those of H2A and H2B have not been conserved for long evolutionary periods (19; Zweidler, unpublished observation).

In Table 14.1 we have distinguished between sequence differences among nonallelic forms within the same species and among analogous (allelic) forms in different species. Most of the allelic variation occurs near the ends of the polypeptides, in regions involved in nonspecific histone–DNA interactions (2), whereas nonallelic variation seems to involve consistently the central regions of the polypeptides, which are involved in the formation of histone complexes (20). To a certain extent the latter correlation may be biased because of the methods necessary for detecting nonallelic forms.

Standard methods for resolving proteins detect differences in charge and size and are not able to distinguish proteins differing in neutral amino acid substitutions. Even primary structure differences in charged amino acids are difficult to detect in the case of the histones because of the high degree of charge variation caused by post-translational modifications. Nonallelic forms of histones coexpressed in the same species were therefore only detected after the development of a new electrophoretic technique that is able to separate proteins on the basis of hydrophobic properties (21–22). This method preferentially detects differences in the hydrophobic regions of proteins and is extremely sensitive for changes in regions that have a high probability to form α-helices. The amino acid substitutions known to change the hydrophobic properties of the histones are underlined in Table 14.1.

2. RESOLUTION AND CHARACTERIZATION OF NONALLELIC FORMS OF CORE HISTONES

2.1. Polyacrylamide Gel Electrophoresis in Presence of Nonionic Detergents (PAGEND)

The most useful technique for the detection, quantitation, and isolation of nonallelic forms of core histones is polyacrylamide gel electrophoresis in presence of nonionic detergents. When appropriate concentrations of nonionic detergents are added to acid–urea gel systems

(25, 26), the electrophoretic mobility of most proteins, including the core histones, is differentially reduced (21–22). The detergent effect is most likely due to mixed micelle formation between the detergent and the hydrophobic regions of the denatured proteins (22), since it occurs primarily above the critical micelle concentration of the detergent (Zweidler, unpublished observation). The protein–detergent interactions are highly dependent on the urea concentration (more so than the critical micelle concentration of the detergent), making it more efficient to vary the urea concentration than the detergent concentration to change the resolution of different components (22).

PAGEND has been shown to be able to resolve proteins differing in a single conservative substitution (10). Such extreme sensitivity to small changes in hydrophobic properties has its drawbacks. Thus if a protein has been exposed to oxidizing agents, PAGEND resolves the different forms produced by partial oxidation of methionine residues (22, 27). Since methionine sulfoxides in proteins are difficult to reduce, methionine oxidation must be prevented by keeping the protein fractions in presence of an excess of a reducing agent throughout isolation and storage (19, 20). In addition, polyacrylamide gels polymerized with an oxidizing catalyst, such as ammonium persulfate, contain a high level of reactive groups that can bind or modify proteins. For best results polacrylamide gels should be aged over night to reduce the number of reactive groups and to improve the uniformity of the gel matrix. Furthermore, the remaining persulfate ions should be removed from the gels by preelectrophoresis and the uncharged or immobile reactive groups inactivated by electrophoresis of a scavenging agent (for instance cysteamine or 2-methylthiopropylamine) ahead of the protein sample.

2.2. High Resolution Analysis of Core Histones

PAGEND resolves histone components that occur in very small relative amounts. For accurate quantitation of such minor components, the histone fraction should be essentially free not only of oxidation artefacts but also of nonhistone proteins and degradation products (19). To a certain extent this can be achieved by direct two-dimensional electrophoresis of acid soluble proteins of whole cells and tissues

(28). In most cases, however, histone analysis requires partial purification of chromatin, nucleosomes, or at least nuclei. A procedure for purifying core histones applicable to most animal cells and tissues is shown in Table 14.2.

For optimal resolution of all core histone components, it is necessary to perform electrophoretic analysis of the same sample under at least

TABLE 14.2. Method for Isolating Core Histones

Cells or Tissues Quick-Frozen and Stored at −70°C

1. Without thawing homogenize in 10 volumes of ice-cold buffer A (10 mM K-Tris-maleate pH 7.4, 50 mM glycine, 5 mM $MgCl_2$, 1% thiodiglycol, 20 μg/ml aprotinin) using a Waring blendor with foam adapter or an Ultraturrax with nonfoaming probe for 30–60 sec at 4–5000 rpm.
2. Filter the homogenate through 100-μ and 20-μ nylon screen.
3. Centrifuge at 200 g for 5 min through an equal volume of buffer A plus 5% sucrose and 0.25% Triton X-100.
4. Remove the supernatant.

Nuclear Pellet

1. For certain tissues, such as liver, repeat the homogenization and centrifugation until the supernatant is clear.
2. Suspend the nuclear pellet in 5 volumes of buffer A. Add with rapid stirring 3 times the volume of 200 mM NaCl, 100 mM $MgCl_2$.
3. Centrifuge at 2000 g for 5 min through an equal volume of buffer A plus 5% sucrose, 150 mM NaCl, 75 mM $MgCl_2$.
4. Reextract the pellet once more with buffer A plus 150 mM NaCl, 75 mM $MgCl_2$, centrifuge and pool the supernatants which contain the soluble nuclear nonhistone proteins and the outer histones.

Extracted Nuclear Pellet

1. Wash the nuclear pellet by suspension in 10 volumes of buffer B (10 mM K-Tris-maleate pH 7.4, 1 mM $CaCl_2$, 1% thiodiglycol, 20 μg/ml aprotinin) followed by centrifugation at 2000 g for 5 min.
2. Suspend the pellet in 5 volumes of buffer B at room temperature, digest with 40 U/ml of micrococcal nuclease for 10 min and centrifuge at 2000 g for 5 min.
3. Extract the deoxyribonucleoprotein complexes from the nuclei by homogenizing the pellet in 20 volumes of buffer B plus 1 M NaCl, followed by centrifugation at 40,000 g for 5 min.

TABLE 14.2. (*continued*)

Chromatin Supernatant
Precipitate acid insoluble material by adding HCl to 0.35 N and remove the precipitate by centrifugation at 2000 g for 5 min.
Core Histone Supernatant
1. Collect the histones by precipitation with 12% TCA and centrifugation at 10,000 g for 10 min.
2. Wash the precipitate consecutively with (*a*) 5% TCA, 1% thiodiglycol, (*b*) acetone plus 0.4% conc. HCl, 1% TDG, (*c*) acetone–1% thiodiglycol.
3. Dissolve in 5% thiodiglycol for storage or 8 M urea, 100 mM dithiothreitol, 5% thiodiglycol for electrophoretic analysis.

two different conditions (19, 20). H2A histones, which have a very high affinity for nonionic detergent, are best resolved in 7.5–8.5 M urea and 6mM Triton X-100 (Figure 14.1) whereas H3 and H2B histones are best resolved in 5–6 M urea and 6 mM Triton X-100 (Figure 14.2*a*). The two systems can be combined into a single composite gel (Figure 14.2*b*) where all the histones are first separated in a 10% polacrylamide gel portion containing 8 M urea and 6 mM Triton X-100 before entering the main portion of the gel containing 12% acrylamide, 6 M urea, 6 mM Triton X-100. While qualitative comparisons are most efficiently performed on thin slab gels (Figure 14.1), quantitation is done more efficiently using cylindrical gels (Figures 14.2 and 14.3) because the increased path length gives a better signal-to-noise ratio and a larger dynamic range. In our experience, the most reliable and most accurate method for quantitating proteins in acrylamide gels is to stain with Amido Black and to scan the gels at 300 nm (19) because (1) Amido Black forms very stable complexes with all proteins; (2) Amido Black has a high molar absorptivity at 300 nm which is relatively insensitive to the environment and to fading; (3) the absorbance per microgram of protein for Amido Black complexes of all four core histones is essentially identical (Table 14.3).

2.3. Characterization of Individual Core Histone Components

It is possible to identify and partially characterize individual electrophoretic components by labeling tissue culture cells with different

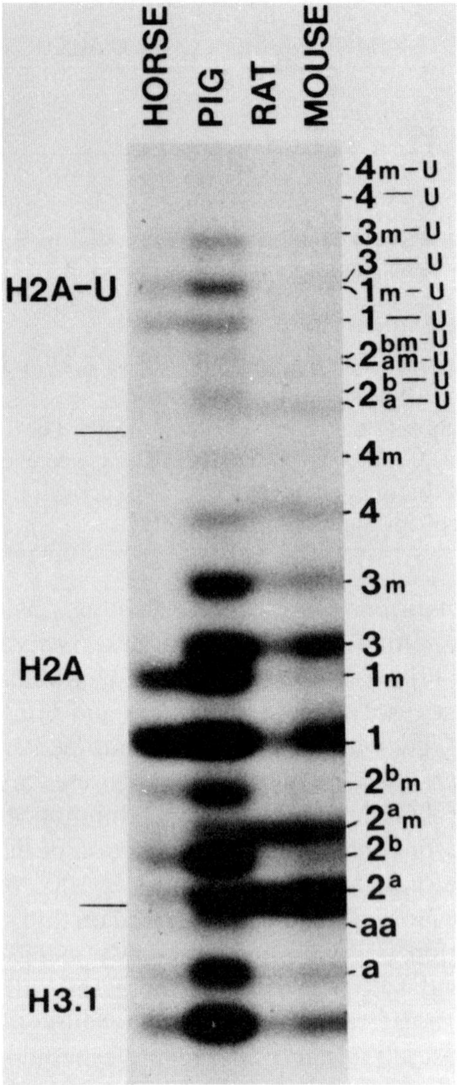

FIGURE 14.1. High-resolution analysis of mammalian histones by PAGEND. Core histones were isolated by the method described in table 4.2 from the liver of adult horse, pig, rat, and mouse and resolved by electrophoresis for 51 hours at 220 V in a 300 × 150 × 1.5 mm, 10% polyacrylamide slab gel containing 5% acetic acid, 7.5 M urea, 6 mM Triton X-100. The proteins were stained with Serva Blue R. The H2A variants 1, 2^a, 2^b, 3, 4 are described in the text; m = post-translational charge modification due to acetylation or phosphorylation (see Figure 3), U = covalent complexes with ubiquitin. Note the different ratios of H2A.2^a and H2A.2^b in the different species and the absence of H2A.3 in horse.

H^3-amino acids and separating tryptic peptides of individual bands in polyacrylamide gels (29). In most cases it is necessary, however, to isolate individual components in relatively large amounts by a combination of ion exchange chromatography and preparative electrophoresis (10, 13) to allow accurate amino acid analysis, peptide analysis, and sequencing.

2.3.1. H2A Variants. Most of the core histone components resolved by PAGEND are not primary structure variants but post-translationally modified forms. For instance, the H2A histones of most mammals can be resolved into about 20 electrophoretic components (Figure 14.1). Labeling with ^{32}phosphate or ^{3}H-acetate (in presence of cycloheximide) reveals the existence of two similar sets of bands displaced by about 3% in electrophoretic mobility (Figure 14.3). The slower set of bands represents the phosphorylated or acetylated forms of the components of the faster set. In addition, the slower half of the H2A bands actually is a repeat of the faster half of the H2A bands at about one-tenth the amount, and represents the covalent complexes between H2A histones and ubiquitin Figures 14.1 and 14.3; References 13, 30). Thus there are only five apparently unmodified forms of H2A. Of these, all except H2A.4 have identical molecular weights as determined by SDS gel electrophoresis (31). Electrophoresis of the purified components in acid–urea gels, according to Panyim and Chalkley (26), shows that H2A.1 and H2A.2^a have an identical charge as well and must therefore be separated in PAGEND on the basis of differences in detergent binding. H2A.2^b has an electrophoretic mobility equivalent to about one-half charge less than H2A.2^a (calibrated with acetylated or phosphorylated forms of H2A.2^a), whereas unmodified H2A.3 essentially comigrates with acetylated or phosphorylated H2A.1, in presence and absence of detergent. Amino acid analysis and microfingerprinting (10) showed that the detergent affinity difference between H2A.1 and H2A.2 correlates with a Leu/Met substitution at position 51, while the charge differene between H2A.1 and H2A.3 is due to a His/Glu substitution at position 124 or 125, and the partial charge difference between H2A.2^a and H2A.2^b is due to a deletion of one of the histidine residues at position 124/5 (32). Although there are additional differences in the amino acid sequences between these variants (20), they evidently do not influence either the molecular weight, the charge, or the detergent affinity. These observations demonstrate that none of the electrophoretic systems, including PAGEND, can resolve all possible primary structure variants, and that the microheterogeneity in the primary structure of the core histones could be more extensive than apparent from the analysis of intact proteins. In addition to these somatic H2A proteins, which occur in all tissues of all mammals, some species express spermatocyte-specific H2A proteins (33; Zweidler, unpublished observation) which have not been well characterized yet. H2A.4 is dramatically increased in the spermatocytes of many mammals (33, 34) and was originally characterized as a spermatocyte-specific H2A variant (H2A.S) of the mouse (20, 34).

The mammals, as well as many other species, express a minor histonelike protein, M1 (13, 14; also called H2A.Z, 28), which seems to share some sequence homology with H2A (28) but is clearly distinct from H2A histones with respect to metabolic properties (35), evolutionary conservation (36), distribution in the nucleus (37), and histone complex formation (Zweidler, unpublished observation). We

FIGURE 14.2. (*A*) Change in the histone variants with age in the mouse liver. Total histones were isolated from the livers of ICR mice from one week before birth (−1) to 30 weeks after birth, and resolved on 30-cm-long cylindrical polyacrylamide gels (12% T) containing 5% acetic acid, 6 M urea, and 6 mM Triton X-100. The proteins were stained with Amido Black. T = adult mouse thymus standard which has an identical histone composition as embryonic liver (lane 2). Variants (V) are as described in text. Note the increase in the relative amounts of H2A.2, H2A.3, H3.3, H1°, and H2B.1 with age, starting one week after birth. (*B*) Core histones (upper gel) and outer histones (lower gel) of different tissues of 12-month-old ICR mice, isolated as described in table 2 and resolved by electrophoresis in composite 35-cm-long cylindrical polyacrylamide gels containing a 10.5-cm-long upper portion with 10% polyacrylamide, 8 M urea, 6 mM Triton X-100, and a 24.5-cm-long lower portion with 12% polyacrylamide, 5.75 M urea, 6 mM Triton X-100. Only the lower portion of the gel is shown. Staining is with Amido Black. For nomenclature see text. Lanes: 1 = calf thymus; 2 = mouse thymus; 3 = mouse spleen; 4 = intestinal mucosa; 5 = mammary gland; 6 = lung; 7 = heart; 8 = Kupffer cells of the liver; 9 = salivary glands; 10 = kidney; 11 = purified polyploid hepatocytes. The tissues were arranged in decreasing order of their rate of replication in the adult. Note the decrease in the relative amounts of H2A.1, H3.1, H3.2, H2B.2 proportional to the rate of replication and the corresponding increase in H2A.3, H2A.2, H3.3, H2B.1, and H1°.

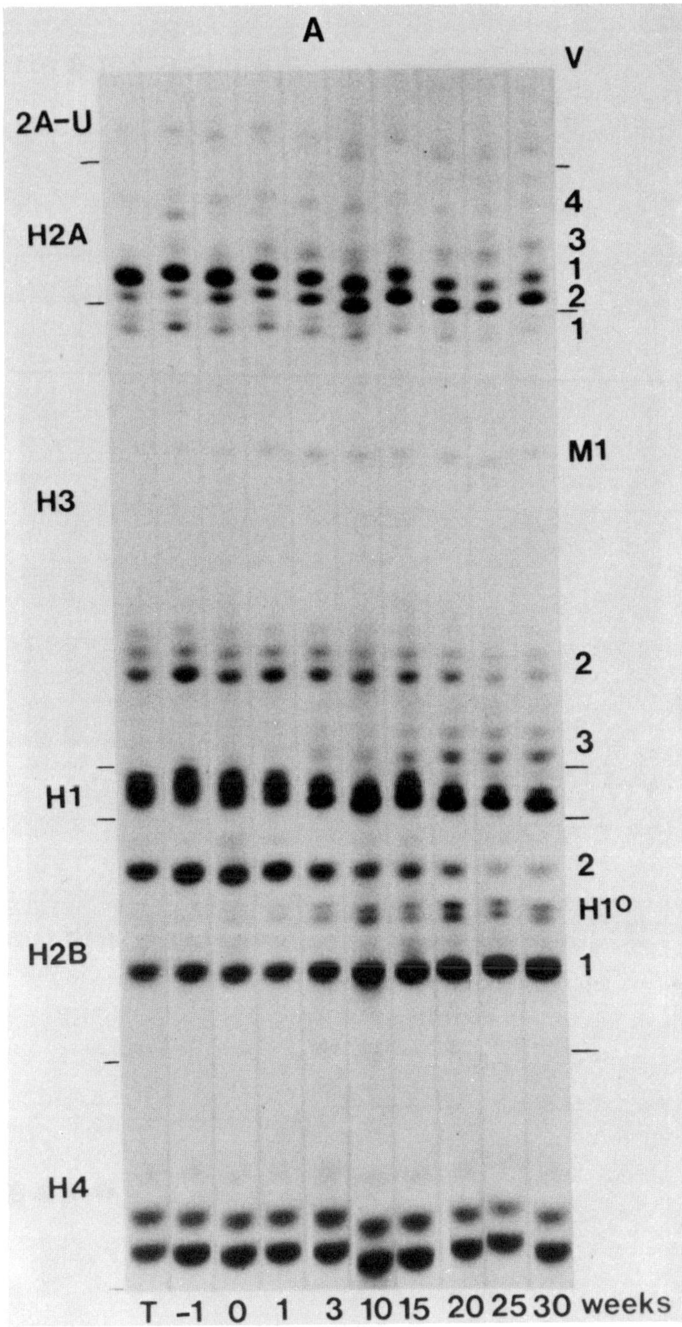
A
V
2A-U
4
H2A
3
1
2
1
M1
H3
2
3
H1
2
H1O
H2B
1
H4
T -1 0 1 3 10 15 20 25 30 weeks

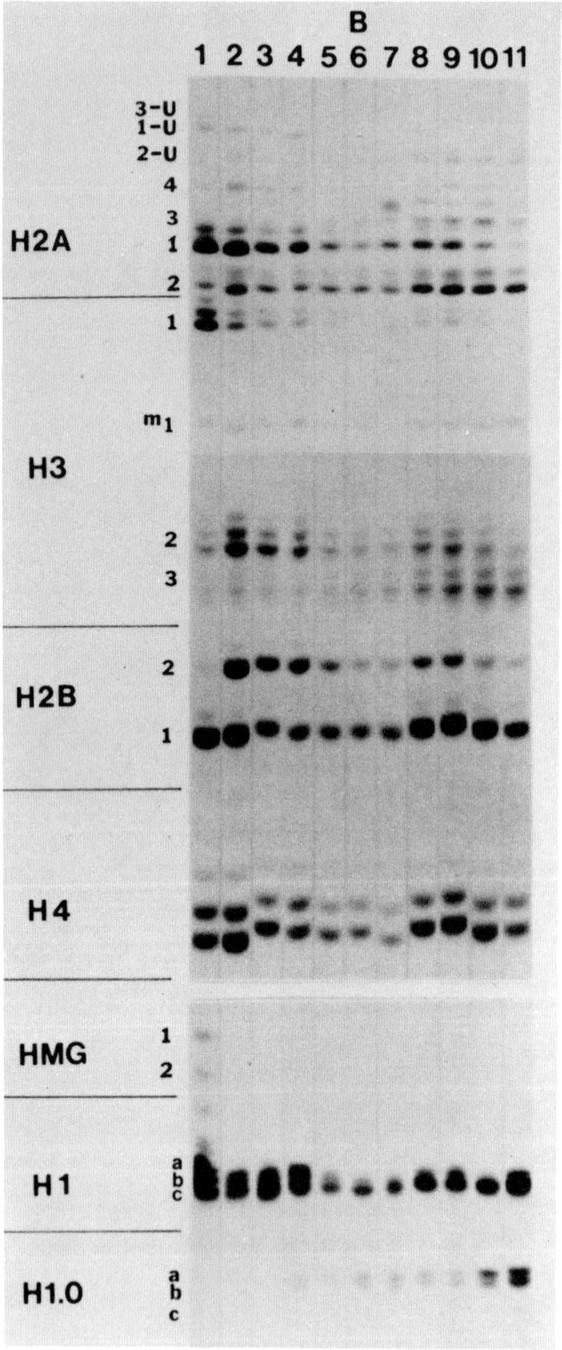
B
1 2 3 4 5 6 7 8 9 10 11
3-U
1-U
2-U
4
3
H2A
1
2
1
m1
H3
2
3
H2B
2
1
H4
HMG
1
2
H1
a
b
c
H1.0
a
b
c

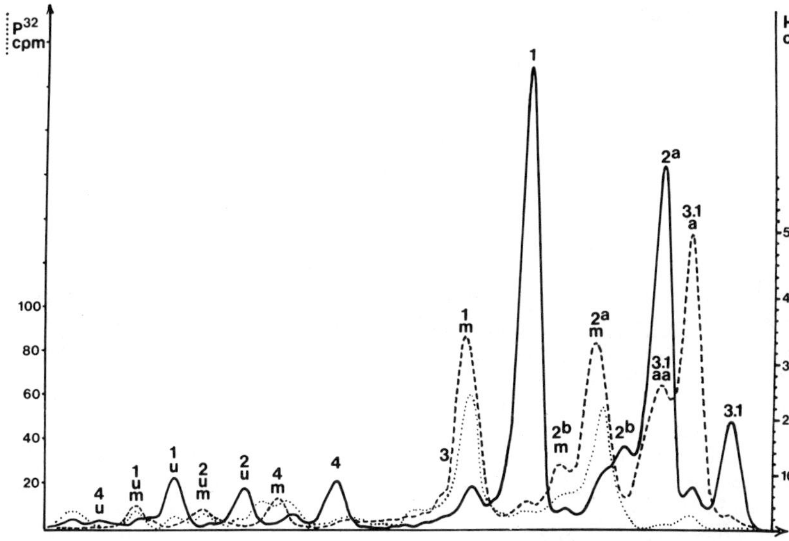

FIGURE 14.3. Identification of charge modification in H2A histones. The core histones of murine erythroleukemia cells labeled with either ^{32}P-phosphate or ^{3}H-acetate (in presence of cycloheximide) wre isolated as described in table 14.2 and electrophoresed as in Figure 14.1, but on 30-cm-long cylindrical polyacrylamide gels 4 mm in diameter (from left to right). The proteins were stained with Amido Black 10B and scanned at 300 nm (solid profile). The gels were then sliced into 1-mm slices and the radioactivity determined after digestion in H_2O_2 (dotted profile = ^{32}P, dashed profile = ^{3}H-acetate). Nomenclature is as in Figure 14.1 and text. Note the low level of *N*-terminal acetylation of H2A.1, demonstrating effective inhibition of histone synthesis. The phosphorylated and the acetylated forms essentially comigrate about 3% slower than the unmodified species. H2A.2^b migrates between H2A.2^a and its modified form.

prefer to regard M1 as a distinct histone species until it can be shown that it is functionally related to H2A.

2.3.2. H2B Variants. In contrast to H2A, only one unmodified form of H2B can usually be detected in somatic tissues of mammals. Where there are additional H2B forms, they appear to be species or family specific. The mouse expresses two somatic H2B variants with identical molecular weight and charge which differ only in a Ser/Gly substitution at position 75 (10) and the affinity for nonionic detergents (Figure 14.2). In addition, the mouse expresses a prominent spermatocyte-specific H2B variant (20) which has a lower apparent molecular weight,

TABLE 14.3. Histone Quantitation with Amido Black[a]

Histone	Area/Microgram × 10^{-3}	$\sigma n - 1$	n
H2A	1701	±156	11
H2B	1722	±78	12
H3	1670	±116	9
H4	1683	±183	10
H1	1520	±130	5
H5	1917	±64	5

[a] Histones were isolated by CMC chromatography and preparative electrophoresis from calf thymus, mouse liver, and chicken erythrocytes (10, 13) and quantitated by amino acid analysis. Aliquots of the same fractions were subjected to electrophoresis by PAGEND, stained with 0.2% Amido Black 10B (Merck) in methanol:water:acetic acid = 5:5:1, destained by diffusion in 10% isopropanol, 1% acetic acid, sealed into Pyrex glass tubes, and scanned at 300 nm with a Gilson UV monitor. The recorder signal was filtered with a Spectrum noise filter calibrated to give a 1 V signal at 10 D, and integrated with a Columbia Supergrator-2 at a scanning speed of 0.5 mm/sec. No differences were found between the histones of different species or different histone variants.

a lower positive charge, and a higher detergent affinity and differs from the somatic H2B.1 in a deletion at the *N*-terminus and at least seven conservative amino acid substitutions involving hydrophilic residues (20, 34). Both of these additional H2B variants seem to occur in all mouse species (Zweidler and Potter, unpublished observation) but not in the rat or other mammals. Other mammals, however, contain different testes-specific H2B variants (38).

2.3.3. H3 Variants. All mammals express the same three somatic forms of H3, each of which is resolved into several charge-modified forms due primarily to acetylation (Figure 14.2). Variants H3.1 and H3.2 differ only in a Cys/Ser substitution at position 96, whereas variant H3.3 differs from H3.2 in the substitution of Ile–Gly for Val–Met at positions 89–90 (10). Variant H3.1 seems to be mammal specific, whereas variants H3.2 and H3.3 appear to be coexpressed in identical form in all vertebrates studied so far (13; Zweidler, unpublished).

In addition to the three somatic H3 variants, the mouse also expresses a spermatocyte-specific variant H3.S, which differs from the somatic variants in several positions in the *C*-terminal half (20, 34). The same H3.S variant is expressed in all wild mouse species studied (Zweidler and Potter, unpublished), as well as in the rat (33), but not in other mammals.

3. DIFFERENTIAL EXPRESSION OF HISTONE VARIANT GENES

3.1. Changes in Histone Variant Composition of Somatic Tissues during Development

In the sea urchin there are dramatic changes in the expression of different H2A, and H2B, and H1 variants during early embryogenesis (18, 39, 40). In the chicken (19) there is an increase in the complexity of the histone complement due to the appearance of a second variant each of H2B and H3 during somite formation. In contrast to the situation in sea urchin, both the early and the late appearing variants of the chicken embryo continue to be expressed in all tissues throughout life. The relative amount of the late H2B and H3 variants of the chicken remains very low in embryos and rapidly dividing adult tissues but increases in nondividing adult tissues. This is most dramatic for the late H3 variant (identical to mammalian H3.3), which eventually becomes the predominant H3 variant in the liver and kidney of older animals. The two chicken H2A variants are expressed in essentially constant amounts at all stages of development and in all tissues.

In the mouse (20, 41), all somatic variants are expressed very early in development, and no significant quantitative changes of the histone complement occur from the neurula stage (about one day after implantation) until birth. After birth the proportions of the different histone variants are maintained at embryonic levels in those tissues that produce short-lived differentiated cells (hemopoetic tissues, intestinal mucosa, etc.) and therefore continue to divide at relatively high rates in the adult. In contrast, dramatic changes in histone variant proportions occur in those tissues that produce long-lived differentiated cells and therefore have very low rates of cell division in the adult (liver, kidney, etc.). In these nondividing cells those variants that are most prominent in embryonic tissues decrease in relative amount while at the same time a set of minor embryonic variants increases to become the predominant adult variants (Figure 14.2).

These observations suggest that the change in histone variant proportions are related to growth characteristics and have led to the hypothesis (20) that there are at least two classes of somatic histone genes, namely, those dependent on DNA replication for their expression (replication variants = H2A.1, H2B.2, H3.1, H3.2) and therefore occur in large amounts in rapidly dividing cells, and those expressed even in absence of DNA replication and therefore accumulate in nondividing cells by replacement incorporation (replacement variants = H2A.2, H2A.3, H2B.1, H3.3, H1°).

A detailed study of the relationship between the rate of replication and the histone variant changes was undertaken using highly purified hepatocyte nuclei of mice raised under controlled conditions (Zweidler, in preparation). Figure 14.4 shows the decay of the putative replication dependent variants H2A.1, H2B.2, and H3.1 + H3.2, as well as the rate of replication as measured by the labeling index (L) after continuous infusion of H3-thymidine for 10 days. The postnatal growth of the liver occurs in three distinct phases (42): (1) rapid cellular growth phase in the first month, when the number of diploid nuclei increases more than tenfold; (2) slow expansion growth phase in the next few months, when the average ploidy of hepatocytes increases from 2C to more than 6C; (3) an adult stage where the total amount of DNA in hepatocytes and the labeling index reach an essentially constant level, reflecting a steady state between cell death and spontaneous regeneration at about 0.1% per day. The decay of the putative replication variants shows at least two distinct phases: (1) a rapid decrease during cellular and expansion growth which occurs at different rates for the different histone species (apparent half-lives of 50 days for H2A.1, 75 days for H2B.2, and 150 days for H3.1 + 2); (2) a slow decrease after about 4–6 months of age, which is similar for all three histone species (apparent half-life of 360 days, corresponding to a turnover of ca. 0.2% per day). H2A.1 and H2B.2 appear to reach a minimum plateau at about 10% of their respective histone species at about 12–14 months of age. This could reflect an equilibrium between the rate of input of replication variants due to slow tissue turnover (ca. 0.1% per day) and the turnover of these histones. H3.3 evidently never reaches this equilibrium during the lifetime of the mouse. Since the input due to replication stays essentially constant at a very low level in the adult hepatocytes, the continued decrease

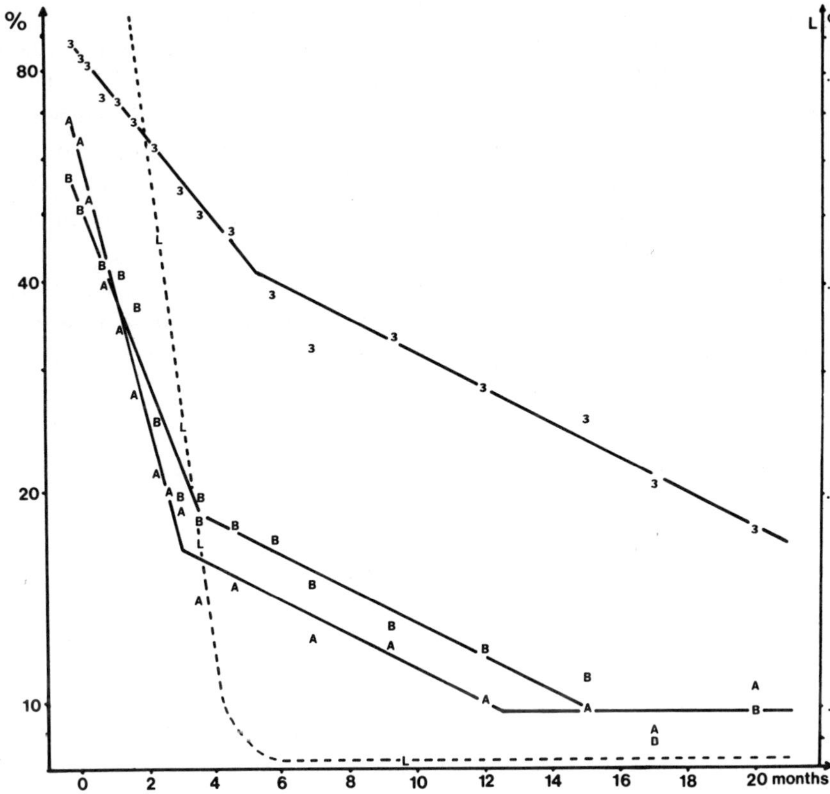

FIGURE 14.4. Decay of (A) H2A.1, (B) H2B.2, and (3) H3.1 + 2 in purified mouse hepatocytes with age. The relative amounts of the core histones were analyzed on gels similar to those shown in Figure 14.2*B* and expressed as percentage of the respective histone species on a logarithmic scale. The rate of replication, expressed as percent nuclei starting DNA synthesis per day (L%, right-hand scale), was determined by autoradiography of isolated liver nuclei of animals infused with a total of 0.05 mCi of ^{3}H-thymidine over a 10-day period. Each point represents the average of 2–12 independent determinations. Note the fast early phase with different apparent half-lives for the three histone types starting around birth, the slower late phase with similar apparent half-lives for the three histones after the rate of replication has reached a plateau at about 0.1%/day, and the plateau reached by H2A.1 and H2B.2 at 12–14 months.

in H2A.1, H2B.2, and H3.1 + 2 must be due to replacement of these variants independent of DNA replication. The rapid decrease in H2A.1 + H2B.2 during the first phase could be due either to a higher rate of replacement or to an altered relative rate of synthesis of H2A and H2B variants for the last few rounds of replication in terminally differentiated hepatocytes. We believe that the latter possibility is more likely because the change in variants starts in the first few weeks after birth when the rate of growth is still very high, and therefore most of the histone protein present at any given time is recently synthesized. In addition, a similar phenomenon is observed in murine erythroleukemia cells after induction of terminal differentiation (35).

3.2. Histone Variant Metabolism in Differentiating Murine Erythroleukemia Cells (35)

Certain cell lines derived from Friend leukemia virus transformed erythroblasts can be induced with a variety of agents to undergo terminal differentiation *in vitro* (44). Within three days after exposure to hexamethylenebisacetamide (HMBA) or dimethylsulfoxide (DMSO), essentially all the cells become irreversibly committed to stop dividing after about five more cell cycles. The system mimics many aspects of normal erythropoesis (43). During the time of induction the growth rate decreases rapidly from a very high level characteristic of uninduced cells to a low level characteristic for induced cells. During this time of decreasing growth rate, the proportions of H2A, H2B, and H1 variants change in a fashion similar to hepatocytes in young mice. Thus the relative amounts of H2A.1, H2B.2 decrease rapidly to about 30% of their respective histone species, while the relative amounts of H2A.2 and H2B.1 increase correspondingly (Figure 14.5).

This change in variant proportions can be accounted for almost entirely by changes in the relative rates of synthesis of these variants (Figure 14.5). The change occurs while the rate of replication (input of new histones) is still high, and does not seem to change further after a constant low level of overall histone synthesis is reached by day 4.

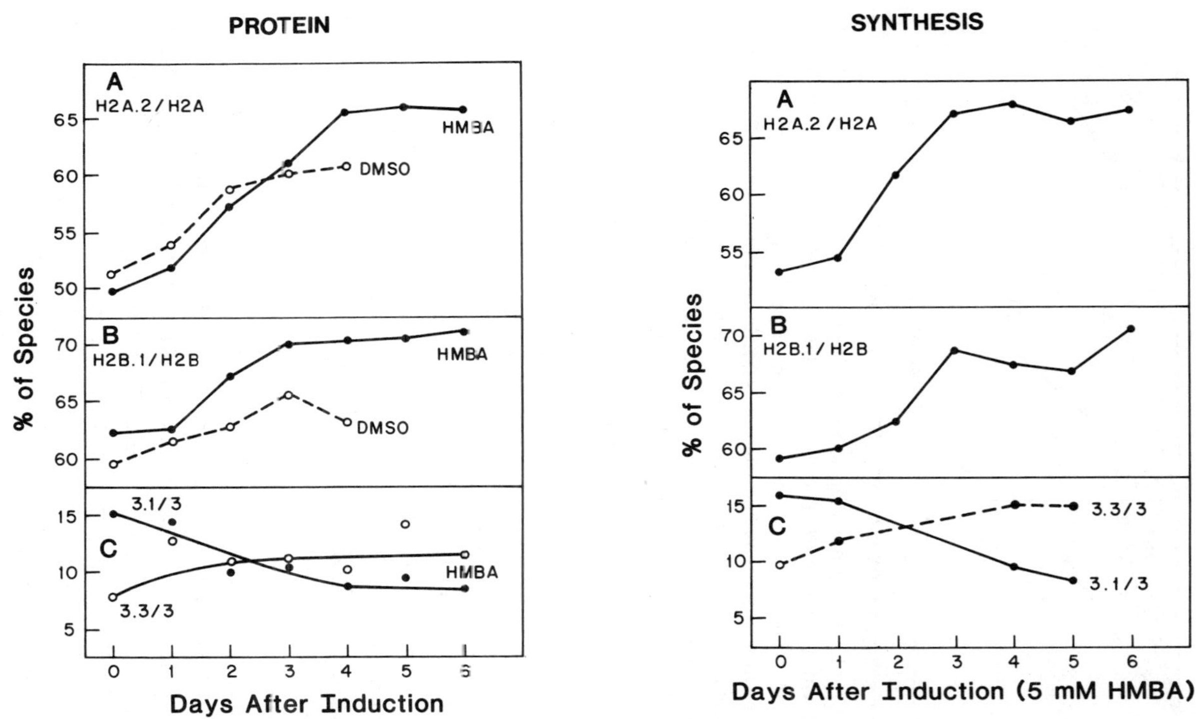

FIGURE 14.5. Changes in the relative amounts and the relative rates of synthesis of different histone variants after induction of terminal cell differentiation in a Friend leukemia virus transformed mouse erythroblast cell line with 5 mM hexamethylene–bisacetamide (HMBA) or 1.5% dimethylsulfoxide (DMSO). Note that the changes in amounts and in rates of synthesis are essentially parallel, indicating that the change in variant proportions is primarily due to a change in the relative rate of synthesis. (From Reference 35.)

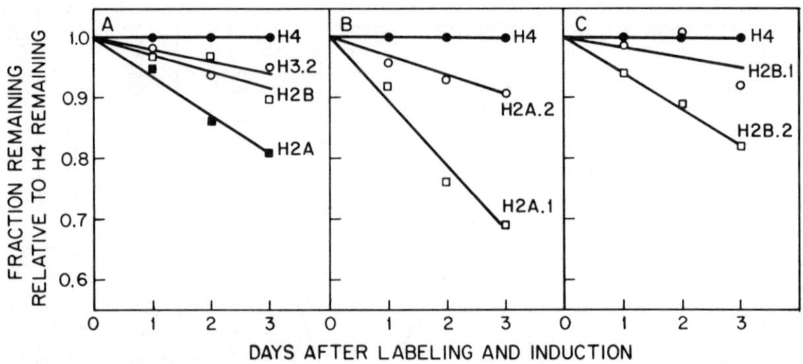

FIGURE 14.6. Turnover of different histones relative to H4 in murine erythroleukemia cells induced with 1.5% DMSO. The cells were labeled with a mixture of ^3H amino acids before induction and chased with cold medium during induction. Note that the different histone species have different relative rates of turnover (*A*). Within the same histone species (*B* and *C*), the replication-dependent variants (H2A.1, H2B.2) have a higher rate of turnover than the replication-independent variants (H2A.2, H2B.1). Similar results were obtained with uninduced cells and with cells induced by HMBA (35).

Histone replacement could not account for the changes in the H2A and H2B variant proportions since the overall rate of turnover is low (Figure 14.6) and since the difference in the relative rates of turnover between the variants is only slightly increased after induction (35). The observation of differential turnover of variants of the same histone species even in rapidly dividing cells implies some form of biological recognition of different variants or their associated nucleosomes. It confirms directly the postulate for asymmetric histone variant replacement made on the basis of histone variant changes in nondividing cells *in vivo* (20). The changes in the relative rate of synthesis of histone variants after induction of terminal differentiation in MEL cells are similar to those observed in hepatocytes of young animals. Both in the MEL system and in the hepatocytes the changes in H3 variant proportions are very slow, indicating that the mechanisms governing changes in H3 variants may be different from those in H2A and H2B variants. The complexity of the system is most clearly demonstrated in the regenerating liver, where both induction and inhibition of histone synthesis can be studied *in vivo*.

3.3. Histone Variant Changes during Mouse Liver Regeneration (20; Figure 14.7)

After surgical removal of about 60% of the liver in adult mice, essentially all the remaining, normally nondividing hepatocytes undergo a synchronous round of replication leading to a doubling of the ploidy, the size of each cell, and therefore the size of the remaining lobes of the liver (45, 46). At the time of DNA replication, the amounts of H2A.1, H2A.2, H2B.1, H2B.2, H3.1, H3.2 all increase in parallel with the amount of DNA, indicating that the synthesis of all these variants is induced together with DNA synthesis (Figure 14.7). The amounts of the variants H2A.3, H3.3 (and H1.0) are not increasing significantly during DNA replication and are therefore not induced together with DNA synthesis. After completion of DNA synthesis, the amounts of H2A.1, H2B.2, H3.1, and H3.2 decrease again, while those of H2A.2, H2B.1 continue to increase, and H2A.3, H3.3, and H1.0 start to increase. The minor histone H2A.4 and M1 stay constant throughout the course of liver regeneration, that is, they follow the change in DNA. On the basis of these observations, one can classify the histone variants into four groups which differ in their relationship to DNA replication.

3.3.1. Strictly Replication-Dependent Variants. H2A.1, H2B.2, H3.1, and H3.2 are induced at the beginning of DNA synthesis and repressed at the end of DNA synthesis.

3.3.2. Partially Replication-Dependent Variants. H2A.2 and H2B.1 are induced at the beginning of S phase but not completely repressed at the end of S phase.

3.3.3. Replication Independent Variants. H2A.3, H3.3, and H1.0 are neither induced nor repressed in conjunction with DNA synthesis and may be continuously expressed at very low levels or expressed only in nondividing cells.

3.3.4. Noncompeting Minor Histones. H2A.4 and M1 occur in somatic tissues in similar amounts relative to DNA regardless of changes in other histone variants.

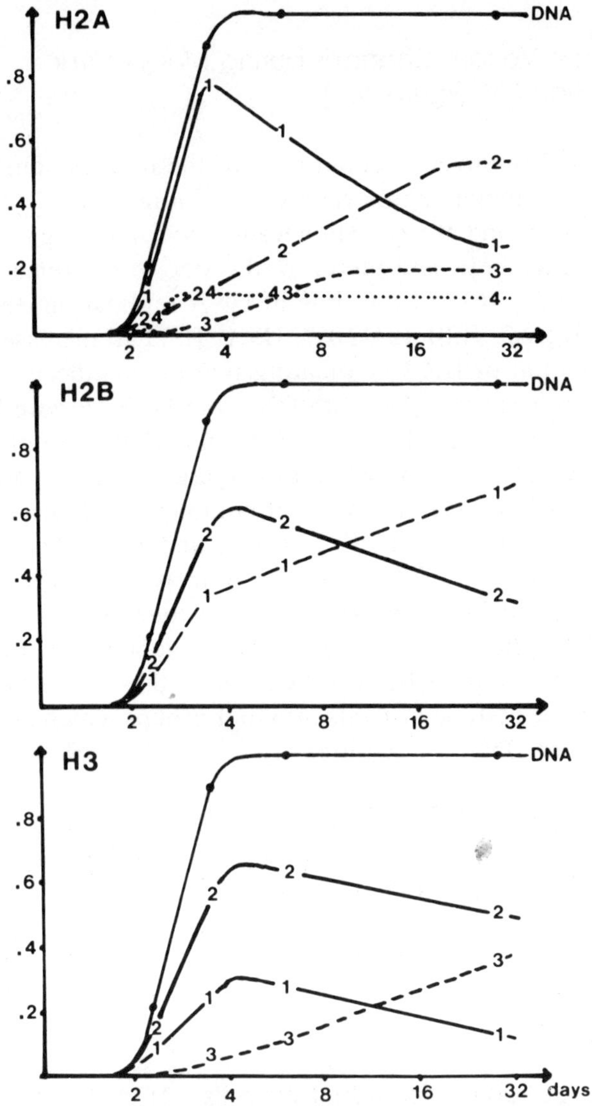

FIGURE 14.7. Changes in the newly synthesized DNA and histones after partial hepatectomy in 4-month-old mice. Total liver DNA doubled between 2 and 4 days after hepatectomy, which was reflected in an increase in the average ploidy of hepatocyte nuclei from approximately 4C to 8C. The absolute amount of different histone variants minus the amount present at 30 hours after hepatectomy was taken as the newly synthesized amount and plotted against time in days (log scale) after hepatectomy. Note the four different types of variant responses in H2A. See text for interpretation.

3.4. Spermatocyte-Specific Changes in Histone Variants

Although histone replacement in absence of DNA replication has only recently been described for somatic cells (20), it is a well-known phenomenon in sperm maturation (47). In addition to the extensive late changes in sperm chromatin, there is an earlier phase in spermatogenesis when more subtle changes in the histone complement occur. Thus during the long meiotic prophase of the mouse, a set of spermatocyte-specific variants of H2A, H2B, H3, and H1 is incorporated into chromatin in absence of replicative DNA synthesis—but at a time of repair DNA synthesis possibly associated with recombination events (33, 34). At the same time, the minor somatic histone variant H2A.4 increases dramatically in relative amount. In this system H2A.4, H2A.S, and H3.S increase during early meiotic prophase, while H2B.S and H1.S only appear in late prophase (pachytene).

4. GENETICS OF HISTONE VARIANT POLYMORPHISMS

The histone variants described so far are coexpressed in inbred mouse strains and are therefore nonallelic by genetic definition. We have recently discovered several electrophoretic histone variants that behave like typical allelic polymorphisms. In a *Mus musculus domesticus* colony established from a wild population by Dr. M. Potter we found a polymorphism of the spermatocyte specific variants H1.S and H2B.S (Zweidler and Potter, in preparation). Heterozygotes express equal amounts of the two forms of both variants, homozygotes either one or the other. This indicates that these histones are coded by single genes. In test crosses with inbred mouse strains, H1.S and H2B.S always cosegregate (0/50 recombinants), indicating that the two genes are closely linked. Subsequently we discovered the H1.S polymorphism in inbred mouse strains. While C57BL/6 strains have the slower form, all other strains tested have the faster form. We also discovered a polymorphism in the replication-dependent somatic histone variant H2A.1 which occurs only in the ICR line of BALB/c. A backcross analysis between C57BL/6 ICR and BALB/c ICR showed that the H1.S and the H2A.1 genes are not linked (Figure 14.8).

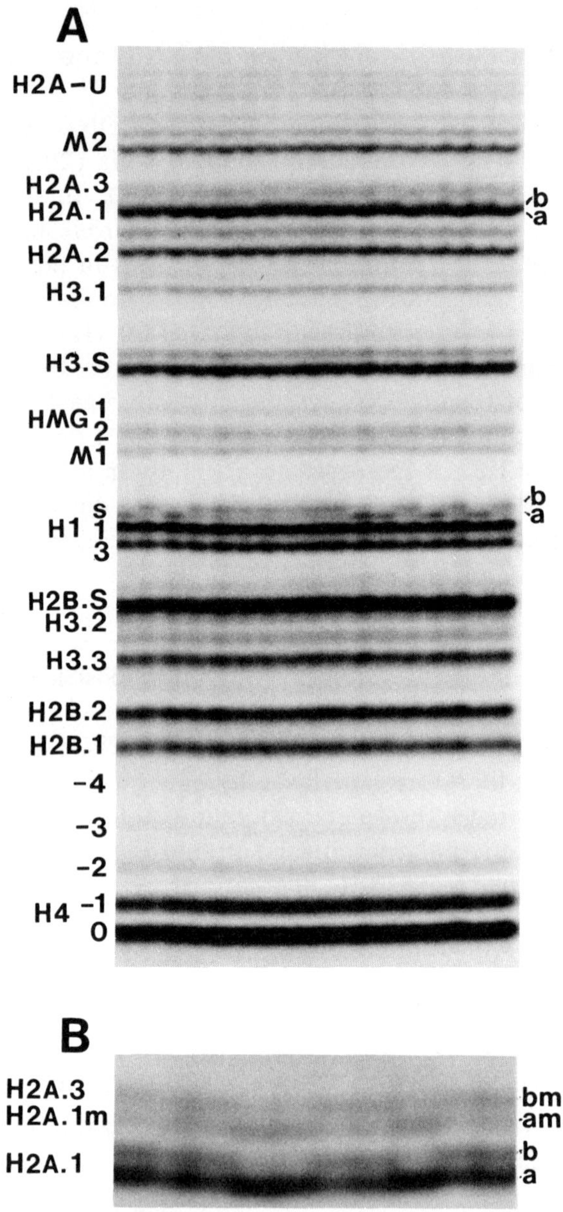

A
H2A-U
M2
H2A.3
H2A.1
H2A.2
H3.1
H3.S
HMG 1 2
M1
H1 s 1 3
H2B.S
H3.2
H3.3
H2B.2
H2B.1
-4
-3
-2
H4 -1 0
b
a
b
a
B
H2A.3
H2A.1m
H2A.1
bm
am
b
a

The H2A.1 polymorphism has some unusual features. Thus in testes of the inbred mouse strain BALB/c ICR the two forms occur in a ratio of 2:1, indicating that there are three active H2A.1 genes in this tissue. In heterozygotes (e.g., BALB/c ICR × C57BL/6 F1) the ratio is always 5:1. In backcrosses with the BALB/c ICR parent, half the offspring have a ratio of 2:1 and half a ratio of 5:1, as expected for a three-gene system. Analysis of a variety of somatic tissues of BALB/c ICR animals revealed to our surprise that the ratio of the two H2A.1 forms varied from 2:1, in the testes, to 5:1 in liver, kidney, and lung, and approximately 10:1 in thymus and spleen. We interpret these observations to indicate that there are at least 11 H2A.1 genes arranged in three distinct domains of 5, 3, and 3 genes, which are induced sequentially depending on demand. Thus rapidly dividing cells with an S-phase duration of approximately 6 hours (e.g., blood cells; 48) use all three domains; slowly dividing cells with an S phase of about 12 hours (e.g., liver, kidney in young animals; 48) only use two domains of three genes; while during the premeiotic S phase of spermatocytes, which lasts about 24 hours (48), only one domain of three genes is induced. This would explain the change in the relative rate of synthesis of H2A.1 of terminally differentiated cells (Figures 14.4 and 14.5).

In view of these results we have reanalyzed the other histones for potential polymorphisms and believe that H2A.2^a and H2A.2^b (Figures

FIGURE 14.8. Segragation of polymorphic forms of H2A.1 and H1.S in 17 individual mice of a backcross between BALB/c ICR × C57BL/6 ICR F1 and BALB/c ICR. Total testes histones were extracted from isolated nuclei and resolved by PAGEND electrophoresis in a 30-cm-long, 1.5-mm-thick 10% polyacrylamide slab gel containing 7.5 M urea and 6 mM Triton X-100. Samples were layered in adjacent slots formed by the teeth of a teflon comb across the surface of the slab gel. Nomenclature of the different variants (left side) is according to Reference 20. M2 = H2A.4 in Figure 14.7. H2A.1^a = wild type from BALB/c and C57BL/6; H2A.1^b = mutant specific for BALB/c ICR. H1.S^a = wild type from BALB/c; H1.S^b = mutant form from C57BL/6. Note that the two genes segregate independently. (*A*) Analysis of all the histone components after electrophoresis for 6600 V/hour. (*B*) Expanded region containing H2A.1 after electrophoresis for 12,000 V/hour showing more clearly the backcross segregation into homozygotes with an H2A.1^a/H2A.1^b ratio of 2:1 (same as the BALB/c ICR parent) and heterozygotes with a ratio of 5:1, as expected if there were three active H2A.1 genes in spermatocytes, one of which is mutated in BALB/c ICR. m = modified form. H2A.1^bm comigrates with H2A.3.

14.1 and 14.3), represent a polymorphic system, since the ratio of the two forms varies widely in different species (Figure 14.1) but stays constant in all tissues and throughout development in the same species. Thus the H2A.2 genes might form a single domain of six genes with an H2A.2^a/H2A.2^b ratio of 5:1 in mice, 2:4 in pig and horse.

5. SUMMARY, CONCLUSIONS, OUTLOOK

In inbred BALB/c ICR mice we can distinguish 10 types of H2A, 3 types of H2B, and 4 types of H3 variants that differ in either their primary structure and/or the regulation of their expression. On the basis of gene dosage considerations we can account for 20 H2A genes, which is similar to the number of genes estimated by DNA hybridization techniques (49). This extensive divergence of mammalian histone genes, both in their structural and their regulatory regions, would not have been expected in view of the high degree of evolutionary conservation and the small degree of histone heterogeneity observed in nonmammalian animals. The increased degree of divergence of mammalian histone genes most likely reflects changes in the organization of this multigene family. In most lower animals the histone genes are arranged in tandem repeats of compact sets containing one each of the five histone species (50), a configuration that allows the stoichiometric regulation of the histone gene number by unequal crossover mechanisms (51). Tandem repeats seem to be resistant to divergence of individual genes due to the same mechanism. The tandemly repeated histone gene arrangement is probably most favorable for species that have relatively unprotected stages in early embryonic development and therefore select for individuals with a high maximal rate of chromosome replication, cell division, and growth. In mammals, which protect the early stages of development, the need for a high maximal rate of cell division and therefore for maintaining the tandem repeat arrangement of histone genes probably diminished. Thus in the mouse tandem repeats are no longer present, the histone genes are not arranged in sets containing one gene each for the five histones, and the individual histone genes are separated by long stretches of intervening DNA (52).

The primary need for histone synthesis is for packaging newly synthesized DNA into nucleosomes during the S phase of the cell cycle. In general, the synthesis of histones probably must be repressed in absence of DNA synthesis to prevent an accumulation of free histones which may be toxic. In yeast, histone synthesis is regulated by induction at the beginning of S phase and by a separate mechanism of repression at the end of DNA synthesis (53). A typical histone gene is therefore strictly replication dependent. Certain organisms may have evolved the need for histone synthesis in absence of DNA replication for specific purposes. Thus in sea urchin and other species it may be an advantage to synthesize and store histone proteins or mRNAs in the oocyte for use in the reorganization of the zygote chromatin [replacement of sperm-specific proteins, CS-histones (40)] and during the rapid cleavage divisions of early embryos [α-variants (18)]. Oocyte-specific histones may not have been described for mammals because they are difficult to detect, or because they are not needed in the slowly developing mammalian embryo. Well-known examples of replication-independent histone synthesis for replacement of old histones are the sperm-specific histones or protamines (47). In addition, many species have a preceding phase of replacement incorporation of spermatocyte-specific histone variants during the meiotic prophase which is characterized by increased DNA repair as a consequence of recombination events (54). In somatic cells, replication-independent histone synthesis may be necessary for histone turnover and nucleosome repair in long-lived nondividing cells (20). This may be especially important for warm-blooded large animals such as birds and mammals which have long life spans and a limited total cell population. Low rates of histone synthesis in absence of DNA replication could be achieved by either making the repression at the end of S phase leaky, or by making a small number of histone genes insensitive to both the induction and the repression signals during the cell cycle. Both of these possibilities seem to have been realized in the mammalian system (partially replication-dependent, leaky histone variants = H2A.2, H2B.1; replication-independent variants = H2A.3, H3.3).

The functional significance of histone variants is difficult to assess because the function of histones in general can only be tested indirectly on the basis of their interaction with each other and the DNA to form nucleosomes. An alternative test for functional significance would be the correlation between changes in the regulatory region and the structural region of individual variant genes. Thus H3.3 occurs in identical form in all vertebrates examined and is always a minor component in dividing cells but accumulates in nondividing cells (19, 20, 35). In the case of H2A variants, similar types of variants differing a Leu/Met substitution at position 51 occur in the sea urchin (55), chicken (13), and the mammals (10). However, the two forms have not been continuously preserved in parallel. Thus fishes, frogs, and snakes seem to lack $H2A^{met}$, and turtles lack $H2A^{leu}$ (Zweidler, unpublished observation). Furthermore, even where the two types of variants occur simultaneously, they are not always expressed under the same circumstances. Thus in sea urchin, $H2A^{met}$ is an oocyte and early embryo-specific variant while several forms of $H2A^{leu}$ are late variants (18). In the chicken the two forms are coexpressed in a constant ratio in all tissues and all developmental stages (19). In the mouse the $H2A^{met}$ (H2A.2) is partly replication dependent, one $H2A^{leu}$ form (H2A.1) is strictly replication dependent, while another $H2A^{leu}$ form (H2A.3) is replication independent (20). All somatic H2B variants, and spermatocyte-specific variants of H2A, H2B, and H3, where they occur, are generally species specific. Thus only in the case of H3.3/H3.2 is there a correlation between changes in the structural and the regulatory region, and therefore reasonably convincing evidence for functional divergence. However, even in the histone variant cases that do not show conservation of pattern of expression, it cannot be excluded that specific functional properties have evolved more recently, for example, among the mammals.

Even if there were no functional significance to the histone variants, their study is important and valuable as a tool for studying the expression and the metabolism of different members of this multigene family. Information gained on the protein level is an important prerequisite for investigations into the regulation of transcription at the gene level. The mammalian histone genes could well be one of the most useful gene systems for such studies because members of the same highly preserved protein species are responsive to different induction and repression signals during the cell cycle or to differentiation-specific signals in germ cells.

ACKNOWLEDGMENTS

We thank Drs. Sam Franklin, Michael Urban, and Gregory Grove for their outstanding contributions; Judy Mamas, Mary Ann Houck, Patrick Zweidler, and Regula Zweidler for their invaluable technical assistance; Dr. M. Potter for supplying us with material from a large number of wild mice. This work was supported by U.S. Public Health Grants CA-15135, CA-06927, CA-09035, and RR-05539 and by an appropriation from the Commonwealth of Pennsylvania.

REFERENCES

1. I. Isenberg, *Ann. Rev. Biochem.* **48,** 159 (1979).
2. J. D. McGhee and G. Felsenfeld, *Ann. Rev. Biochem.* **49,** 115 (1980).
3. D. G. Searcy and R. J. DeLange, *Biochem. Biophys. Acta* **609,** 197 (1980).
4. B. Laine, D. Kmiecik, P. Sautiere, G. Biserte, and M. Cohen-Solal, *Eur. J. Biochem.* **103,** 447 (1980).
5. P. A. Temussi, *J. Theor. Biol.* **50,** 25 (1975). G. R. Reeck, E. Swanson, and D. C. Teller, *J. Mol. Evol.* **10,** 309 (1978).
6. R. J. DeLange, D. M. Fambrough, E. L. Smith, and J. Bonner, *J. Biol. Chem.* **244,** 5669 (1969).
7. C. V. C. Glover and M. A. Gorovsky, *Proc. Natl. Acad. Sci. USA*, **76,** 585 (1979).
8. L. Patthy, E. L. Smith, and J. Johnson, *J. Biol. Chem.* **248,** 6834 (1973).
9. W. F. Brandt and C. von Holt, *Eur. J. Biochem.* **121,** 501 (1982).
10. S. G. Franklin and A. Zweidler, *Nature* **266,** 273 (1977).
11. J. A. Hooper, E. L. Smith, K. R. Sommer, and R. Chalkley, *J. Biol. Chem.* **248,** 3275 (1973).
12. W. F. Brandt and C. von Holt, *Eur. J. Biochem.* **46,** 407 (1974).
13. M. K. Urban, S. G. Franklin, and A. Zweidler, *Biochemistry* **18,** 3952 (1979).
14. A. Zweidler, *Life Sci. Res. Rep.* **4,** 187 (1976).
15. J. W. Wallis, L. Hereford, and M. Grunstein, *Cell* **22,** 799 (1980).
16. D. C. Allis, C. V. C. Glover, J. K. Bowen, and M. A. Gorovsky, *Cell* **20,** 609 (1980).
17. B. Laine, P. Sautiere, and G. Biserte, *Biochemistry*, **15,** 1640 (1976).
18. L. Cohen, K. M. Newrock, and A. Zweidler, *Science* **190,** 994 (1975).
19. M. K. Urban and A. Zweidler, *Dev. Biol.* **95,** 421 (1983).

20. A. Zweidler, *Dev. Biochem.* **15,** 47 (1980).
21. A. Zweidler and L. Cohen, *Fed. Proc.* **31,** 926 (1972).
22. A. Zweidler, *Methods Cell Biol.* **17,** 223 (1978).
23. P. N. Lewis and E. M. Bradbury, *Biochim. Biophys. Acta* **336,** 153 (1974).
24. P. D. Van Helden, W. N. Strickland, M. Strickland, and C. Von Holt, *Biochim. Biophys. Acta* **703,** 17 (1982).
25. R. A. Reisfeld, U. J. Lewis, and D. E. Williams, *Nature* **195,** 281 (1962).
26. S. Panyim and R. Chalkley, *Arch. Biochem. Biophys.* **130,** 337 (1969).
27. C. R. Alfageme, A. Zweidler, A. Mahowald, and L. H. Cohen, *J. Biol. Chem.* **249,** 3729 (1974).
28. W. M. Bonner, M. H. P. West, and J. D. Stedman, *Eur. J. Biochem.* **109,** 17 (1980).
29. M. H. P. West and W. M. Bonner, *Biochemistry* **19,** 3238 (1980).
30. I. L. Goldknopf and H. Busch, *Proc. Natl. Acad. Sci. USA* **74,** 864 (1977).
31. U. K. Laemmli, *Nature* **227,** 680 (1970).
32. T. Hayashi, Y. Ohe, H. Hayashi, and K. Iwai, *J. Biochem.* **88,** 27 (1980).
33. P. K. Trostle-Weige, M. L. Meistrich, W. A. Brock, K. Nishioka, and J. W. Bremer, *J. Biol. Chem.* **257,** 5560 (1982).
34. A. Zweidler and S. G. Franklin, *Int. Congr. Biochem.* **11,** 46 (1978) and in preparation.
35. G. W. Grove and Zweidler A. *J. Cell Biol.* **95,** 81 (1982) and in preparation.
36. R. S. Wu, D. Nishioka, and W. M. Bonner, *J. Cell Biol.* **93,** 426 (1982).
37. C. D. Allis, Y. S. Ziegler, M. A. Gorovsky, and J. B. Olmsted, *Cell* **31,** 131 (1982).
38. A. Shires, M. P. Carpenter, and R. Chalkley, *Proc. Natl. Acad. Sci. USA* **72,** 2714 (1975).
39. K. M. Newrock, L. H. Cohen, M. B. Hendricks, R. J. Donelly, and E. S. Weinberg, *Cell* **14,** 327 (1978).
40. G. Childs, R. Maxson, and L. Kedes, *Dev. Biol.* **73,** 153 (1979).
41. A. Zweidler, M. Urban, and P. Goldman, *Miami Winter Symp.* **15,** 531 (1977) and in preparation.
42. R. Tsanev, *Results Prob. Cell Differ.* **7,** 197 (1975).
43. C. Friend, W. Scher, J. G. Holland, and T. Sato, *Proc. Natl. Acad. Sci. USA* **68,** 378 (1971).
44. P. A. Marks and R. A. Rifkind, *Ann. Rev. Biochem.* **47,** 419 (1978).
45. J. James, M. Schopman, and Delfgaauw, P. *Exp. Cell Res.* **42,** 375 (1966).
46. E. G. Bade, Sadnik, I. L., C. Pilgrim, and W. Maurer, *Exp. Cell Res.* **44,** 676 (1966).
47. A. J. Louie, E. P. M. Candido, and G. H. Dixon, *Cold Spring Harbor Symp. Quant. Biol.* **38,** 803 (1974).

48. *Biology Data Book*, 2nd ed., FASEB Vol. **1,** 95 (1972).
49. K. V. Anderson and P. A. Lengyel, *Cell* **21,** 717 (1980).
50. C. C. Hentschel and M. L. Birnstiel, *Cell* **25,** 301 (1981).
51. K. D. Tartof, *Ann. Rev. Genet.* **9,** 355 (1975).
52. D. B. Sittman, I. M. Chiu, C. J. Pan, R. H. Cohn, L. H. Kedes, and W. F. Marzluff, *Proc. Natl. Acad. Sci. USA* **78,** 4078 (1981).
53. M. A. Osley and L. Hereford, *Proc. Natl. Acad. Sci. USA* **79,** 7689 (1982).
54. Y. Hotta, A. C. Chandley, and H. Stern, *Chromosoma* **62,** 255 (1977).
55. W. F. Brandt, W. N. Strickland, M. Strickland, L. Carlisle, D. Woods, and C. von Holt, *Eur. J. Biochem.* **94,** 1 (1979).

CHAPTER 15

THE H1 SUBTYPES OF MAMMALS: Metabolic Characteristics and Tissue Distribution

RONALD W. LENNOX AND LEONARD H. COHEN
The Institute for Cancer Research
7701 Burnholme Avenue
Fox Chase
Philadelphia, Pennsylvania

CONTENTS

1. INTRODUCTION
2. H1 COMPLEMENTS OF DIVIDING AND NONDIVIDING SOMATIC MOUSE CELLS AND TISSUES
3. H1 SUBTYPE CHANGES DURING SPERMATOGENESIS
4. *O*-PHOSPHORYLATION OF H1 SUBTYPES
5. RELATIONSHIPS BETWEEN MOUSE H1 SUBTYPES AND THOSE OF OTHER MAMMALS
6. DISCUSSION AND CONCLUSIONS
REFERENCES

1. INTRODUCTION

The early studies of Cole and his collaborators showed that histone H1 could be resolved chromatographically into multiple subfractions containing proteins differing in primary structure (1–4). They found that the relative amounts of such subfractions vary from tissue to tissue within an organism (1–5) and, moreover, obtained evidence of differential synthesis of some of the H1 forms in mouse mammary gland during hormonally induced growth and differentiation (6–9). Their studies established the existence of H1 subtypes and suggested that some of the subtypes are functionally distinct from others.

Further progress in understanding the biological roles of H1 subtypes has been hampered by difficulties in discerning distinctive features by which each subtype might be identified. For example, the total number of subtypes in any one species is not known because it is not clear that the chromatographic subfractions of different tissues represent the same set of proteins. Nor has it been possible to identify homologous subtypes in different mammalian species except in the cases of H1° (10, 11) and H1t (12–14), two proteins with sufficiently distinctive properties to be identified with relative certainty. These problems are compounded by the fact that H1 molecules in dividing cells contain varying numbers of phosphate groups that affect their chromatographic and electrophoretic properties and thwart unambiguous resolution and identification of the subtypes.

Recently, resolution of mouse H1 into eight unphosphorylated and 15 phosphorylated components has been achieved by two-dimensional electrophoresis in long gels (15–17). It appears likely that these represent almost all of the H1 components except the superphosphorylated forms present in cells in mitosis (18–20). The resolving power offered by this system has permitted a number of questions to be answered: What is the distribution of the subtypes and their phosphorylated forms among different cells and tissues? How do such distributions arise? Do the various subtypes have distinguishing features more likely to be related to function than is chromatographic or electrophoretic behavior? Can such features be used to identify homologous subtypes in different tissues and in different mammalian species?

The answers to these questions, which are reviewed here, permit a unified view of mammalian H1 subtypes. They also provide some insights into the differences in function among the various subtypes.

2. H1 COMPLEMENTS OF DIVIDING AND NON-DIVIDING SOMATIC MOUSE CELLS AND TISSUES

In this section we describe changes in the H1 complement that occur when mouse somatic cells stop dividing, and we discuss the differences in the proportions of the various forms of H1 among different somatic tissues.

A number of different types of dividing mouse cells contain five of the H1 subtypes, a–e, and one or more phosphorylated forms of each (15, 16) (Table 15.1). However, different cell types have different proportions of the various subtypes. For example, an embryonal carcinoma cell line, F9, and large thymocytes, have very little H1e, whereas 3T3 cells and mouse embryo fibroblasts (MEF) have substantial levels of this subtype. Since F9 cells and thymocytes have extremely short G_1 phases, these observations suggest the possibility that H1e is synthesized principally in G_1. One mouse cell line we have examined, PYS, lacks H1b, and another, HR9, lacks both H1b and H1d (15). Similarly, Hohmann and his collaborators have shown that the Chinese hamster cell line, CHO, lacks two of the chromatographic subfractions present in two other lines, DON and V-79 (21).

In addition to these five subtypes, some somatic tissues contain H1°, first detected by Panyim and Chalkley (10, 11), who showed that H1° is absent from thymus and proposed that it is an H1 subtype peculiar to nongrowing tissue. More recently it was shown that H1° accumulates in tissue culture cells whose growth has been arrested (22–24). The greater resolution offered by two-dimensional gel electrophoresis has now made it possible to demonstrate that, in addition to the accumulation of H1°, a number of other alterations in the H1 complement take place in cells after confluence and the cessation of division. First, all phosphate groups are removed from H1 within 2–3 days. Second, H1a declines in amount, and after 10–14 days can no longer be detected. Third, with longer times the amount of

TABLE 15.1. Metabolic and Evolutionary Properties of H1 Subtypes[a]

	H1a	H1b	H1c	H1d	H1e	H1t	H1°
Number of phosphorylated forms observed[b]	1	5	2	2	1	3	1
Maximum number of phosphate groups per mol[b]	1	3	1	2	1	3	1
Synthesized in dividing cells[c]	+	+	+	+	+	+	—
Synthesized in nondividing cells[d]	?	?	+	+	+		+
Metabolic stability[e]	L	I	L	H	H		H
Evolutionary stability[e]	L	I	L	H	H	L	H

[a] Data are from references (15, 16, 17, 26, 43, 45).
[b] These values for subtypes a-e and H1° apply to interphase cells and probably do not include mitotic superphosphorylated states. The values for H1e are minimum values because phosphorylated forms of H1e and H1d overlap.
[c] H1t is synthesized in spermatocytes, in meiotic prophase after the completion of DNA replication.
[d] The question marks indicate that the synthesis observed in confluent fibroblasts may be in dividing cells.
[e] H, high; I, intermediate; L, low. These distinctions are only with respect to the other subtypes.

H1b declines also. These decreases in H1a and H1b, which are accompanied by increases in the relative amounts of H1c and H1e as well as by the appearance and accumulation of H1°, have been observed in embryo fibroblasts (both rat and mouse) (16, 25) and also in 3T3 cells (Figure 15.1).

These alterations in H1 complement in nondividing cells are brought about by alterations in the pattern of H1 subtype synthesis coupled with the selective removal of certain H1 subtypes from chromatin. In dividing mouse embryo fibroblasts, the five H1 subtypes (H1a–H1e) are synthesized in proportions similar to the proportions of the subtypes present in the chromatin. H1 continues to be synthesized in confluent cells, but at a much lower rate; furthermore, the subtypes are synthesized in quite different proportions. The differences in synthetic rates can explain the alterations that occur in the amounts

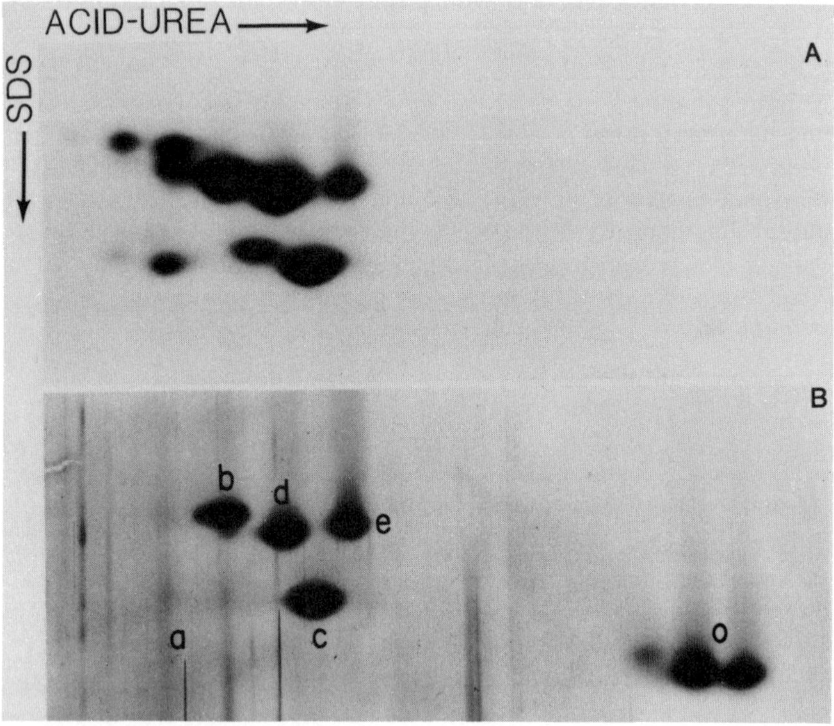

FIGURE 15.1. Two-dimensional gel electrophoresis of H1 proteins of dividing (*A*) and nondividing (*B*) 3T3 cells. Electrophoresis in acid–urea gels was the first dimension, in SDS gels the second.

of the subtypes present in chromatin only if the subtypes differ in metabolic stability as well. Thus the synthesis of both H1a and H1b fall to negligible rates in nondividing cells, and yet the loss of H1a is much faster than that of H1b. Also, the rates of synthesis of H1e and H1° are much lower than that of H1c, and yet the amounts of these two subtypes increase markedly relative to H1c. These results (16) show that H1a and H1c are both much less stable in nondividing cells than are the other subtypes. The H1c content of the chromatin is maintained despite this instability by its high rate of synthesis. H1a and H1c appear to be metabolically unstable in dividing cells as well, since Pehrson and Cole have observed a high rate of turnover in an electrophoretic component containing these two subtypes in both dividing and nondividing cells (26).

Changes similar to those occurring in nondividing cultured cells can also be seen in tissues of the intact mouse during and after periods of nondividing cell formation (16; Figure 15.2, table 15.2). In this case, since nondividing cell populations can be observed over much longer periods of time, and more extensive changes are seen. Tissues of one-week-old mice resemble those of dividing cultured cells in containing the five subtypes, H1a–H1e, and phosphorylated forms of each. In the adult, tissues with few dividing cells, such as liver, kidney, and lung, lack H1a and H1b almost entirely. They also lack all phosphorylated forms of H1. The loss of H1a from these tissues is almost complete by one month after birth and precedes that of H1b, which is almost complete by four months.

The loss of H1b from these tissues occurs principally during a time when there is very little cell division (27–35) and must therefore occur in nondividing cells in the absence of DNA replication. These observations, together with the results obtained with cultured cells (discussed previously) and lymphocytes (discussed later) indicate that both H1a and H1b are removed from the chromatin of nondividing cells. The traces of H1a and H1b still present at four months are probably confined principally to dividing cells and to cells of the immune system.

Concomitant with the decline in the amounts of H1a and H1b, there are marked increases in the amounts of H1e and H1°. There is also a small decrease in the amount of H1d and a small increase in that of H1c. The proportions of the H1 subtypes that are still present in significant amount in these tissues at four months—H1c, H1d, H1e, and H1°—differ somewhat from tissue to tissue: in liver, H1e is the major subtype, with H1c and H1d present in smaller but similar amounts; in kidney, H1e and H1c are present in almost equal amounts and there is less H1d; in lung, H1e is the most prominent subtype and H1d the least prominent. H1° is most prominent in liver and least prominent in lung; in all three tissues it represents 20–30% of the total H1 protein. We do not know whether the tissue differences seen at four months persist in aging mice.

In contrast to these three tissues, the thymus gland of the adult mouse retains levels of subtypes a–e similar to those present at one week and has no H1°, the only change during the first four months of age being a diminution of phosphorylated forms (16; Figure 15.2).

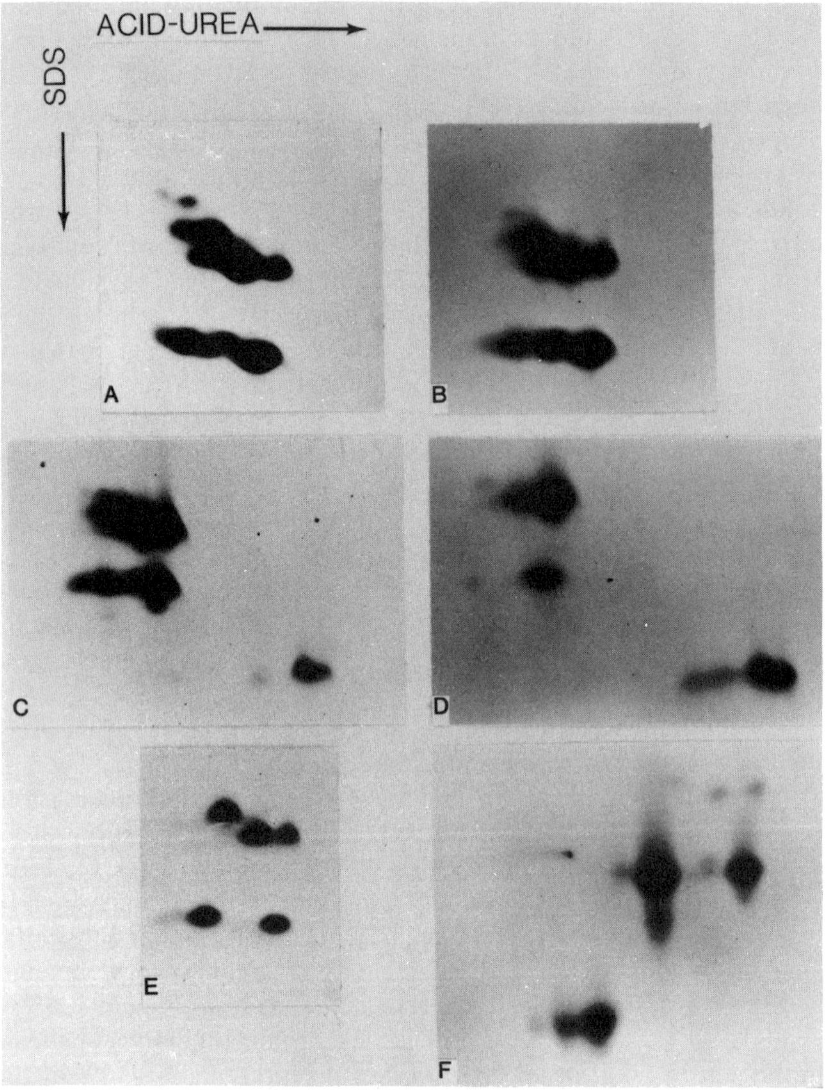

FIGURE 15.2. Two-dimensional gel electrophoresis of H1 fractions prepared from tissues of mice aged 1 week (A, C, E) and 4 months (B, D, F). Tissues are thymus (A, B), liver (C, D), and testis (E, F).

TABLE 15.2. Developmental Characteristics of H1 Subtypes[a]

	H1a	H1b	H1c	H1d	H1e	H1t	H1°
Present in dividing cells and tissues of young animals	+	+	+	+	+		±
Present in nondividing nonlymphoid cells of adults			+	+	+		+
Present in nondividing lymphocytes and macrophages[b]	(+)	+	+	+	+		
Present in prepachytene spermatocytes[c]	+	?	+	?	?		
Present in pachytene spermatocytes and spermatids	+		+			+	
Unnecessary for mitotic cell division[c]		+		+	?		+

[a] Data are from references 15–17, 45.
[b] Long-lived lymphocytes may eventually lose H1a.
[c] The extremely low levels of H1b, d, e, and ° present in the testis may be restricted to nondividing somatic elements (°, d, and e) or type a spermatogonia (b, d, and e).

The presence of H1a and H1b in the thymus cannot be due entirely to the dividing cell population, which, although substantial, constitutes only 10% of the cells of this tissue (36–38). In large part, it is because the other 90% of the cells, although nondividing, are principally newly formed cells (the small thymocytes) which stay in the thymus for only 2–4 days (36–41). Losses of H1a and H1b and accumulation of H1e do eventually occur in these nondividing cells too, as was demonstrated by the observation that peripheral lymphocytes have lower levels of H1a and H1b than do the small thymocytes from which they are derived (16). Thus lymphocytes undergo many of the changes observed in other nondividing cells, although it is not yet known if the loss of H1a and H1b is ever complete. However, lymphocytes, as well as macrophages (unpublished data), differ from other nondividing cells in that they lack H1° (16). This difference in chromatin composition might be of significance in regard to the ability of such cells to commence rapid cell division in the course of an immune response.

The H1 complement of the spleen of the 1-week-old mouse resembles that of the thymus. In the adult mouse, however, the spleen is quite different from the thymus and other somatic tissues; it contains no phosphorylated forms of H1, a reduced level of H1a, an increased level of H1e, and very little H1° (16), as expected of a tissue enriched in macrophages and lymphocytes.

The observations on the differences in H1 complement and patterns of subtype synthesis between growing and nongrowing cells can be summarized as follows (15, 16): (1) Most dividing cells both in culture and *in vivo* contain H1a–H1e and lack H1°. (2) All subtypes present in dividing cells are phosphorylated during interphase. (3) Phosphorylated forms of H1 are usually restricted to dividing cells. (4) Nondividing cells synthesize the H1 subtypes at lower rates and in different proportions than do dividing cells. In particular, they do not synthesize H1a or H1b. (5) Different H1 subtypes have different metabolic stabilities, H1a and H1c being more unstable than H1b, d, and e. (6) Because of the alterations in patterns of synthesis and the differing stabilities of individual subtypes, the H1 complements of nonlymphoid nondividing cells consist principally of H1c, H1e, and H1° with traces of H1d and little or no H1a or H1b.

A number of conclusions can be drawn from these observations. (1) With the exception that lymphoid cells do not contain H1°, the subtypes in somatic tissues are not tissue specific, nor are the alterations following the cessation of cell division tissue specific. (2) Phosphorylations of H1 are not essential for the production or maintenance of diversity in chromatin structure in nondividing cells. (3) H1b, H1d, and H1° are not essential for cell division. (4) The various subtypes have unique metabolic properties that are responsible, at least in part, for producing the changes in H1 subtype ratios that occur in somatic tissues during growth, and for the differences in subtype ratios between the thymus and other somatic tissues in the adult mouse. The metabolic differences among subtypes also include differences in degree and types of phosphorylation that they undergo (see Section 4).

The metabolic differences among the murine H1 subtypes are summarized in Table 15.1 and the developmental differences in Table 15.2. As described in Section 5, these differences allow the identification of homologues of the murine H1 subtypes in other mammals.

3. H1 SUBTYPE CHANGES DURING SPERMATOGENESIS

The H1 complement of the testis of the very young rat or mouse resembles that of somatic tissues at the same age, containing subtypes a-e in similar amounts (17, 42, 43). In the adult rat or mouse, however, the testis has a high level of H1t—a subtype not found at all in somatic tissues—and a similarly high level of H1a, which, as noted in the preceding section, is only a minor component of adult somatic tissues. It has a smaller but substantial amount of H1c, and very low levels of H1b, H1d, H1e, and H1° (Figure 15.2) (17, 42–45). Recently, strong evidence has been obtained that all of the testis H1 subtypes except H1t are the same proteins as those present in somatic tissues (17).

As first shown by Seyedin and Kistler, the H1 composition of chromatin changes during spermatogenesis (43) and the H1 complements of the most abundant germ cells are unlike that of any somatic cell. By examining cell populations enriched for specific stages, Bucci et al. (45) showed that practically all of the cells of the testis other than the somatic cells have high levels of H1a, whereas H1t does not begin to be deposited in the chromatin until the pachytene stage. Subsequently, during the early spermatid stages, all of the histones are replaced by other basic proteins (46). Bucci et al. (45) concluded that H1a, b, c, d, and e are all synthesized in spermatogonia and early spermatocytes. They also concluded that in pachytene spermatocytes H1t replaces subtypes b, c, d, and e. However, the degree of purity of the cell fractions obtained was not sufficient to exclude the possibility that spermatogonia and prepachytene spermatocytes already lack all subtypes but b and c. Recent studies of the mouse testis indicate that in that species the prepachytene spermatocytes, and possibly spermatogonia as well, contain almost exclusively H1a and H1c, and that the principal subtype replaced by H1t during the pachytene stage is H1a (44).

In view of their results, Bucci et al. (45) proposed that H1a has a role associated with DNA replication, and that H1t plays roles that facilitate those chromatin functions occurring during the late stages of spermatogenesis. The notion that H1a is important for replication is consistent with the observed loss of this subtype from nondividing cells, described in the preceding section. However, it does not explain the fact that H1a is by far the predominant subtype in the chromatin of prepachytene germ cells but is only a small fraction of the total H1 protein in the chromatin of rapidly dividing somatic cells. An appealing hypothesis is that H1a and H1c allow extended chromatin conformations (25) that, in prepachytene spermatocytes, are required throughout the chromosomes for the precise pairing necessary for meiotic recombination (17), and that H1t is involved in subsequent chromosome condensations.

4. O-PHOSPHORYLATION OF H1 SUBTYPES

Early work by Stocken and his colleagues showed that H1 prepared from tissues actively engaged in cell division contains a substantial amount of phosphorus (47–49). Since that time, the occurrence of multiple *O*-phosphorylations of H1 in dividing cells has been demonstrated in cultured cells by several authors (15, 16, 18–21, 50–53), and recent work has indicated that similar phosphorylations occur in dividing cells in the intact animal in a number of tissues (15, 16). Nondividing cells, whether in culture or *in vivo*, do not contain any *O*-phosphorylated H1 (15, 16) except after stimulation by certain hormones, when a very small amount of H1 becomes phosphorylated at residue ~37 (54, 55). Hence *O*-phosphorylations of H1 are not essential for maintaining diverse ground state chromatin conformations, including active and inactive chromatin, but appear to participate in temporary alterations in chromatin structure that occur during the cell cycle and during activation of hormonally inducible genes.

In dividing cells, the phosphorylation state of H1 depends on the phase of the cell cycle. During interphase the H1 molecules contain small numbers of phosphate groups, generally between zero and four (19, 53). During mitosis (18, 19, 53) and correlated closely with mitotic chromosome condensation (53), they acquire three or four additional phosphate groups, which are then removed during chromosome decondensation in telophase.

The interphase phosphorylations of H1 follow a rather complex schedule. During most of the G_1 phase there is essentially no active phosphorylation, although a significant amount of the H1 present has retained one or more of the phosphate groups acquired previously, presumably during the preceding S phase. In HeLa cells, for example, a significant proportion of H1 molecules retain one or more phosphate groups throughout most of G_1 (19), and in the intact mouse, small thymocytes, which are recently formed nondividing cells, contain substantial amounts of phosphorylated H1 (16). If cells do not reenter S phase, all phosphate groups are lost from H1 (53, 16); however, it is not known if or when all previously acquired phosphate groups are removed in cells that continue to divide.

Active phosphorylation of H1 begins at about the time of onset of DNA replication (48, 50, 51). Recent work with HeLa cells on the phosphorylation sites of H1 indicates that most of the S-phase phosphorylation occurs at two sites within the highly cationic *C*-terminal half of the H1 molecule (20). These two S-phase phosphorylations are very different from one another. One occurs even if the onset of replication has been blocked by hydroxyurea, whereas the other fails to occur unless replicated DNA is allowed to accumulate. It was proposed that the first kind of phosphorylation is in preparation for replication and that the second kind occurs behind the replication forks, possibly maintaining replicated chromatin in a conformation in which DNA cannot be replicated again.

HeLa cell H1 has been resolved into two subfractions (19, 20), and these two types of S-phase phosphorylation have been shown to occur in both. In one of the subfractions they are the only phosphorylations that occur during the S phase, whereas in the other subfraction several additional phosphorylations occur in both the *N*-terminal and *C*-terminal regions, at relatively low rates in comparison with the two principal ones. The intramolecular sites of these minor S-phase phosphorylations are all major sites of phosphorylation during mitosis. It is thus evident that cells can exert subtype-selective control over the schedule of phosphorylation at several sites in the H1 molecule.

More recent studies, carried out with mouse cells and employing two-dimensional electrophoresis, have revealed clearly that dividing cells contain phosphorylated forms of all of their H1 subtypes (Figure 15.3). This includes H1° (56, 16) and H1t (17) when they are present. However, as can be seen in Figure 15.3, the subtypes differ markedly during interphase with respect to both the proportion of molecules phosphorylated and the number of phosphate groups per molecule (15). Figure 15.3 also reveals that different subtypes undergo qualitatively different types of phosphorylations. Some subtypes, such as H1b and H1c, undergo two kinds of phosphorylation, one that has no effect on the migration of the protein in SDS gels, and one that results in lowered mobility. The latter type of phosphorylation was first detected in *Drosophila* and its effect on mobility shown to be due to an effect on protein conformation (57, 58). In mammals, as far as we can tell, it is restricted to only one of the phosphorylation sites in the H1 molecule. Because of its conformational effects, it has been referred to as *c*-phosphorylation to distinguish it from the more common type of H1 phosphorylation undergone by all subtypes.

It is clear from the data described above that phosphorylation sites differ in a number of respects, including cell cycle schedule of phosphorylation, effects of their occupancy on the conformation of the molecule, and distribution among H1 subtypes. In some instances subtypes have been shown to lack a particular phosphorylation site: one rabbit thymus subtype lacks the seryl residue at position ~37 that is the site of hormone-induced phosphorylation in rat liver H1 (54–55), and one HeLa cell H1 subtype lacks the seryl residue at position ~17 that is one of the sites of phosphorylation during S phase and mitosis (20). These differences provide one type of mechanism for the subtype-selective control of H1 phosphorylation, and it appears that cells have other mechanisms as well.

5. RELATIONSHIPS BETWEEN MOUSE H1 SUBTYPES AND THOSE OF OTHER MAMMALS

Cole, Chalkley, and their collaborators showed that H1 proteins exhibit a high degree of variability from species to species when examined by column chromatography or one-dimensional gel electrophoresis (1–5). This divergence has prevented the detection of subtype homologies among various species by these methods. The metabolic

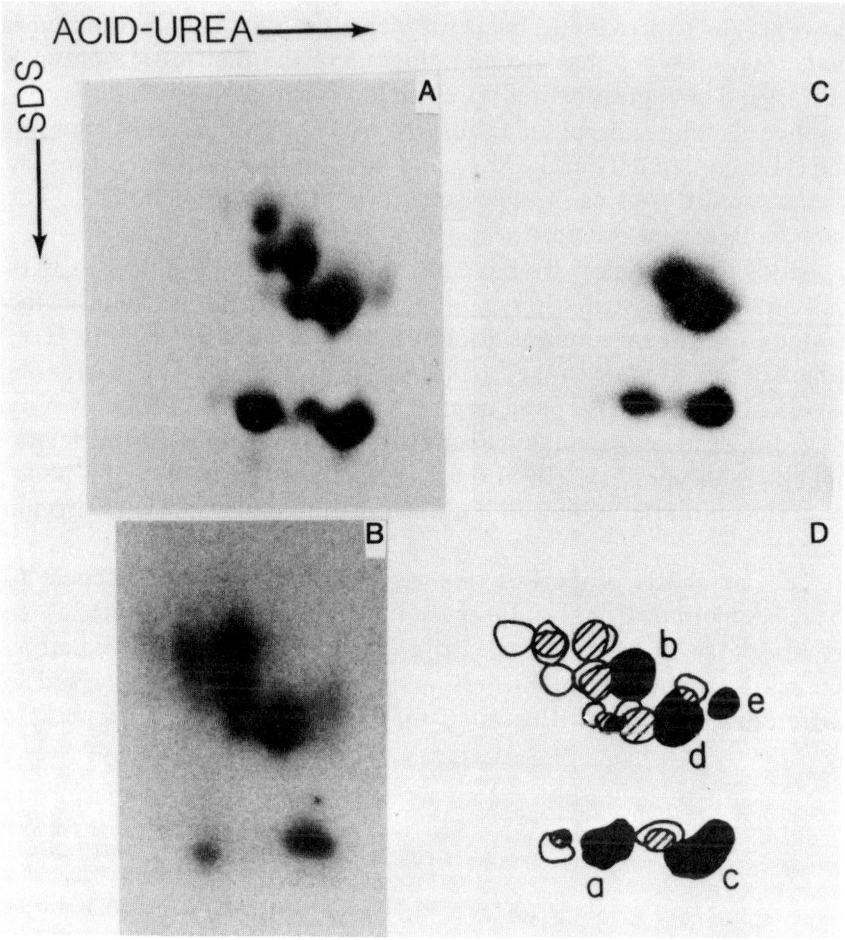

FIGURE 15.3. Two-dimensional gel electrophoresis of H1 proteins of F9 teratocarcinoma cells and identification of phosphorylated forms. (*A*) [^{3}H]lysine labeled H1; (*B*) ^{32}P-labeled H1; (C) [^{3}H]lysine labeled H1 treated with alkaline phosphatase before electrophoresis; (*D*) tracing of an autoradiogram of [^{3}H] lysine-labeled H1 superimposed on an autoradiogram of ^{32}P-labeled H1. Regions labeled only with ^{3}H are black, those only with ^{32}P are white, and regions labeled with both isotopes are hatched. Comparison of panels (*C*) and (*D*) shows that the only components remaining after alkaline phosphatase treatment are the five major spots not labeled with ^{32}P, representing the subtypes a–e.

and developmental properties described in the preceding sections, however, have provided a basis for identifying, in a number of mammals, H1 subtypes that are homologous to the mouse H1 subtypes (25, 59). The identification relies on the fact that the metabolic differences among subtypes manifest themselves in differences between the H1 complements of dividing and nondividing cells in culture, as well as in differences in the distribution of subtypes among tissues such as liver and thymus.

All of the mammals we have examined contain H1 subtypes with cell and tissue distributions similar to those seen in the mouse: the thymus or spleen contains proteins we can identify as H1a-H1e, whereas the liver or kidney has little or no H1a or H1b, increased levels of H1e and H1c, and contains H1°. In addition, as shown by Seyedin et al. (44), many mammalian species contain the testis-specific subtype, H1t. In at least the rat and the mouse, the phosphorylation patterns that distinguish different H1 subtypes are the same.

The metabolic properties that we have examined thus appear to have been quite strongly conserved in mammals in evolution (25). In contrast, the structures of the different H1 subtypes, as revealed by their electrophoretic mobilities, appear to have been conserved to different degrees, as illustrated diagrammatically in Figure 15.4.

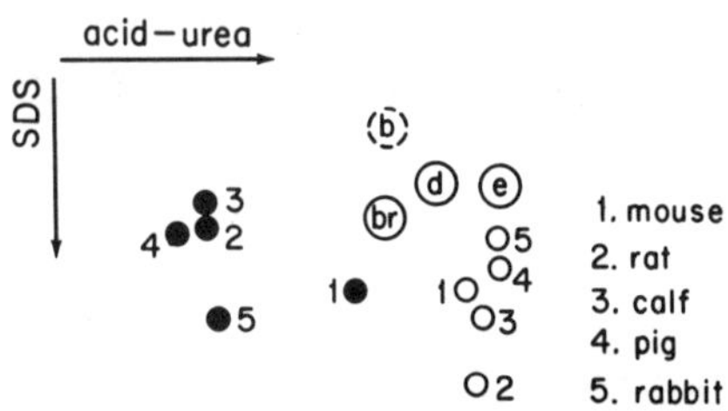

FIGURE 15.4. Drawing showing relative mobilities of H1 proteins from mammals in two-dimensional gel electrophoresis. The large open circles labeled d and e indicate the positions of subtypes d and e of all species. The circle labeled br indicates the position of rabbit H1b; the dotted circle labeled b indicates the positions of subtype b from the other species. Subtypes c are indicated by small open circles and subtypes a by small closed circles; the species identification of each of these subtype spots is given by the number adjacent to it. In mixtures, the subtype b spot was frequently oddly shaped, as though the individual spots overlapped but did not coincide exactly.

Subtypes d and e of all of the species examined are indistinguishable. Subtype b shows a small degree of variation: those of mouse, pig, and cow comigrate in both electrophoretic dimensions and can barely be resolved from that of rat, while that of rabbit is well separated from all of the others. Subtypes a and c are highly variable, differing considerably in every mammal examined, including the three rodents (25).

Thus the use of metabolic criteria allows one to codify the H1 subtypes despite changes in their structures (25).

6. DISCUSSION AND CONCLUSIONS

The work reviewed here establishes criteria by which the various mammalian H1 subtypes may be identified and compared. Seven (or eight if the two forms of H1° are counted separately) have now been identified. They differ from each other in a number of metabolic, developmental, and evolutionary properties. Thus they are synthesized at different relative rates in dividing and nondividing cells, they exhibit different metabolic and evolutionary stabilities, they differ with respect to the number and cell cycle schedule of their phosphorylations (see Tables 15.1 and 15.2), and their relative abundance varies from tissue to tissue. Thus far, all mammalian species examined have the same number of H1 subtypes with similar sets of characteristics.

The observed differences in metabolic stability among the subtypes, together with the changes in their synthetis rates when cells stop dividing, explain to a large extent the changes in the H1 complement that occur during development in tissues such as liver, kidney, and lung. They also explain some of the differences, in the adult, between the H1 complements of these tissues and that of the thymus, in which most of the nondividing cells that are formed leave the gland before alterations in the H1 composition of the chromatin occur. They do not, however, explain the observed differences in subtype ratios among liver, kidney, and lung of the adult animal or the absence from lymphocytes and macrophages of H1°, a protein of very high metabolic stability. Nor do they account for the overwhelming abundance in early spermatocyes of H1a, which is present in only modest amounts in dividing somatic cells and, moreover, is removed from the chromatin of nondividing somatic cells before any other subtype. It would appear, therefore, that the H1 composition of chromatin may be subservient to cell function rather than a simple consequence of metabolic properties inherent in the subtypes themselves.

Although the properties described here are not functions, they are probably related to function, and they provide insights into possible functional differences among the subtypes. The exceptionally low metabolic and evolutionary stabilities of H1a and H1c compared to those of other subtypes, for example, suggest that these two subtypes are less involved than the others in macromolecular interactions that hinder their dissociation from chromatin or constrain their freedom to evolve, as if they were situated in more open chromatin conformations than are the other subtypes. As pointed out earlier (Section 3), this possibility is consistent with the fact that H1a and H1c are by far the most abundant subtypes in the chromatin of prepachytene spermatocytes.

Because of such considerations, we have proposed (15, 25) that the H1 subtypes differ in their effects on chromatin conformation such that a given subtype or combination of subtypes produces a chromatin conformation that is different from those produced by other subtypes or combinations thereof. Perturbations from these various "ground state" chromatin conformations during replication and during hormone-induced gene expression require phosphorylations of H1 molecules. In cells traversing the cell cycle, the structure of every portion of the chromatin must be altered during a part of the S phase to allow replication of its DNA. The best explanation for the fact that the number of sites in the H1 molecule that are phosphorylated during the S phase is different for different subtypes is that the number of phosphorylations required to make the DNA accessible to DNA polymerase is related to the number of steps required to unfold the chromatin. Thus H1 subtypes that are present in highly folded chromatin ground states may have to undergo more phosphorylations than those that are present in simpler ground states. The fact that H1a and H1c undergo fewer phosphorylations than other H1 subtypes is thus consistent with our conclusion that they are in more open chromatin structures.

In most instances the structural differences between transcribed and nontranscribed chromatin cannot depend on phosphorylations of H1, since nondividing cells do not contain any detectable O-phosphorylated forms of H1. However, H1 proteins associated with genes that are inducible through a cAMP-dependent mechanism apparently undergo phosphorylation at a single site when the genes are induced (54). The absence of this particular phosphorylation site from some H1 subtypes represents a potential means for limiting such hormone-dependent changes to selected regions of the genome.

It has been suggested by Newrock et al. (60) and Weintraub et al. (61) that histone subtypes may become distributed nonrandomly in chromosomes by mechanisms operating during cell division and differentiation. It was pointed out by these two groups that if such distributions were different in different cells, histone subtypes could play a role in cell differentiation even if individual subtypes were not tissue specific. Two important assumptions of this hypothesis are that different cell types have different subtype ratios and that there are functional differences among the subtypes. The results obtained with mammalian H1 substantiate these assumptions. However, direct evidence that the subtype arrangement in chromatin is nonrandom or that subtypes have different effects on chromatin conformation is still to be obtained.

Two relatively unexplored aspects of H1 subtypes are the structural bases for their functional differences and the regulation of their synthesis. The control of H1 gene expression is apparently complex, since a number of different patterns of H1 synthesis have been detected, depending on the tissue or cell type and on whether the cells are dividing or not (see Sections 2 and 3 and Tables 15.1 and 15.2.). As for structure, only one tentative sequence of a mammalian H1 protein has been published (68). However, partial sequence data for a few H1 proteins is available (19, 20, 65–70), and these indicate that the central region of H1 is much more conserved, both among mammalian species (68) and among subtypes within a species (20), than are *N*-terminal and *C*-terminal regions but that the central region of H1° differs substantially from that of other mammalian H1 proteins (69, 70). At present, although a number of mammalian histone genes have been isolated, no mammalian H1 genes have been cloned (62–64). Such genes would provide valuable information concerning H1 subtype structures and would be useful in understanding the regulation of H1 gene expression.

The work reviewed here provides strong support for the view that each H1 subtype has unique functions, and should stimulate further efforts to determine what these functions are and their importance in chromatin function. H1 genes will also be of interest as an example of a family of genes whose members have had different degrees of freedom to evolve.

ACKNOWLEDGMENTS

Work from the laboratory was supported by grants GM-24019, CA-06927, and RR-05539 from the National Institute of Health and an appropriation from the Commonwealth of Pennsylvania.

REFERENCES

1. M. Bustin and R. D. Cole, *J. Biol. Chem.* **243,** 4500–4505 (1968).
2. J. M. Kinkade, Jr., *J. Biol. Chem.* **244,** 3375–3386 (1969).
3. J. M. Kinkade, Jr. and R. D. Cole, *J. Biol. Chem.* **241,** 5790–5797 (1966).
4. J. M. Kinkade, Jr. and R. D. Cole, *J. Biol. Chem.* **241,** 5798–5805 (1966).
5. S. Panyim, D. Bilek, and R. D. Chalkley, *J. Biol. Chem.* **246,** 4206–4215 (1971).
6. R. H. Stellwagen and R. D. Cole, *J. Biol. Chem.* **243,** 4656–4668 (1968).
7. R. H. Stellwagen and R. D. Cole, *J. Biol. Chem.* **244,** 4878–4887 (1969).
8. P. Hohmann and R. D. Cole, *Nature* **223,** 1064–1066 (1969).
9. P. Hohmann and R. D. Cole, *J. Mol. Biol.* **58,** 533–560 (1971).
10. S. Panyim and R. D. Chalkley, *Biochem. Biophys. Res. Commun.* **37,** 1042–1049 (1969).
11. S. Panyim and R. D. Chalkley, *Biochemistry* **8,** 3972–3979 (1969).
12. A. Shirer, M. P. Carpenter, and R. D. Chalkley, *Proc. Natl. Acad. Sci. USA* **72,** 2716–2718 (1975).
13. W. S. Kistler and M. E. Geroch, *Biochem. Biophys. Res. Commun.* **63,** 378–384 (1975).
14. S. M. Seyedin and W. S. Kistler, *Biochemistry* **18,** 1371–1375 (1979).

15. R. W. Lennox, R. G. Oshima, and L. H. Cohen, *J. Biol. Chem.* **257,** 5183–5189 (1982).
16. R. W. Lennox and L. H. Cohen, *J. Biol. Chem.* **258,** 262–268 (1983).
17. R. W. Lennox and L. H. Cohen. To be submitted to *Dev. Biol.* (1983).
18. P. Hohmann, R. A. Tobey, and L. R. Gurley, *J. Biol. Chem.* **251,** 3685–3692 (1976).
19. K. Ajiro, T. W. Borun, and L. H. Cohen, *Biochemistry* **20,** 1445–1454 (1981).
20. K. Ajiro, T. W. Borun, S. D. Shulman, G. M. McFadden, and L. H. Cohen, *Biochemistry* **20,** 1454–1464 (1981).
21. D. J. Wilkinson, B. G. Shinde, and P. Hohmann, *J. Biol. Chem.* **257,** 1247–1252 (1982).
22. J. Pehrson and R. D. Cole, *Nature* **285,** 43–46 (1980).
23. J. A. D'Anna, R. A. Tobey, and L. R. Gurley, *Biochemistry* **19,** 2656–2671 (1980).
24. C. Pieler, G. R. Adolf, and P. Swetly, *Eur. J. Biochem.* **115,** 329–333 (1981).
25. R. W. Lennox, *J. Biol. Chem.* **259,** in press (1984).
26. J. R. Pehrson and R. D. Cole, *Biochemistry* **21,** 456–460 (1982).
27. S. Leto, G. C. Kokkonen, and C. H. Barrows, Jr., *J. Gerontol.* **31,** 149–154 (1976).
28. S. Leto, G. C. Kokkonen, and C. H. Barrows, Jr., *J. Gerontol.* **31,** 144–148 (1976).
29. C. H. Barrows, Jr. and G. Kokkonen, *Growth* **42,** 71–85 (1978).
30. B. Schultze and W. Oehlert, *Science* **131,** 737–738 (1960).
31. R. Baserga and W. E. Kisieleski, *J. Natl. Canc. Inst.* **28,** 331–339 (1962).
32. J. D. Simnett and A. G. Heppleston, *Lab. Invest.* **15,** 1793–1801 (1966).
33. I. Y. R. Adamson, D. H. Bowden, M. G. Cote, and H. Witschi, *Lab. Invest.* **36,** 26–32 (1977).
34. I. L. Cameron, *Texas Rep. Biol. Med.* **28,** 203–248 (1970).
35. S. M. Tilghman and A. Belayew, *Proc. Natl. Acad. Sci. USA* **79,** 5254–5257 (1982).
36. M. Matsuyama, M. Wiadrowski, and D. Metcalf, *J. Exp. Med.* **123,** 559–576 (1966).
37. D. Metcalf and M. N. Wiadrowski, *Cancer Res.* **26,** 483–491 (1966).
38. B. J. Bryant, *Eur. J. Immunol.* **2,** 38–45 (1972).
39. K. Borum, *Cell Tissue Kinet.* **6,** 545–552 (1973).
40. L. E. Feinendegen, A. J. Heiniger, G. Friedrich, and E. P. Cronkite, *Cell Tissue Kinet.* **6,** 573–585 (1973).
41. R. G. Scollay, E. L. Butcher, and I. L. Weissman, *Eur. J. Immunol.* **10,** 210–218 (1980).

42. S. M. Seyedin and W. S. Kistler, *Biochemistry* **18,** 1376–1379 (1979).
43. S. M. Seyedin and W. S. Kistler, *J. Biol. Chem.* **255,** 5949–5954 (1980).
44. S. M. Seyedin, R. D. Cole, and W. S. Kistler, *Exp. Cell Res.* **136,** 399–405 (1981).
45. L. R. Bucci, W. A. Brock, and M. L. Meistrich, *Exp. Cell Res.* **140,** 111–118 (1982).
46. T. C. Rodman, S. D. Litwin, M. Romani, and G. Vidali, *J. Cell Biol.* **80,** 605–620 (1979).
47. M. G. Ord and L. A. Stocken, *Biochem. J.* **107,** 403–410 (1968).
48. W. S. Stevely and L. A. Stocken, *Biochem. J.* **110,** 187–191 (1968).
49. R. A. Buckingham and L. A. Stocken, *Biochem. J.* **117,** 157–160 (1970).
50. D. B. Marks, W. K. Paik, and T. W. Borun, *J. Biol. Chem.* **248,** 5660–5667 (1973).
51. L. R. Gurley, R. A. Walters, and R. A. Tobey, *J. Cell Biol.* **60,** 356–364 (1974).
52. L. R. Gurley, R. A. Walters, and R. A. Tobey, *J. Biol. Chem.* **250,** 3936–3944 (1975).
53. L. R. Gurley, J. A. D'Anna, S. S. Barham, L. L. Deaven, and R. A. Tobey, *Eur. J. Biochem.* **84,** 1–15 (1978).
54. T. A. Langan, *Proc. Natl. Acad. Sci. USA* **64,** 1276–1283 (1969).
55. T. A. Langan, S. C. Rall, and R. D. Cole, *J. Biol. Chem.* **246,** 1942–1948 (1971).
56. J. A. D'Anna, L. R. Gurley, R. R. Becker, S. S. Barham, R. A. Tobey, and R. A. Walters, *Biochemistry* **19,** 4331–4341 (1980).
57. M. Blumenfeld, *Biochem. Genet.* **17,** 163–166 (1979).
58. P. C. Billings, J. W. Orf, D. K. Palmer, D. A. Talmage, C. G. Pan, and M. Blumenfeld, *Nucl. Acids Res.* **6,** 2151–2164 (1979).
59. R. W. Lennox and L. A. Abrams, *J. Cell. Biol.* **95,** 83a (1982).
60. K. M. Newrock, C. R. Alfageme, R. V. Nardi, and L. H. Cohen, *Cold Spring Harbor Symp. Quant. Biol.* **42,** 421–431 (1978).
61. H. Weintraub, S. J. Flint, I. M. Leffak, M. Groudine, and R. M. Grainger, *Cold Spring Harbor Symp. Quant. Biol.* **42,** 401–407 (1978).
62. D. B. Sittman, I. M. Chiu, C. J. Pan, R. H. Cohn, L. H. Kedes, and W. F. Marzluff, *Proc. Natl. Acad. Sci. USA* **78,** 4078–4082 (1981).
63. F. Sierra, A. Lichtler, F. Marashi, R. Rickles, T. Van Dyke, S. Clark, J. Wells, G. Stein, and J. Stein, *Proc. Natl. Acad. Sci. USA* **79,** 1795–1799 (1982).
64. F. Sierra, A. Leza, F. Marashi, M. Plumb, R. Rickles, T. Van Dyke, S. Clark, J. Wells, G. S. Stein, and J. L. Stein, *Biochem. Biophys. Res. Commun.* **104,** 785–792 (1982).

65. S. C. Rall and R. D. Cole, *J. Biol. Chem.* **246,** 7175–7190 (1971).
66. G. M. T. Jones, S. C. Rall, and R. D. Cole, *J. Biol. Chem.* **249,** 2548–2553 (1974).
67. L. W. Liao and R. D. Cole, *J. Biol. Chem.* **256,** 3024–3029 (1981).
68. R. D. Cole, in P. Tso, Ed. *The Molecular Biology of the Mammalian Genetic Apparatus*, Elsevier/North Holland, 1977, pp. 93–104.
69. B. J. Smith, J. M. Walker, and E. W. Johns, *FEBS Lett.* **112,** 42–44 (1980).
70. J. R. Pehrson and R. D. Cole, *Biochemistry* **20,** 2298–2301 (1981).

CHAPTER 16

ORGANIZATION AND EXPRESSION OF HUMAN HISTONE GENES

GARY S. STEIN
FELIPE SIERRA
MARK PLUMB
FARHAD MARASHI
LISA BAUMBACH
Department of Biochemistry and Molecular Biology
University of Florida College of Medicine
Gainesville, Florida

JANET L. STEIN
NADINE CAROZZI
KEITH PROKOPP
Department of Immunology and Medical Microbiology
University of Florida College of Medicine
Gainesville, Florida

CONTENTS

1. INTRODUCTION

2. STRUCTURE AND ORGANIZATION OF HUMAN HISTONE GENES
   - 2.1. The General Organization of Human Histone Genes
   - 2.2. Specific Human Histone Coding Sequences
   - 2.3. Putative Regulatory Sequences

3. THE REGULATION OF HISTONE GENE EXPRESSION
   - 3.1. Are Histone Proteins Synthesized in a Cell Cycle Specific Manner?
   - 3.2. Is the Cellular Abundance of Histone mRNA under Cell Cycle Control?
     - 3.2.1. Is the Abundance of Histone mRNA Temporally Coupled to DNA Replication?
     - 3.2.2. Are Histone mRNAs Periodically Synthesized during the Cell Cycle?
   - 3.3. Is There More Than One Type of Regulation of Histone Gene Expression?
   - 3.4. Can Histone Gene Expression Be Uncoupled from DNA Replication?

4. RELATIONSHIP BETWEEN HISTONE GENE EXPRESSION AND DNA REPLICATION

5. CONCLUSION

REFERENCES

1. INTRODUCTION

In this chaper we summarize several of the experimental approaches we have been taking to examine human histone genes. The structure and organization of human histone genes is discussed, particularly within the context of the putative relationships of specific regions of the genes to their expression. Approaches to assessing the levels at which control of histone gene expression resides are also considered. Results are presented that suggest: (1) Human histone genes are a family of moderately reiterated sequences with variations in the structure, organization, and possibly in the regulation of the various copies. (2) At least 15 different, though not necessarily all, human histone genes are coordinately expressed during the S phase of the cell cycle and appear to be temporally and functionally coupled with DNA replication. (3) There are both transcriptional and post-transcriptional components to the regulation of those histone genes expressed in conjunction with DNA replication.

2. STRUCTURE AND ORGANIZATION OF HUMAN HISTONE GENES

2.1. The General Organization of Human Histone Genes

Several years ago when we initially screened a λCh4A human gene library [constructed by Lawn, et al. (1)] and isolated clones containing genomic human histone sequences, our expectation for the organization of the human histone genes was a simple tandem repeat such as had been observed for *Drosophila* (2) and "early" sea urchin histone genes (3, 4), that is, a repeated unit of approximately 7 kb in length containing one of each of the five histone genes (H2A, H2B, H3, H4, and H1), with a nontranscribed spacer sequence between each mRNA coding region. We anticipated this type of organization because it was known at the time that the human histone genes were moderately reiterated (approximately 40 copies per haploid genome for the sequences coding for each of the five histone proteins) (5), and *in situ* hybridization studies, in which human chromosomes were hybridized with radiolabeled human mRNA (6) or sea urchin histone cRNA (7), suggested clustering of the histone genes.

However, it soon became apparent that although the human histone genes were clustered, they were not organized in the form of a simple tandem repeat (8). Rather, as shown in the restriction maps in Figure 16.1, the human histone gene clones we have isolated and characterized exhibit at least four types of arrangements with respect to restriction sites and the representation and order of coding sequences; each of these arrangements is totally distinguishable from the others. In all the clones we have studied, at least two different histone genes are found adjacent to each other (within a few kilobases). Clones λHHG 5, λHHG 55, and λHHG 415 (Figure 16.1*a*) each contain one of each of the four core histone genes, H2A, H2B, H3, and H4. On the other hand, clones λHHG 6, λHHG 17, and λHHG 22 each contain two genes coding for each of the inner core histones, H3 and H4 (Figure 16.1*a*). Clone λHHG 415, which was isolated by screening the human library with a chicken genomic H1 histone DNA probe (9), contains an H1 coding region, as well as at least one of each of the other histone genes (10). This indicates that in humans at least some H1 histone genes are clustered together with the core histone gene sequences.

Two other laboratories have reported the isolation and characterization of human histone genes. Shown in Figure 16.1*b* are maps of those clones isolated by Heintz et al. (11) and Clark (9) in which the organization of histone genes differs from those in our clones. Thus far, a total of nine different types of human histone gene clusters have been reported (clones λHh2 and λHh7 isolated by Heintz et al. appear to be identical to clones λHHG 39 and λHHG 17, respectively). It is likely that the clones identified to date are not a representative sample of the entire spectrum of possible arrangements of the histone genes in the human genome because the selection was strongly biased toward clones containing H4 histone genes. Heintz et al. used a cloned human H4 probe to select their clones (11), whereas most of our clones were isolated using a probe containing chicken H3 plus H4 coding sequences (provided by Dr. J. R. E. Wells). Clark used a probe containing sequences for all four core histone genes from chicken (H2A, H2B, H3, and H4) and selected two human histone gene clones, neither of which contains H4 coding sequences (9). Interestingly, there is no obvious overlap between the various types of clusters that have been isolated.

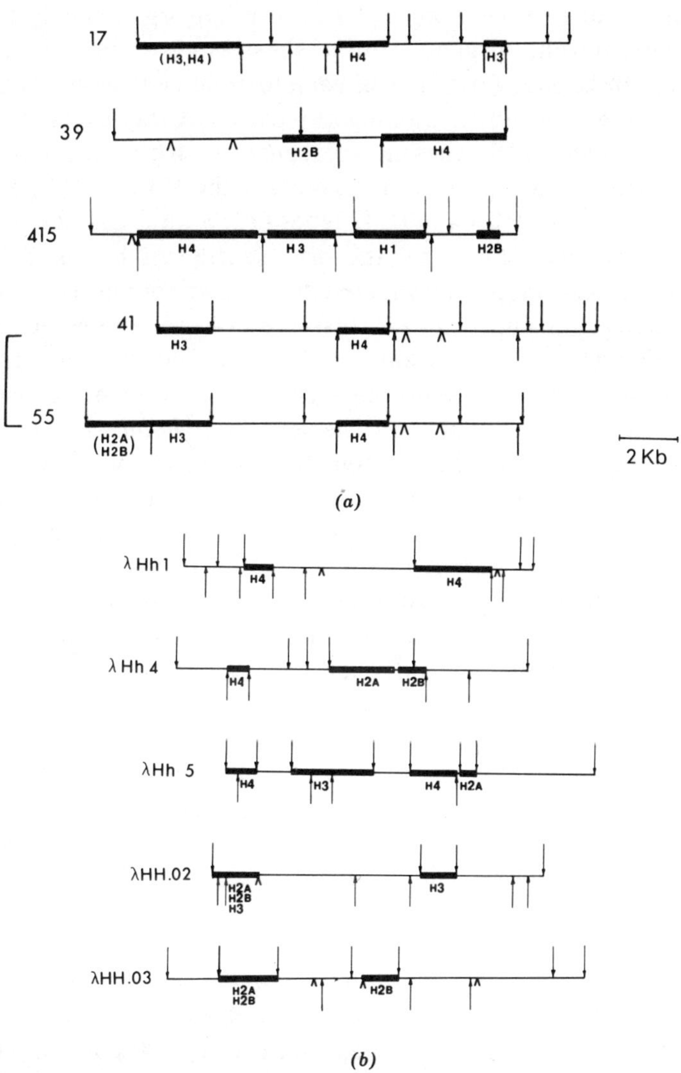

FIGURE 16.1. Restriction maps of cloned human histone genes. (*a*) λHHG clones isoalted by Sierra et al. (8). λHHG clones were mapped with respect to EcoRI (↓), Hind III (↑), and Bam HI (^) restriction endonucleases. The bar at the bottom of the figure indicates the scale. Boxes indicate restriction fragments that hybridize with heterologous histone gene probes. The locations of histone coding regions within the λHHG phage have been confirmed by hybrid selection—*in vitro* translation. (*b*) Clones isolated in other laboratories. λHh clones were isolated by Heintz et al. (11), λHH clones by Clark and Wells (9).

Analysis of histone genes in several other organisms has shown that a polymorphic organization of these genes is prevalent in eucaryotes. In yeast, two copies of each histone gene have been found per haploid genome (12). Cloning of these DNA fragments has shown that the H2A and H2B genes are adjacent to each other with divergent transcriptional polarity (12, 13) and that the H3 and H4 genes are not adjacent to the H2A and H2B genes (12; Smith, Chapter 1). More recent work has shown that the organization of histone genes in vertebrates is far more complicated than shown for yeast. In *Xenopus laevis*, the genes appear to be clustered (14, 15; and see Chapter 8 by Van Dongen et al.), and some clusters are tandemly repeated (16); however, more than one gene order has been found, each one associated with a different H1 variant (14). The newt *Notophthalamus viridescens* has homogeneous 9 kb clusters containing one each of the five histone genes; however, these clusters are not arranged in tandem repeats, but are separated by up to 50 kb of unrelated DNA sequences (17, 18; see Chapter 9 by Stephenson). Chicken and mouse histone genes, like human, have a polymorphic organization with no obvious simple tandem repeats (see Chapter 10 by Harvey and Wells, and Chapters 11 and 12 by Engel, Marzluff, and Graves.)

Of course, the possibility that tandem repeats exist in higher vertebrates cannot be formally excluded based on the available information. It is conceivable that some of the clones that have been studied represent fragments derived from a larger tandem repeat. Detailed studies of genomic histone gene organization (in progress) should resolve whether any of the histone gene clusters observed in higher vertebrates (chicken, mouse, and human) are tandemly repeated.

It has been known for several years that the genomes of higher eukaryotes contain, in addition to single-copy sequences, large amounts of DNA of moderate to high repetition (19, 20). In most eukaryotes repeated sequences approximately 300 nucleotides long have been found interspersed throughout the genome (21, 22). In humans the Alu DNA family is predominant among these reiterated sequences, with a repetition frequency of approximately 3×10^5 per haploid genome (23). Experiments in which EcoRI-restricted λHHG phage DNA was hybridized to an Alu I DNA sequence cloned into pBR322 (generously provided to us by Dr. S. Weissman) have shown that the histone genes are interspersed with several members of the Alu family of repetitive DNA sequences, as well as with other DNA sequences whose transcripts are represented in the polysomes of HeLa S3 cells (24). Of particular interest is the location of Alu sequences in the regions flanking several H4 histone genes (P. Kroeger, M. Plumb, J. Stein, and G. Stein, unpublished observations).

The last point we consider with regard to the general organization of human histone genes is some of the experimental evidence for sequence variations in the different copies of the H4 histone genes, despite the stringent evolutionary conservation of the amino acid sequence of the H4 histone proteins. Histone protein amino acid sequences are well conserved evolutionarily; however, variant proteins have been detected for all histones except H4 (25).

Our first indication of the polymorphic nature of human H4 histone genes preceded the availability of cloned genomic histone sequences and was based on direct analysis of the H4 histone mRNAs (26). By electrophoretically fractionating H4 histone mRNAs, initially on the basis of size under denaturing conditions and subsequently on the basis of charge and secondary structure under nondenaturing conditions, Litchler et al. identified at least seven different H4 histone mRNA species in S phase HeLa S3 cells. Presumably, these different species of H4 mRNA are all functional, as they were isolated from polysomes. *In vitro* translation of the individual mRNA species in a wheat germ system gives rise to acid-soluble proteins that comigrate with HeLa cell H4 histone proteins in an acetic acid–urea polyacrylamide gel. In addition, tryptic peptide mapping of *in vitro* translation products from several of the H4 histone mRNAs gave patterns that were indistinguishable from one another or from *in vivo* synthesized HeLa H4 histone mRNA species, all of which code for identical or very similar polypeptides (26, 27).

Two lines of experimental evidence indicate that the various H4 histone mRNAs are encoded in individual genes. Initially we demonstrated differences in the two-dimensional "fingerprints" of the T1 ribonuclease oligonucleotides from the seven H4 histone mRNA species. The oligonucleotides obtained from the smaller molecular weight H4 mRNAs were not subsets of those generated from the larger H4 mRNAs, which rules out a simple product-precursor relationship. The results are similar to Grunstein's findings for H4 histone mRNA fractions from sea urchin (28). More direct evidence that the human H4 histone mRNAs are independently encoded comes from experiments in which various H4 histone mRNA species were assigned to specific H4 histone genes. By hybridizing a mixture of all the HeLa H4 histone mRNAs to various cloned genomic human H4 histone DNA sequences, we were able to demonstrate that each H4 histone gene formed an S1 nuclease-resistant hybrid with only one H4 mRNA. Furthermore, where two H4 histone genes are present in the same cluster, as in clones λHHG6 and λHHG17, different H4 histone mRNA sequences are encoded in the adjacent H4 genes (Figure 16.1*a*). We have initiated studies directed toward the identification, isolation, and characterization of multiple forms of other core histone mRNAs—H2A, H2B, and H3 species. We have been able to identify at least four distinct H2A, H2B, or H3 histone mRNA species in several human cell lines, including normal diploid, transformed, and tumor cells.

The obvious question that arises is the biological significance of multiple forms of H4 histone mRNAs that are genetically encoded and serve as templates for the synthesis of apparently identical histone polypeptides. Our current thinking encompasses a working model with the following components: All histone genes in a cluster are coordinately controlled, with their expression being modulated by common regulatory sequences and/or regulatory molecules. Only a subset of the reiterated histone genes is expressed in any given cell at any given time, with variations in those histone genes (gene clusters) expressed in different cells and/or in different biological circumstances. Selection of clusters to be transcribed would be based on a requirement for a histone H2A, H2B, H3, or H1 subspecies where differences in the mRNAs and in the histone proteins have been observed. The H4 genes expressed would be predicated on location within a cluster containing coding sequences for a specific H2A, H2B, H3, or H1 histone subspecies—with the possibility that all genes in such a cluster contain similar regulatory sequences.

Alternatively, multiple forms of H4 histone mRNAs in HeLa cells and other human cell lines may have arisen by conservative nucleotide substitutions occurring in multiple genes. For example, two H4 histone genes expressed during the same early developmental stages of the sea urchin *Psammechimus miliaris* show as much as 10.3% divergence in the nucleotide sequences of the H4 coding region (29, 30), yet both code for identical H4 proteins.

2.2. Specific Human Histone Coding Sequences

To facilitate detailed analysis of specific human histone genes and their flanking regions, we have subcloned essentially all EcoRI fragments from our eight genomic histone clones (λHHG) into pBR322. Several of these subclones have been further characterized by restriction endonuclease mapping and DNA sequencing.

DNA sequences are available (Figure 16.2) for the H4 histone gene present in λHHG41 (pF0108A), and partial sequences have been determined for the H3 genes in λHHG17 (pST519) and λHHG41 (pF0535) and for the H2b genes in λHHG39 (pTN521) and λHHG55 (pFF435D). Because amino acid sequence data are available for only the H2B histone protein of humans (31), the nucleotide sequences of the human H3 and H4 histone genes have provided the first direct information pertaining to the primary structures of these proteins.

Except for the H2B gene present in λHHG55 (described later), the human histone genes we have sequenced have the capacity to code for typical histone proteins that do not differ greatly from their counterparts in other species. As shown in Table 16.1, the H4 gene in λHHG41 codes for an H4 histone protein that does not differ at all from that of calf (25). Similarly, the regions of the two human H3 genes we have sequenced code for the same amino acid sequence found at the amino terminus of the calf H3 histone protein. In comparing the carboxy terminal 40 amino acids predicted from the nucleotide sequences of the two human H2B genes, we find only 7.5% divergence, consistent with the conservation of the carboxy termini of H2B proteins from other species (25).

Careful examination of the H2B sequences present in λHHG55 (pFF435D) (Figure 16.2) indicates that, despite the extensive homology with other H2B coding sequences, this gene cannot express a functional H2B protein because it contains at least two frameshift mutations, located at the positions coding for amino acids 6 (Ser) and 92 (Arg). Furthermore, several putative regulatory sequences are absent from the flanking regions of this gene (32): (1) While TATA boxes are usually located 65–90 bases upstream from the ATG initiation codon of H2B genes (33), there is none within 130 nucleotides of the AUG of the λHHG55 H2B gene. (2) No translation stop codon occurs at or near the end of the amino acid coding region of the gene. (3) The histone mRNA termination motif ACCA, which character-

```
-240     -230        -220        -210        -200        -190
AATTC TCCCG GGGAC CGTTG CGTAG GCGTT AAAAA AAAAA AAGAG TGAGA GAGGG ACTGA

-180     -170        -160        -150        -140        -130
GCAGA GTGGA GGAGG AGGGA GAGGA AAACA GAAAA GAAAT GACGA AATGT CGAGA GGGCG

-120     -110        -100        -90         -80         -70
GGGAC AATTG AGAAC GCTTC CCGCC GGCGC GCTTT CGGTT TtCAA TCTGG TCCGA TAtCt

-60      -50         -40         -30         -20         -10
CtGTA TATtA CGGGG AAGaC GGtGa CGCtC CGatC GaNcN Nctat CGGGC TCCtG CGGTC

ATG TCC GGC tGt GGa aAG GGC GGA AAG GGC TTA GGC AAA GGT GGC GCT AAG CGC
Met0Ser1Gly Arg Gly Lys5Gly Gly Lys Gly Leu10Gly Lys Gly Gly Ala15Lys Arg

CAC CGC AAG GTC TTG AGA GAC AAC ATT CaG GGC ATC ACC aAG CCT GCC aTT CGG
His Arg Lys20Val Leu Arg Asp Asn25Ile Gln Gly Ile Thr30Lys Pro Ala Ile Arg35

CGT NTA GCT CGG CGT GGC GGC GTT AAG CGG ATC TCT GGC CTC ATT TAC GAG GAG
Arg Leu Ala Arg Arg Gly Gly Val Lys Arg Ile Ser Gly Leu Ile Tyr Glu Glu
                 40                   45                   50

ACC CGC GGT GTG CTG AAa GTG TTC TTG GAG AAT GTG ATT CGG GAC GCA GTC ACC
Thr Arg Gly Val Leu Lys Val Phe Leu Glu Asn Val Ile Arg Asp Ala Val Thr
     55                   60                   65                   70

TAC ACC GAG CAC GCC AAG CGC AAG ACC GTC ACA GCC ATG GAT GTG GTG TAC GCG
Tyr Thr Glu His Ala Lys Arg Lys Thr Val Thr Ala Met Asp Val Val Tyr Ala
              75                   80                   85

CTC AAG CGN CAG GGG AGN aCC CtC TAC GGC TTC GGA GGC TAG GCCGC CGCTC
Leu Lys Arg Gln Gly Arg Thr Leu Tyr Gly Phe Gly Gly Stop
 90                   95                   100     102
                                                  mRNA 3' end
CAGCT TTGCA CGTTT CGATC CCAAA GGCCC TTTTT GGGCC GACCA CTTGC TCAtC CTGAG

GAGTT GGACA CTTGA CTGCG TAAAG TGCAA CAGTA ACGAT GTTGG AAGGT AACTT TGGCA

GTGGG GCGAC AATCG GATCT GAAGT TAACG GAAAG acata accgc
```

(a)

FIGURE 16.2. Nucleotide sequences of several human histone genes. (*a*) The H4 gene present in λHHG 41 (pF0108A) and its flanking regions. Lower case letters indicate residues that were not definitively resolved. An "N" indicates an undetermined nucleotide. The boxes at the 5′ end of the gene indicate the location of the tandem "CAAT" boxes, while the box just preceding the 3′ end of the mRNA coding region indicates the T-hyphenated dyad symmetry. The box located just past these nucleotides indicates the histone-related purine box. Wavy underlines in the 5′ flanking region indicate, in order from 5′ to 3′: (1) 21 base pairs of purines, (2) histone-related GGTCC motif, and (3) TATA box. Horizontal arrows indicate short direct repeats.

istically follows the 3′ hyphenated dyad symmetry regions of histone genes (33) has been replaced by the sequence ACCC.

This nonfunctional H2B sequence poses two questions currently under investigation; the first is the possibility that the gene was modified during the cloning manipulations. We believe this unlikely, as the genetic alterations are so varied, ranging from frameshift mutations to the loss of putative regulatory sequences, while the coding region itself is 99% (190/192 bases) homologous to functional H2B gene sequences. Second, it is of interest to determine whether the H2B sequence is a "pseudogene" located in a cluster (λHHG55) containing other functional histone genes (H2A, H3, and H4, see Figure 16.1). This type of arrangement has been described for both the α- and β-globin gene clusters, where pseudogenes are interspersed with functional members of each of these gene families (34, 35). Alternatively, λHHG55 may represent a histone gene cluster that is completely nonfunctional (a pseudocluster). In this regard it is noteworthy that λHHG41 (see Figure 16.1), an independent isolate analogous to λHHG55, contains two histone genes (H3 and H4) which are apparently

```
       -59
pF0535 AATT CATGC ACGTT TAGGT CCCCG CCGCT GCGCC GGTAA AGCCT GTGTT TTGGT TCGCT
pST519                                                 ACCTT TTCTG TGGTT TTGCC

       1
   535 ATG GCC CGT ACT AAG CAG ACT GCT CGC AAG TCG ACC GGC GGC AAG GCC CCG AGG
   519 ATG GCT CGT ACT AAA CAG ACA GCT CGG AAA TCC ACC GGC GGT AAA GCG CCA CGC
           Ala Arg Thr Lys Gln Thr Ala Arg Lys Ser Thr Gly Gly Lys Ala Pro Arg
                                               10

   535 AAG CAG CTG GCC ACC AAG GCG GCC CGC AAG AGC GCG CCG GCC ACG GGG GGG GTG
   519 AAG CAG CTG GCT ACC AAG GCT GCT CGC AAG
       Lys Gln Leu Ala Thr Lys Ala Ala Arg Lys Ser Ala Pro Ala Thr Gly Gly Val
                20                                  30

   535 AAG AAG CCG CAC CGC TAC CGG CCT GGG ACC GTA GCC CTG CGA GAG ATC CGA CGC
       Lys Lys Pro His Arg Tyr Arg Pro Gly Thr Val Ala Leu Arg Glu Ile Arg Arg
                        40                                      50

   535 TAC CAG AAG TCC ACG GAA CTA
       Tyr Gln Lys Ser Thr Glu Leu
                                60
```

(b)

FIGURE 16.2. (*continued*). (*b*) The 5′ end of the H3 genes present in λHHG 41 (pF0535) and λHHG 17 (pST519).

(*Figure continues on p. 408.*)

functional (see Figure 16.2). Therefore, sequence analyses of the H4 and H3 genes in λHHG55 should clearly distinguish between these two possibilities.

Examination of the histone nucleotide sequence comparison presented in Table 16.1 indicates that there is no more nucleotide divergence between histone genes of different vertebrates than there is between different human histone genes coding for the same (or

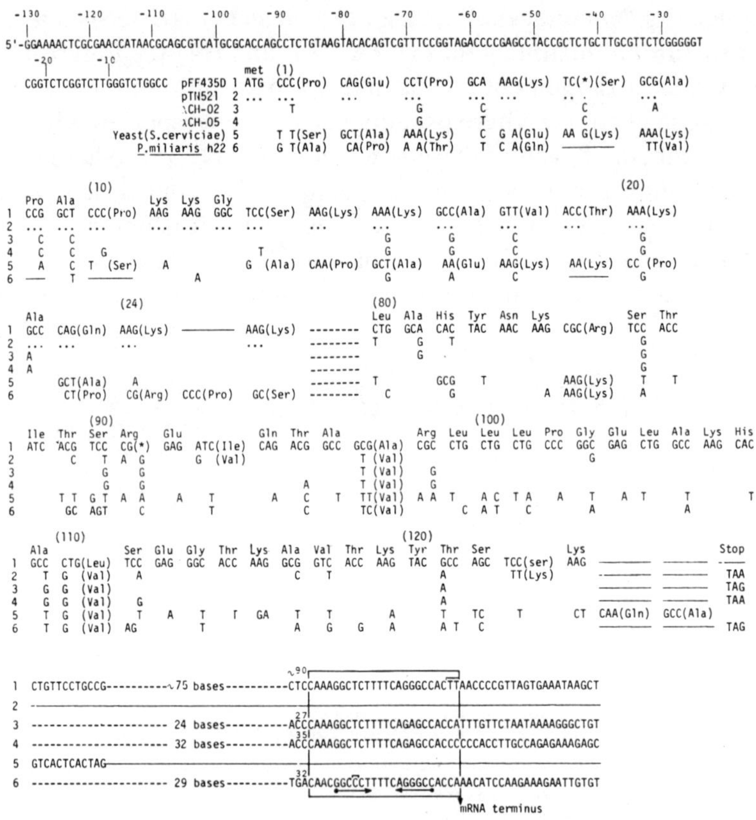

(c)

FIGURE 16.2. *(continued)*. (*c*) Partial sequences of the H2B genes present in λHHG 39 (pTN 521) and in λHHG 55 (pFF435D). Asterisks above nucleotides coding for amino acids 6 and 92 indicate the positions of deletions in pFF435D discussed in the text.

very similar) proteins. This observation is consistent with the identification of multiple histone H4 mRNAs in HeLa cells (26) and with the presence of multiple size classes for core histone mRNAs in both HeLa (36) and other human cells in culture.

The complete H4 nucleotide sequence presented in Figure 16.2 indicates the absence of intervening sequences within the protein coding region of this gene. Similar observations have been made for most histone genes studied so far, with the sole exception of a chicken H3 gene (37). No intervening sequences are apparent in the noncoding region of the H4 mRNA, based on comparison with other H4 histone genes, where the 3′ end of the mRNA is usually located at the ACCA motif immediately following the T-hyphenated dyad symmetry shown in Figure 16.2, and a TATA box is located approximately 50 nucleotides upstream from the ATG initiation codon. The S1 nuclease protection experiments of Lichtler et al. (26) and primer extension analysis carried out by Sierra et al. (38) provide further evidence that the λHHG41 H4 histone gene contains no intervening sequences.

There is only one instance in which two different histone protein coding regions from the same λHHG phage have been sequenced. Data obtained for the H3 and H4 genes in λHHG41 indicate that these two genes are transcribed convergently from opposite strands of DNA.

2.3. Putative Regulatory Sequences

Regions 3′ and 5′ to the mRNA coding sequences of genes transcribed by RNA polymerase II have been postulated to be involved in the control of gene expression. To gain insight into the regulation of human histone gene expression, we have focused our attention on the sequences flanking the H4 gene present in λHHG41. The complete nucleotide sequence has been determined (38; Figure 16.2) and the functional roles of various regions of the gene have been examined.

The 3.1-kb EcoRI fragment from λHHG41 containing the H4 histone gene was subcloned into the Eco RI site of pBR322. Figure 16.3 shows a detailed restriction map of this clone (pF0108) and of a subclone (pF0108A) containing only the 1.8-kb EcoRI/HindIII fragment. pF0108A does not contain the Alu I sequence found in pF0108 (indicated by the white arrow in Figure 16.3*A*).

TABLE 16.1. Nucleotide Sequence Comparison for Different Organisms

Histone	Clone Designation	Literature Reference	Number of Nucleotides Compared	Nucleotide Divergence (%)		Amino Acid Divergence (%)	
				pFF435D	pTN521	pFF435D	pTN521
H2B	pFF435D (human)	32	117	—	17.5	—	7.5
	pTN521 (human)	32	117	17.5	—	7.5	—
	λCH-02 (chicken)	99	116	8.3	10.8	7.5	0
	λCH-03 (chicken)	99	117	10.8	15.0	10	2.5
	λCH-05 (chicken)	99	117	10.2	11.7	7.5	7.5
	(yeast)	13	117	42.0	35.8	7.5	2.5
	(sea urchin *P. miliaris*)	100	117	28.3	24.1	7.5	2.5
				pF0535		pF0535	
H3	pST519 (human)	101	111	18.0		0	
	pHU 2.6 (chicken)	9	78	19.2		0	
	pCH 3dR (chicken)	102	156	14.7		0	
	h 22 (early sea urchin)	29, 30	156	23.0		0	
	λLpH (late sea urchin)	103	156	19.9		0	

				pF0108A	pF0108A
H4	pHU4A (human)	11	193	14.1	3
	mus-hi-1 (mouse)	104	303	13.2	0
	px1 ch4 (*Xenopus*)	105, 106	186	19.3	0
	x1-hi-1 (*Xenopus*)	105, 106	303	18.1	0
	pcx1H4W2 (*Xenopus*)	105, 106	303	17.5	0
	pcx1H4W1 (*Xenopus*)	105, 106	303	18.1	0
	Nv 51	107	138	19.2	0
	(*Notophthalamus*)				
	h 22 (sea urchin)	29, 30	181	20.0	0
	h 19 (sea urchin)	29, 30	303	19.8	0
	pSp2 (sea urchin)	108	302	21.9	0

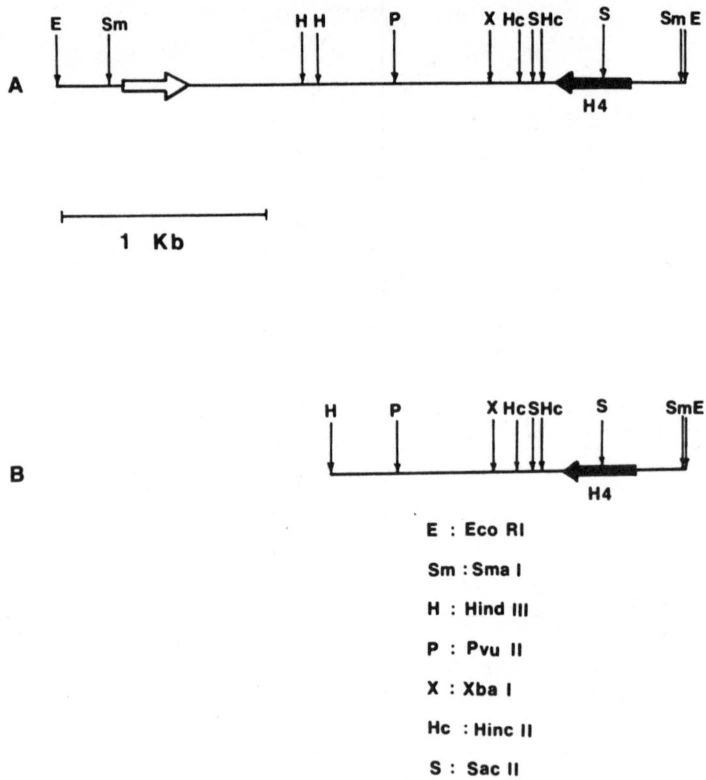

FIGURE 16.3. Restriction maps of λHHG 41 and subclones pF0108 and pF0108A. λHHG 41 is a λCh4a recombinant clone containing H3 and H4 human histone DNA sequences; its isolation and characterization have been described elsewhere (8, 38). pF0108 contains the 3.1-kb EcoRI fragment of λHHG41 (containing the H4 histone gene) inserted into the EcoRI site of pBR322. This 3.1-kb fragment also contains a member of the Alu family of repeated DNA sequences, indicated by the white arrow. The 1.8-kb EcoRI/Hind III fragment containing the H4 gene, but not the Alu sequence, has been subcloned into EcoRI/Hind III digested pBR322 and is designated pF0108A. E = EcoRI; H = HindIII; B = BamHI; Sm = SmaI; P = PstI; X = XbaI; Hc = HincII; S = SacII.

The complete nucleotide sequence of the pF0108A H4 gene, including 5′ and 3′ flanking regions is presented in Figure 16.2. Analysis of the sequences upstream from the ATG translation initiation codon indicates the presence of several sequences that may be involved in some functional or regulatory capacity *in vivo*. For example, the sequence GGTCC which occurs approximately 10 bp upstream from the TATA box is very similar to the GATCC motif found in an analogous position in several sea urchin histone genes (33). Further upstream are two tandem CAAT boxes (indicated in Figure 16.2 by closed boxes) remarkably similar to homology blocks found in most other genes transcribed by RNA polymerase II (39, 40), including several histone genes (33). Although in some cases, for example, the H2A histone gene present in the sea urchin clone h22, two CAAT boxes have been found in tandem arrangement, this homology block has not been observed in other H4 or H1 histone genes (33). Despite the apparent conservation of these homology blocks, it must be emphasized that their functional significance is as yet unknown.

Further upstream from the H4 histone coding region are several nonrandom sequences. Notably, between nucleotides -152 and -174 there is a stretch of 21 nucleotides containing only A and G residues, most usually in the form of the trinucleotide GGA (see Figure 16.2). Similar although not identical stretches have been found in the spacer regions of other histone genes (33, 41), and a role in recombination has been proposed. Finally, several short direct repeats (indicated by horizontal arrows in Figure 16.2) are present in the 5′ flanking region of the H4 gene. It is well documented that direct repeats present in the SV40 and retrovirus genomes act as strong promoters of transcription, both *in vivo* and *in vitro* (42, 43).

The 3′ flanking region of the H4 gene present in pF0108A exhibits many features found in histone genes of other species. The mRNA is most likely terminated *in vivo* at the ACCA motif found a few nucleotides downstream from the hyphenated dyad symmetry, which is characteristic of histone mRNA 3′ ends (33). Interestingly, this region of the RNA can, theoretically at least, form a stem and loop structure, similar to that which appears to be involved in transcription termination by RNA polymerase III (44). Birchmeier et al. (45) found this sequence to be necessary, although not sufficient, for correct termination of transcription of a sea urchin H2A histone gene microinjected into *Xenopus* oocytes. Deletions of the hyphenated dyad symmetry sequence induced readthrough by RNA polymerase II, but reinsertion of the same motif in the middle of an H2B gene did not by itself promote premature termination of transcription at this point. Twelve nucleotides downstream from the ACCA motif there is another histone gene-related motif, characterized mainly by its high A + G content. No specific function has yet been ascribed to these sequences.

To test the functional relevance of these putative regulatory sequences, we used an *in vitro* transcription assay based on the whole HeLa cell lysate described by Manley et al. (46). Preliminary experiments showed that under the appropriate conditions, *in vitro* transcription of EcoRI/Hind III-digested pF0108A DNA produced a correct runoff transcript of 1.8 kb in length (38). This transcript was sensitive to low concentrations of α-amanitin (2μg/ml), implicating the involvement of RNA polymerase II, which has been reported to transcribe human histone genes *in vivo* (47). Primer extension analysis indicated that these transcripts were initiated at the same site used *in vivo*.

This *in vitro* transcription system was used to test a series of deletion mutants constructed from pF0108A, which spanned almost all the 5′ flanking region of the H4 gene but did not include the TATA box (38). The TATA box has been implicated in numerous systems as playing a role in directing the precise site of initiation of transcription by RNA polymerase II (39, 48–50).

Deletion mutants of pF0108A were constructed by BAL-31 nuclease digestion of EcoRI-linearized DNA, followed by the addition of EcoRI linkers and cloning into pBR322. Figure 16.4 shows the sequences upstream from the H4 gene and indicates the end points of the deletions determined for several clones (vertical arrows).

Several of these deletion mutants were selected for further analysis. They were restricted with either HindIII or with HindIII plus EcoRI, and then transcribed in the *in vitro* system. All the clones examined gave rise to an *in vitro* transcript of the expected size (Figure 16.5), including clone pF0108A 5′Δ80, which lacks the direct repeats as well as the "CAAT" boxes previously described. The same deletion mutants exhibited a decreased level of this specific transcript when the DNA template was digested only with HindIII. Apparently, EcoRI cleavage (5′ to the H4 gene) can provide an artificial site of entry for the RNA polymerase, which then begins transcribing at its normal site. Primer extension experiments have shown that the *in vitro* transcripts from clone pF0108A 5′Δ80 are initiated at the *in vivo* 5′ initiation site, as demonstrated by comparison with HeLa cell polysomal RNA (Figure 16.6). We can therefore conclude that no sequences upstream from the TATA box are required for site-specific

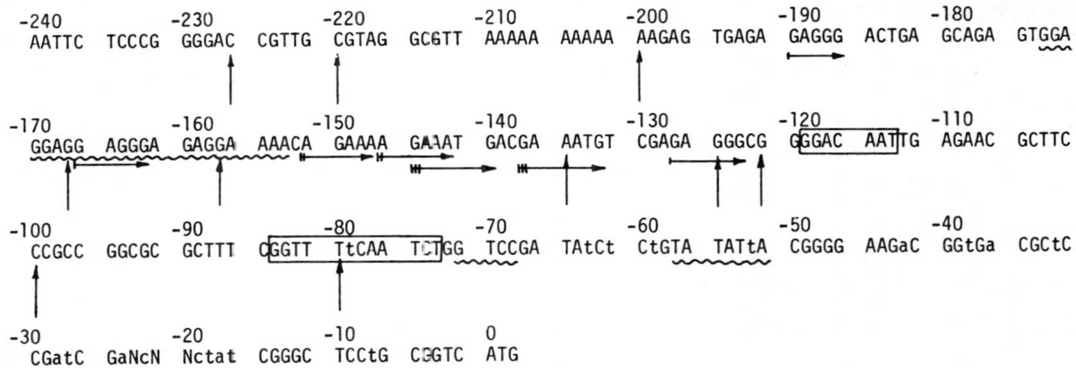

FIGURE 16.4. End points of 5′ deletion mutants of pF0108A. Sequences upstream from the 5′ end of the H4 coding region are shown. Vertical arrows indicate the end points of the different deletion mutants. Nucleotide residues are numbered from the AUG initiation codon and decrease in the upstream direction. The deletion clones are designated by the nucleotide at which the end point of the deletion lies. Deletion mutants were constructed by digestion of EcoRI-linearized pF0108A DNA with BAL-31 nuclease to remove nucleotides located upstream from the H4 gene. Synthetic EcoRI linkers were added, and the DNA was exhaustively digested with EcoRI and Hind III restriction endonucleases and then ligated to calf intestine alkaline phosphatase-treated, EcoRI/HindIII-digested pBR322. Ligation mixes were used to transform *E. coli* strain HB 101.

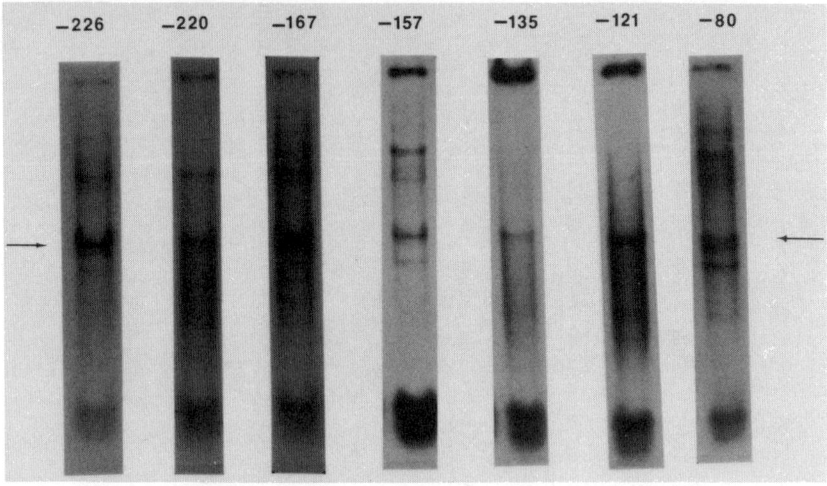

FIGURE 16.5. *In vitro* transcription of deletion mutants constructed from pF0108A. Autoradiograms of 1.5% agarose, 3% formaldehyde gels showing the transcripts obtained from selected deletion mutants. The template DNA was digested with EcoRI and HindIII restriction endonucleases. The arrows indicate the position of the expected readthrough transcript. Numbers above the lanes refer to the position of the deletion end point with respect to the AUG initiation codon.

initiation of *in vitro* transcription of the H4 histone gene present in pF0108A.

To determine if sequences downstream from the 3′ end of the H4 histone gene are required for *in vitro* transcription, our initial approach was to use pF0108 DNA that had been truncated by digestion with several different restriction endonucleases (38). These experiments were performed using double digestions of pF0108 DNA. DNA was first digested to completion with EcoRI restriction endonuclease, thus separating the insert containing the H4 histone gene from vector sequences and eliminating the possibility of interference due to the presence of promoters in the pBR322 vector. The DNA was then cleaved with each of the other restriction endonucleases indicated in Figure 16.3. Transcriptions were performed in parallel in the presence or absence of 2 μg/ml of α-amanitin to identify transcripts produced by RNA polymerase II, the enzyme responsible for transcription of the histone genes *in vivo* (49).

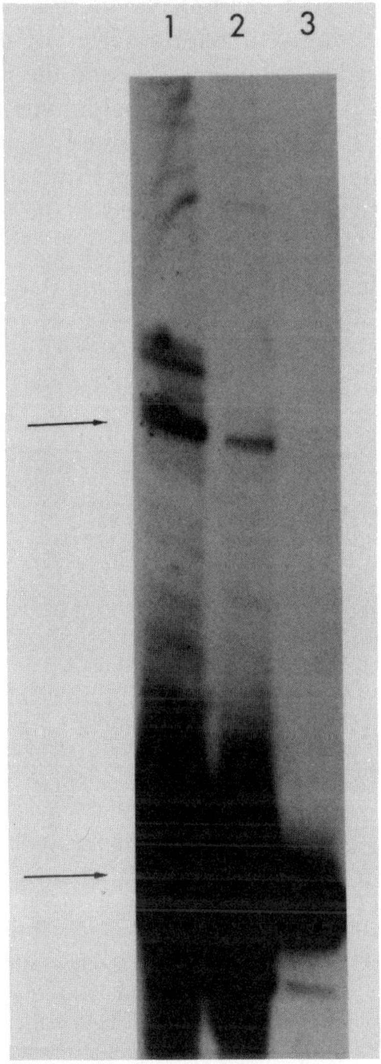

FIGURE 16.6. Primer extension analysis of transcripts generated using pF0108A 5′Δ80 as a template. Autoradiogram of a 10% polyacrylamide/8.3 M urea gel showing the DNA fragments obtained after extension of the 64-bp primer by AMV reverse transcriptase, using different RNA samples as templates. Lane 1: HeLa polysomal RNA. Lane 2: *in vitro* transcripts from pF0108A 5′Δ80. Lane 3: Unhybridized prmer. The positions of the unextended and the extended primers are indicated by arrows.

Figure 16.7*a* shows an example of the type of results obtained when pF0108 DNA digested with a series of different restriction endonucleases was transcribed *in vitro* and the resulting transcripts were analyzed on a 1.5% agarose/formaldehyde gel. The black dots at the left of lanes 1, 3, 5, 7, 9, 10, and 12 indicate the positions expected for the migration of transcripts initiated at the correct 5′ terminus of the H4 gene and terminated at the point of truncation.

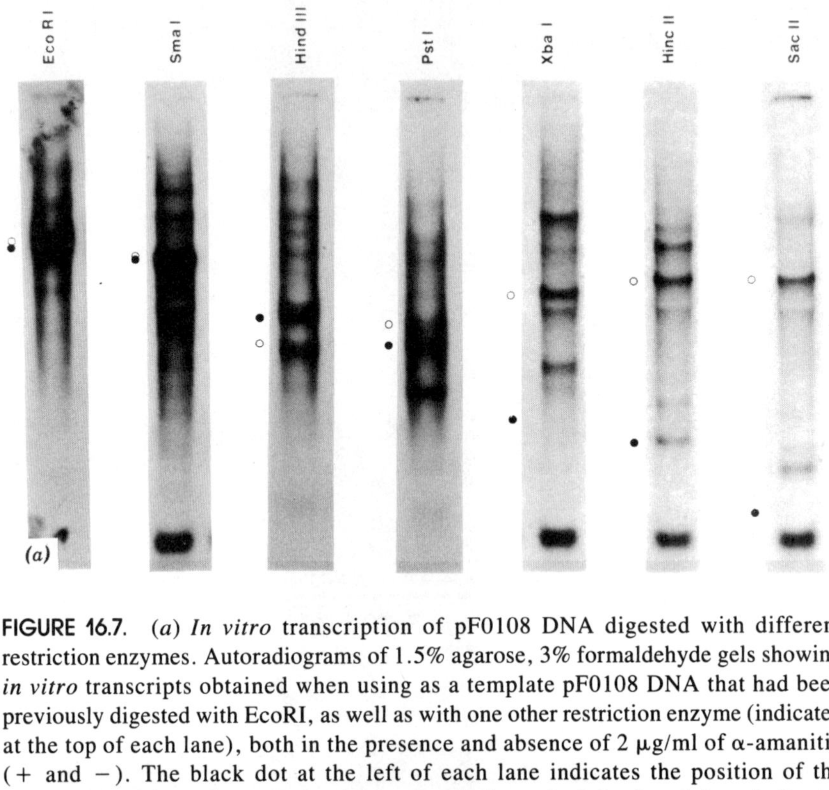

FIGURE 16.7. (*a*) *In vitro* transcription of pF0108 DNA digested with different restriction enzymes. Autoradiograms of 1.5% agarose, 3% formaldehyde gels showing *in vitro* transcripts obtained when using as a template pF0108 DNA that had been previously digested with EcoRI, as well as with one other restriction enzyme (indicated at the top of each lane), both in the presence and absence of 2 μg/ml of α-amanitin (+ and −). The black dot at the left of each lane indicates the position of the expected readthrough transcript. The open circle at the left of each lane indicates the position of the α-amanitin insensitive, insert-dependent transcript discussed in the text. (*b*) Primer extension analysis of *in vitro* transcripts. Autoradiogram of a 10% polyacrylamide/8.3 M urea gel showing the DNA fragments obtained after extension of the 64 bp primer by AMV reverse transcriptase, using different RNA samples as templates. Lane 1: HeLa polysomal RNA. Lanes 2–5: *in vitro* transcripts obtained using as a template pF0108A DNA digested with EcoRI restriction endonuclease. In addition Lane 2: HindIII; Lane 3: PstI; Lane 4: XbaI; Lane 5: HincII. The positions of the unextended and the extended primers are indicated by arrows.

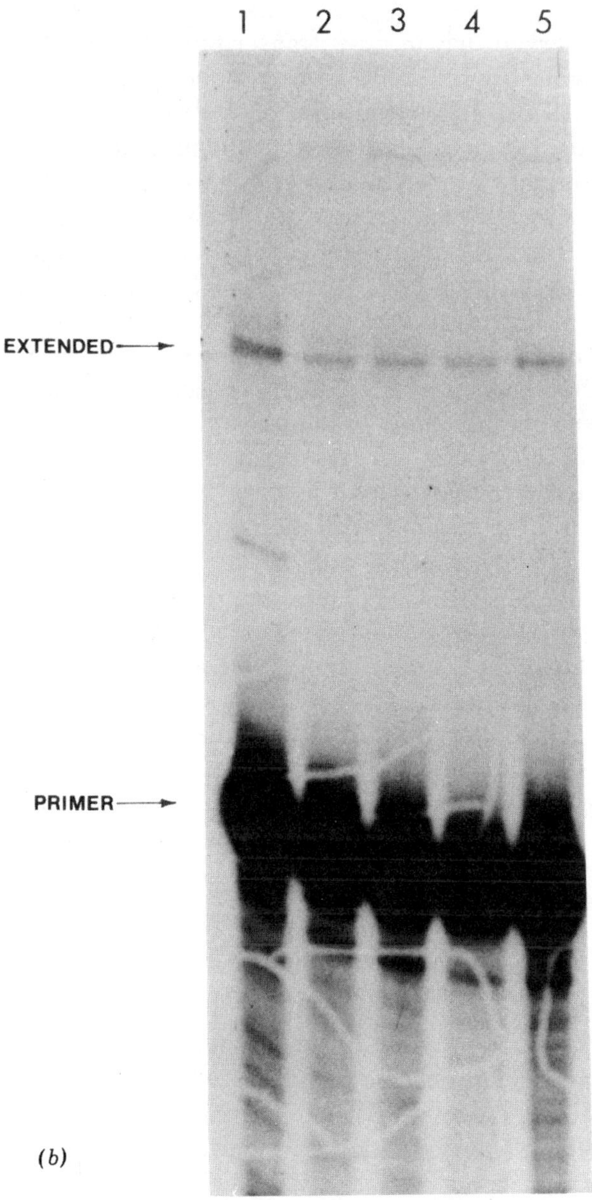

(b)

It can be observed that *in vitro* transcription of pF0108 DNA that had been digested with EcoRl, either alone or in combination with SmaI, HindIII, or PstI restriction endonucleases, gives rise to an α-amanitin sensitive runoff transcript of the predicted size. However, when the template was truncated closer to the 3′ end of the gene (with XbaI, HincII or SacII), no runoff transcript of the appropriate size was observed. A transcript of the predicted size is present among the *in vitro* transcripts obtained using EcoRI/HincII-digested pF0108 as template; however, this transcript is not sensitive to low concentrations of α-amanitin (Figure 16.7) and is also produced when EcoRI/HincII-digested pBR322 DNA is used as a template.

The transcripts indicated by the open circles in Figure 16.7*a* are most likely products of *in vitro* transcription initiated at the Alu DNA sequence present in pF0108 (white arrow in Figure 16.3), since they are not sensitive to low concentrations of α-amanitin [Alu DNA is transcribed *in vitro* by RNA polymerase III (51, 52)] and are not produced when pF0108 DNA is used as a template.

The *in vitro* transcription of truncated pF0108 and pF0108A DNA templates shows that, under appropriate conditions, specific RNA polymerase II transcripts as long as 2.8 kb can be produced. However, removal of sequences within or downstream from the 3′ end of the H4 histone gene abolished the production of transcripts of the expected size. Analysis of these transcription products on 5% polyacrylamide gels containing 8.3 M urea failed to resolve smaller, specifically terminated transcripts (data not shown). These results suggested that sequences located at the 3′ end of the gene might enhance *in vitro* initiation and/or elongation of transcription of the H4 histone gene present in pF0108.

To further explore these possibilities, a series of primer extension experiments (53, 54) was designed to determine if initiation of transcription had indeed occurred in cases where no runoff transcript was detected. A 64-bp AluI/HhaI fragment, containing those nucleotides coding for amino acids 17 through 38 (see Figure 16.2), was used as a primer to assay for specific initiation of transcription. If accurate initiation occurred, this DNA primer would be elongated to a molecule of approximately 160 nucleotides in length. Figure 16.7*b*, lane 1, shows the extended primers obtained with HeLa polysomal RNA. Two bands are resolved, indicating that more than one of the multiple HeLa H4 histone mRNAs (26) has sufficient sequence homology with the 64-bp fragment from pF0108A to form stable hybrids; differences in the 5′ leader probably account for the size heterogeneity observed in the extended primer molecules. Lanes 2–5 shows the extended primers obtained with RNAs transcribed *in vitro* from pF0108A DNA digested with both EcoRI and either Hind III (lane 2), PstI (lane 3), XbaI (lane 4), or HincII (lane 5). In all cases, a single band, comigrating with one of the two extended primers observed with HeLa polysomal RNA, was resolved.

These results confirmed the previous finding that specific and accurate initiation of transcription of the H4 histone gene in pF0108A does occur *in vitro*. Furthermore, specific initiation of transcription was observed with all the templates analyzed, despite the fact that two of these truncated templates (pF0108A DNA digested with restriction endonucleases EcoRI and either XbaI or HincII) gave no detectable α-amanitin sensitive runoff transcript. Although this assay is only semiquantitative, the apparent rate of initiation of transcription was similar in all cases.

In summary, our results seem to indicate that truncating the pF0108A template closer than approximately 800 nucleotides from the 3′ end of the H4 gene either inhibits elongation of transcription or affects transcript stability, since site-specific initiation still occurs at approximately normal rates yet no full-size transcripts are produced.

3. THE REGULATION OF HISTONE GENE EXPRESSION

The analysis of gene expression *in vivo* requires an understanding of the relationship between at least five intracellular processes that may modulate the activity of the functional gene product. These potential regulatory events may be simplistically summarized as follows: (1) transcription; (2) transcript processing (splicing, 5′-capping, and 3′-polyadenylation); (3) transcript translocation to the cytoplasm; (4) translation and post-translational modification; and (5) transcript turnover. Analysis of the cell-cycle specific regulation of histone gene expression has therefore necessitated the study of the temporal relationship between these intracellular events as applied to histone genes in synchronized cells.

3.1. Are Histone Proteins Synthesized in a Cell Cycle Specific Manner?

Histone protein synthesis must be considered in the context of the function of these proteins, so it is worth reemphasizing several properties relevant to an understanding of the regulation of their synthesis: (1) Histone proteins are largely responsible for the organization and structure of the cellular genome. (2) Histone proteins are found complexed to DNA in a relatively invariant molar ratio, and there is reason to believe that a change in this stoichiometry would severely affect a cell's viability. (3) No free histone pools have been identified in higher eucaryotes, indicating that newly-synthesized histones are rapidly sequestered by the nucleus; (4) There is no evidence that histone proteins are derived from precursor translation products. These properties in themselves strongly argue for a coupling between histone protein synthesis and DNA replication. It has now been well documented that in most eucaryotic cells, the bulk synthesis of histones (that which is required during DNA replication) occurs in the S phase of the cell cycle (55–57). Exceptions include the recent identification of "basal levels" of histone protein synthesis in G1 and $G1_0$ in several mammalian cell lines; these basal levels correspond to approximately 8% of the rate of synthesis observed in S phase (58, 59). Similarly, histone H1 synthesis may occur outside cell cycle control under special circumstances (60), and a constant rate of histone synthesis during the cell cycle has been reported in S49 mouse lymphoma cells (61; Coffino et al., Chapter 13).

Early studies with synchronized HeLa cells and with nondividing human diploid fibroblasts stimulated to proliferate indicated that the specific activity of nuclear histones was highest when cells were pulse-labeled with ^{3}H- or ^{14}C-labeled amino acids during S phase (56, 62). To eliminate the possibility of cell cycle specific compartmentalization of newly synthesized histone proteins, we recently pulse-labeled G1- and S-phase HeLa cells with ^{35}S-methionine, and analyzed total cellular proteins (Figure 16.8*a*) or 0.4N H_2SO_4-soluble cellular proteins (Figure 16.8*b*) for the presence of radiolabeled histones. HeLa cells in suspension culture were synchronized by double thymidine block (2 mM). After release from the second thymidine block, cells were maintained at a density of 5×10^5/ml for 1 hour.

S-phase cells were incubated for 2 additional hours at 37°C either in the presence or absence of the DNA synthesis inhibitor, cytosine arabinoside (40 μg/ml). Ten milliliters of either S-phase cells or cytosine-arabinoside-treated S-phase cells were centrifuged, washed in spinner salts solution containing 5 μM methionine (37°C), and resuspended in 2.8 ml of low (5 μM) methionine medium (containing cytosine arabinoside when appropriate). ^{35}S-methionine (140 μCi) was added, and cells were pulse-labeled for 45 min at 37°. G_1-phase HeLa cells were obtained by mitotic selective detachment and labeled as described earlier for S-phase cells. Cells were harvested by centrifugation at 1500 × g for 5 min, rinsed with ether, and dried before solubilized in 9.5 M urea–2% (w/v) NP-40–2% (w/v) ampholine–5% (w/v) mercapto-ethanol–0.3 M NaCl–1.0 mg/ml protamine sulfate. ^{35}S-labeled peptides were fractionated in a two-dimensional electrophoretic system in which the first dimension was nonequilibrium pH gradient electrophoresis (NEPHGE) and the second dimension was an SDS-containing 15% (w/v) acrylamide gel. As shown in Figure 16.8, S-phase total cellular histone proteins have a considerably higher specific activity than those isolated from G1 cells (55), an observation consistent with S-phase-specific histone protein synthesis. The synthesis of histone proteins primarily during S phase is also suggested by the very low levels of histone protein synthesis and cytoplasmic histone mRNAs at nonpermissive temperatures in temperature-sensitive cell cycle mutants (63; Hirschhorn, Marashi, Baserga, Stein and Stein, unpublished results).

3.2. Is the Cellular Abundance of Histone mRNA under Cell Cycle Control?

The S-phase-specific synthesis of histone proteins poses the question of whether regulation occurs at the level of translation or whether translation is governed by the availability of histone mRNA. Within this context, several properties of histone gene transcripts must be borne in mind: (1) With the exception of a chicken H3 gene (37) there is no evidence that histone genes contain intervening sequences. As histone mRNAs are generally not polyadenylated, the only documented post-transcriptional processing prior to translocation and translation is 5′-capping (64, 65). (2) Possibly because of the minimal

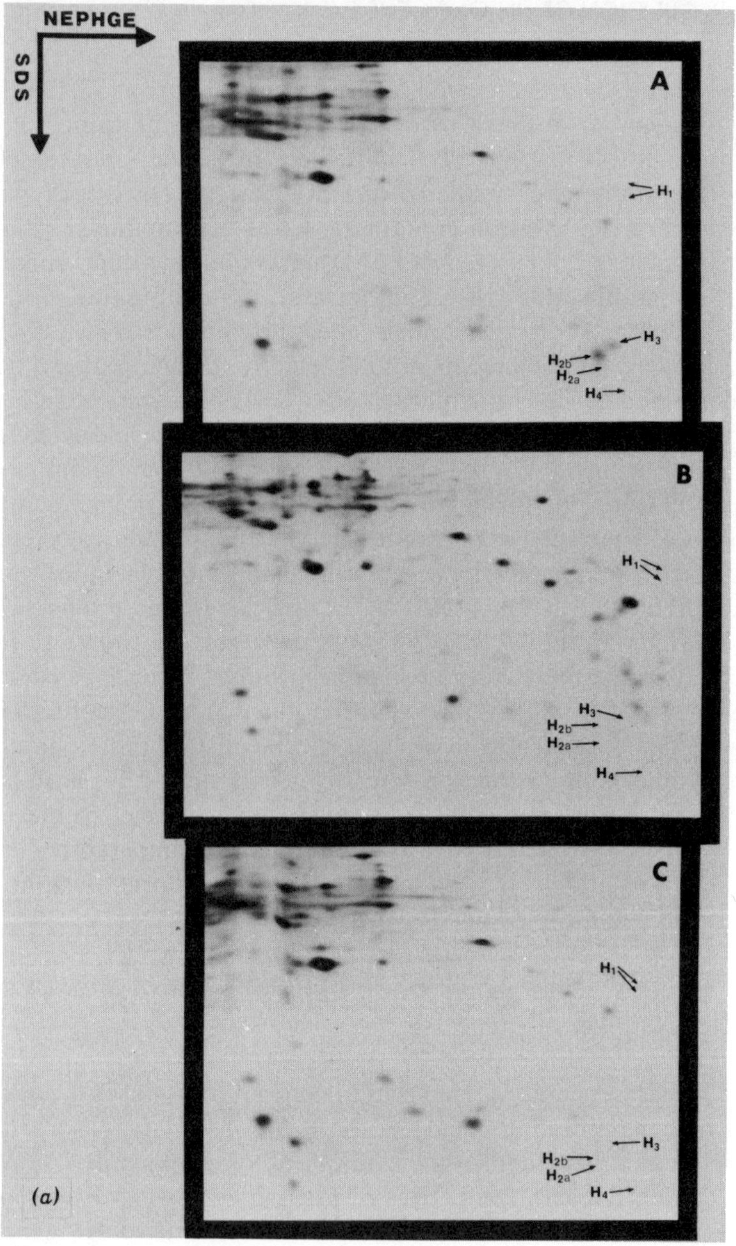

FIGURE 16.8. (*a*) Two-dimensional electrophoretic analysis of ^{35}S-methionine-labeled total cellular proteins A = S phase; B = G_1 phase; c = cytosine-arabinoside-treated S phase. (*b*) Two-dimensional NEPHGE/SDS electrophoresis of acid-extracted nuclear proteins of (A) S-phase, (B) G_1, and (C) cytosine-arabinoside-treated S-phase cells. ^{35}S-methionine-labeled cells were lysed in 10 mM KCl–10 mM Tris–

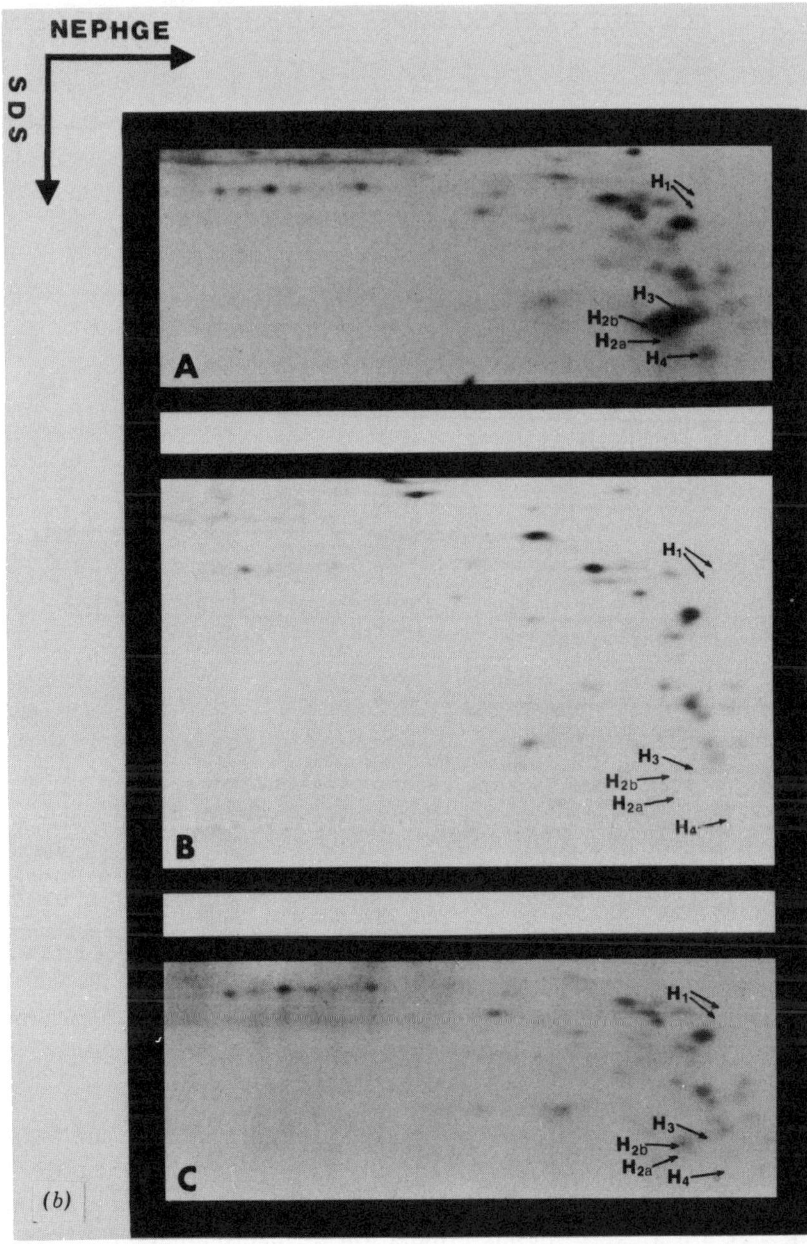

1.3 mM $MgCl_2$ (pH 7.4) containing 0.65% (v/v) Triton X-100, and nuclei were pelleted by centrifugation at 800 × g. Nuclear pellets were extracted with 0.4 M H_2SO_4 for 30 min, and the acid-soluble nuclear proteins were precipitated from the supernatant at −20°C overnight after addition of three volumes of 95% ethanol.

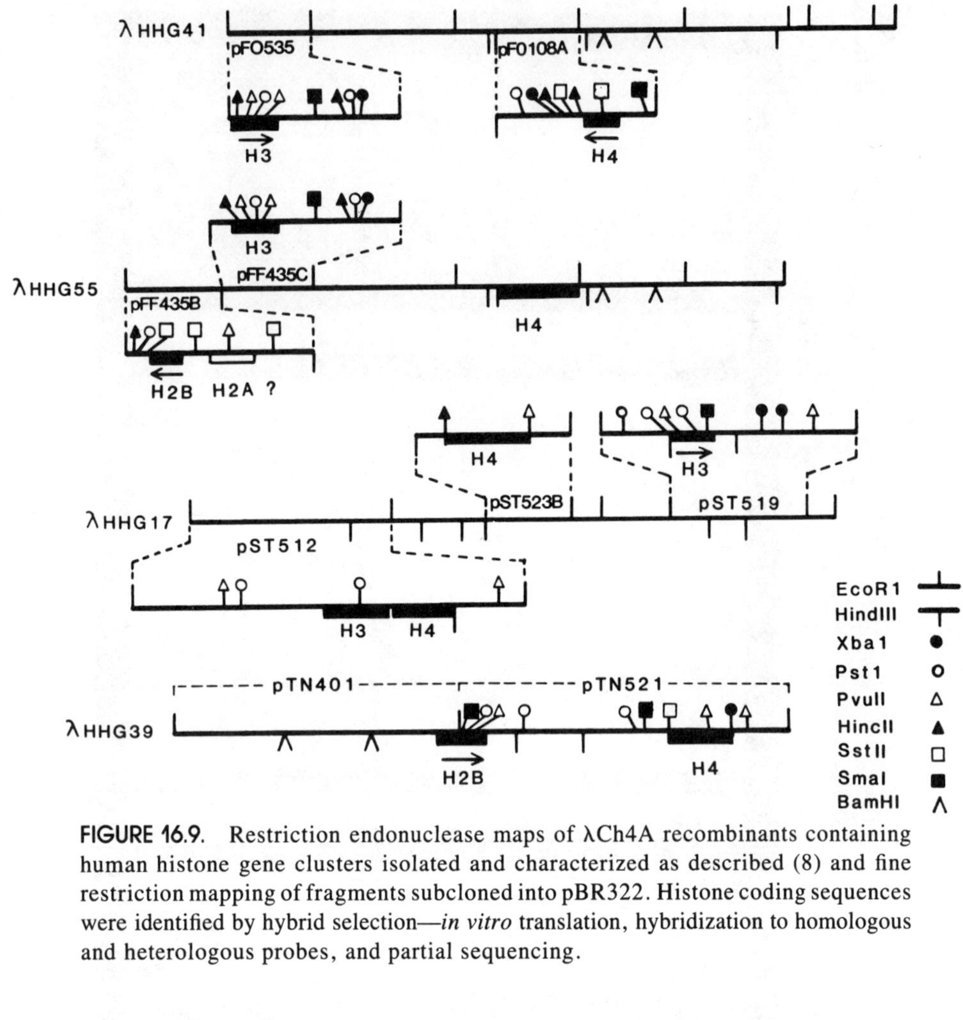

FIGURE 16.9. Restriction endonuclease maps of λCh4A recombinants containing human histone gene clusters isolated and characterized as described (8) and fine restriction mapping of fragments subcloned into pBR322. Histone coding sequences were identified by hybrid selection—*in vitro* translation, hybridization to homologous and heterologous probes, and partial sequencing.

post-transcriptional processing required, histone mRNAs are found only transiently in the nucleus, the bulk of the cellular histone mRNAs being found associated with the polysomes. (3) As alluded to, histone mRNAs are a heterologous population of macromolecules derived from a polymorphic middle repetitive family of genes.

We have used the cloned genomic human histone sequences (Figures 16.1*a* and 16.9) to examine the representation and synthesis of histone mRNAs in synchronized, continuously dividing HeLa S3 cells (36,

66) and in quiescent human diploid fibroblasts stimulated to proliferate (67). This problem necessitated the analysis of multiple histone mRNAs of variable sequence homology to our cloned genomic probes in order to determine whether members of this repetitive family of genes are differentially regulated. We therefore employed an experimental approach that permits the resolution of more than 15 different histone gene transcripts and used the technique to obtain results consistent with the cellular abundance of human histone mRNAs being coordinately regulated at both the transcriptional and post-transcriptional levels during the cell cycle. In agreement with earlier observations from this laboratory and with findings of others (68–73) in which histone mRNAs were analyzed by RNA excess hybridization with homologous histone cDNA, histone mRNA sequences are predominantly synthesized and present during S phase when DNA

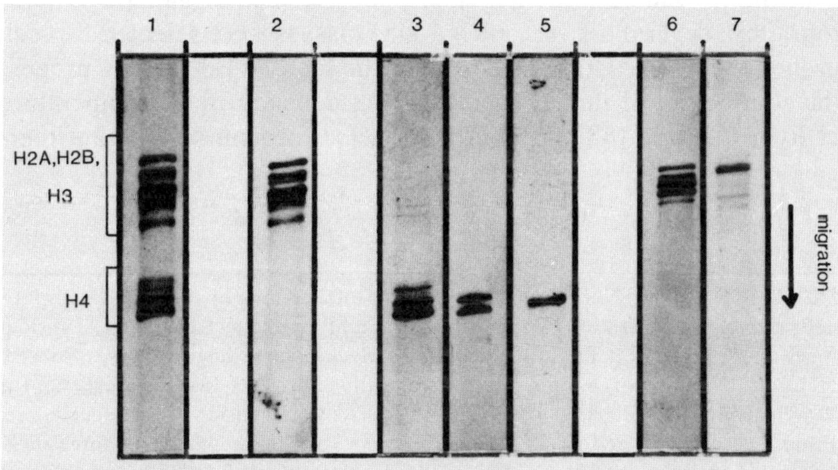

FIGURE 16.10. The effect of temperature on hybrid selection of histone mRNA variants: 150 μg of *in vivo* ^{3}H-uridine pulse-labeled total cellular RNA was hybridized with 50 μg of filter-immobilized cloned human histone sequences. Hybridized RNAs were eluted and resolved by denaturing 6% (w/v) polyacrylamide, 50% (w/v) urea gel electrophoresis, and visualized by fluorography. Lane 1: pF0108A (H4), pFF435B (H2A + H2B); and pFF435C (H3) hybridized at 43°C; lane 2: pFF435B (H2A + H2B) hybridized at 45°C; lanes 3, 4, and 5: pF0108A (H4) hybridized at 43°C, 45°C, and 48°C respectively; and lanes 6 and 7: pFF435C hybridized at 45°C and 48°C, respectively.

replication and histone protein synthesis occur. Recently similar results have also been obtained by Heintz et al. using cloned histone gene probes (74).

3.2.1. Is the Abundance of Histone mRNA Temporally Coupled to DNA Replication? The relationship between histone mRNA levels and DNA replication can be considered only within the limits of the assay used to detect cellular histone mRNA levels. The sequence data presented (Figure 16.2 and Table 16.1) indicate that within the same cell, a sequence divergence of 15–20% between gene transcripts coding for the same or very similar histone proteins is to be expected. Hybridization results emphasize the complexity of the problem. For example, when *in vivo* ^{3}H-uridine-labeled total cellular RNA was hybridized to filter-immobilized plasmid DNAs containing human histone gene sequences and the eluted RNAs were resolved by denaturing polyacrylamide gel electrophoresis and visualized by fluorography, several distinct mRNA size classes were selected by each probe (Figure 16.10). As shown for the histone H3 and H4 probes, the complexity of the resolved mRNAs depends on the temperature of hybridization [43–48°C in 50% (w/v) formamide]. At an inter-

FIGURE 16.11. (Top) Northern blot analysis of total cellular histone mRNA in HeLa cells after synchronization by two cycles of 2 mM thymidine treatment. At various times after release from the second thymidine block, total cellular RNA was isolated, 50 μg of RNA from each preparation were resolved by 1.5% (w/v) agarose, 6% (w/v) formaldehyde gel electrophoresis, and the RNAs were transferred to nitrocellulose. Filters were hybridized at 48°C in 50% (w/v) formamide to one of three nick-translated probes: (1) pFF435C (H3); (2) pFF435(B + C) (H2A, H2B, and H3), and p2.6H (chicken H3 + H4); (3) pF0108A (H4) and hybrids were visualized by autoradiography. (Bottom) The relative abundance of histone mRNAs (●——●, ± S.E.M.) in cells released into S phase after synchronization by two cycles of thymidine treatment and assayed by Northern blot analysis. Hybrids were quantitated for nine northern blots by densitometry and/or liquid scintillation spectrometry, after hybridization with the five histone gene probes as described. Values represent the mean relative abundances of histone H4, H3, or H2A + H2B mRNAs. The rate of DNA synthesis (○——○) was monitored by pulse-labeling 10^6 cells with 0.2 μCi of ^{14}C-thymidine for 20 min and measuring the incorporation of radiolabel into acid-precipitable material.

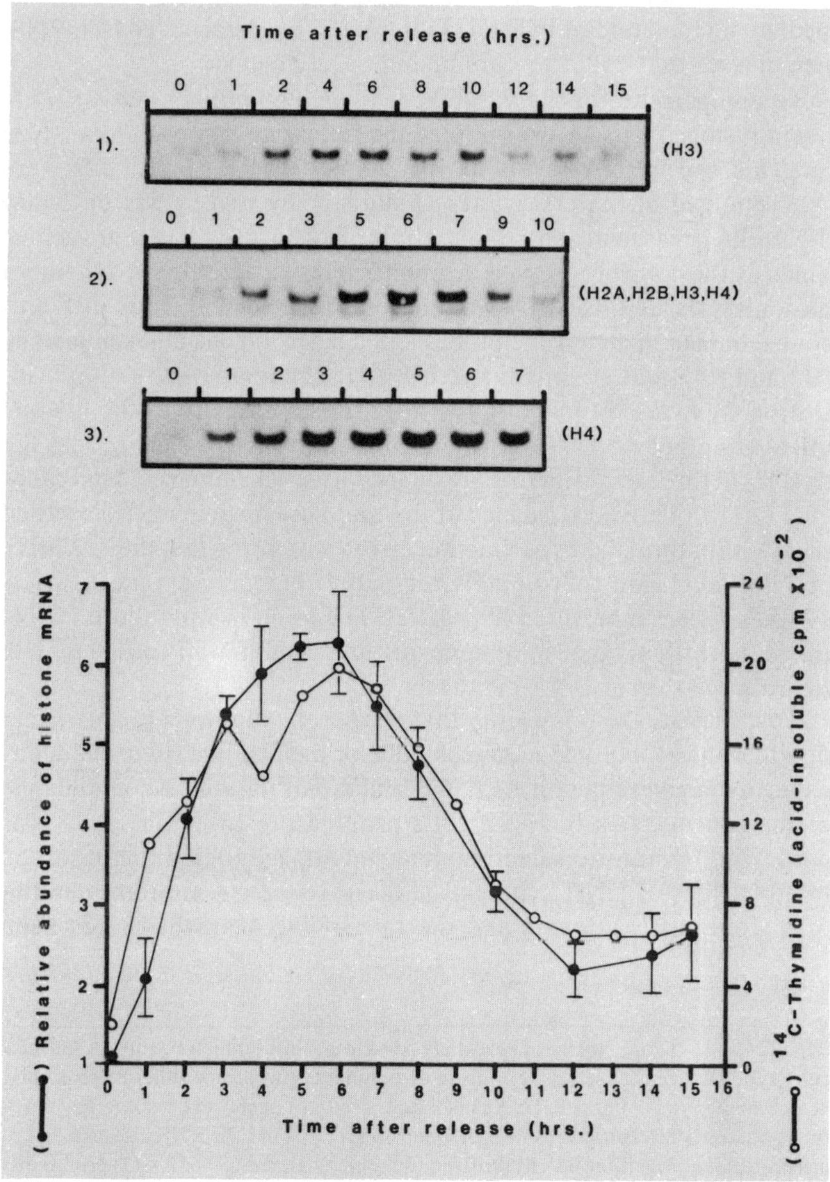

Time after release (hrs.)
0 1 2 4 6 8 10 12 14 15
1).
(H3)
0 1 2 3 5 6 7 9 10
2).
(H2A,H2B,H3,H4)
0 1 2 3 4 5 6 7
3).
(H4)
Relative abundance of histone mRNA
14C-Thymidine (acid-insoluble cpm X10-2)
Time after release (hrs.)

mediate temperature of hybridization (45°C) we detect over 15 mRNA size classes that code for core histone proteins.

To compensate for the variable sequence homology between different histone mRNAs, we adopted the following approach to analyze total histone mRNA levels during the cell cycle: HeLa cells were synchronized at the G1/S phase boundary by two cycles of 2-mM thymidine treatment, and total cellular RNA was isolated at various times as the cells progressed through S phase after release. Northern blot analysis of this RNA was then performed with four different cloned human histone gene probes and a heterologous chicken histone (H3 and H4) gene fragment (p2.6H) (9). The temperature of hybridization was varied (45–50°C) to alter the complexity of the histone mRNAs detected. Three northern blot analyses are shown, and the data from three experiments are quantitatively summarized in Figure 16.11. The relative abundance of the four core histone mRNAs varied only within limits of experimental error, indicating that their relative cellular levels are coordinately regulated. The *in vivo* rates of DNA synthesis were monitored in parallel, and as shown in Figure 16.11, the relative abundance of histone mRNAs is temporally coupled with the relative rate of DNA synthesis.

To eliminate the possibility that temporal coupling was an artifact of either the synchronization technique or the cell line, two analogous experiments were performed. We found that the relative abundance of histone mRNAs in HeLa cells progressing from G1 to S (synchronized by mitotic selective detachment) paralleled the observed relative rates of DNA synthesis (Figure 16.12). A similar coupling of histone mRNA levels and the rate of DNA synthesis has been

FIGURE 16.12. (Top) Northern blot analysis of total cellular histone mRNA in HeLa cells synchronized by selective mitotic detachment. At various times after mitotic detachment, total cellular RNA was isolated, 50 μg of each were resolved by 1.5% (w/v) agarose, 6% formaldehyde gel electrophoresis, and the RNAs transferred to nitrocellulose. The filter was hybridized to nick-translated pST519 (H3) probe, and visualized by autoradiography. (Bottom) Histone mRNA accumulation (●——●, ± S.E.M.) assayed for three Northern blots probed with pST519 (H3), pF0108A (H4), or pFF435B (H2A + H2B), quantitated as described in Figure 16.8*b*. DNA synthesis rates (○——○) were monitored by measuring the incorporation of ^{14}C-thymidine into acid-precipitable material in a 20 min pulse.

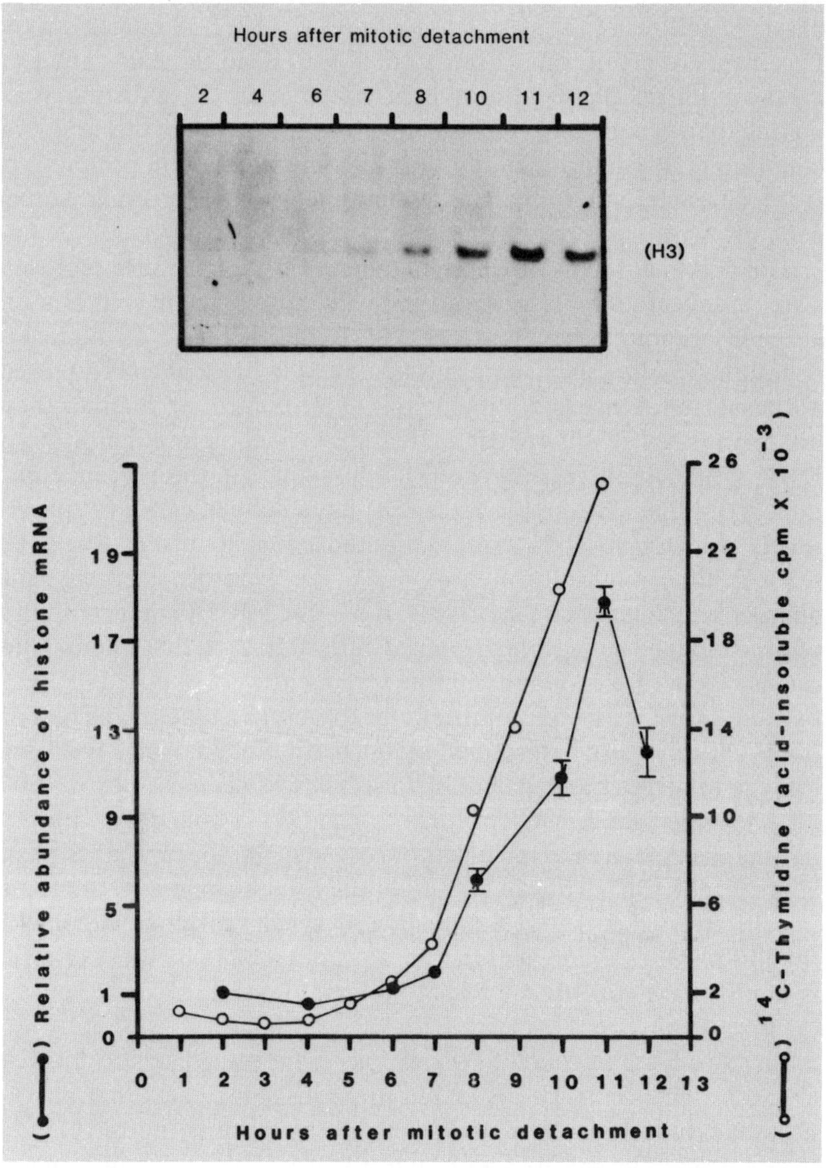
Hours after mitotic detachment
2 4 6 7 8 10 11 12
(H3)
Relative abundance of histone mRNA
^{14}C-Thymidine (acid-insoluble cpm X 10^{-3})
0 1 2 3 4 5 6 7 8 9 10 11 12 13
Hours after mitotic detachment

observed following the stimulation of quiescent WI-38 human diploid fibroblasts to proliferate (Figure 16.13) (67) and in yeast cells progressing into S phase after synchronization in G1 with the yeast mating pheromone (75). These results indicate that the temporal coupling is not determined by the cell synchronization protocol, or is it a characteristic of the HeLa genotype or of transformed human cells grown in culture.

The maximal levels of HeLa histone mRNAs in S phase represent a six- to sevenfold increase compared with those found in cells blocked at the G1-S boundary (Figure 16.11). It was therefore of interest to determine whether the histone mRNA levels in blocked cells represent a "basal" pool that is not under cell cycle control or whether these histone mRNA levels reflect a "leaky" or incomplete inhibition of DNA synthesis in cells at the G1/S boundary in the presence of 2 mM thymidine. Wu and Bonner (58) have defined cellular "basal" levels of histone protein synthesis as those that are insensitive to the DNA synthesis inhibitor hydroxyurea. We therefore used the same approach to determine the origins of histone mRNAs present in thymidine blocked cells. Synchronized cells at 0, 2, 6, or 11 hours after release from the second thymidine treatment were exposed to 1 mM hydroxyurea for 30 min, total cellular RNA was isolated, and histone mRNA levels were determined by northern blot analysis. We found that the relative abundance of histone mRNAs decreased to 30–60% of the levels found in cells blocked at the G1-S boundary with thymidine, irrespective of the levels observed in the appropriate S-phase control cells (36). Thus only 50% of the total histone mRNAs found in cells arrested at the G1/S boundary by synchronization with double thymidine blocks represents a "basal" pool which, by these criteria, is not S-phase specific (~ 5% of maximal levels).

3.2.2. Are Histone mRNAs Periodically Synthesized during the Cell Cycle? The evidence discussed previously indicates that histone protein synthesis is regulated primarily by modulating the cellular levels of histone mRNAs available for translation during S phase. Three possible regulatory mechanisms are invoked by this observation: (1) Histone mRNA synthesis is constitutive in the presence of a variable turnover rate. (2) The rate of histone gene transcription varies in the presence of a constant turnover rate. (3) Both the rates of histone mRNA synthesis and turnover are under cell cycle control. Experiments designed to distinguish among the three are severely complicated in the case of histone mRNAs by the fact that histone mRNAs are only transiently present in the nucleus, and by the assumption that histone mRNA turnover occurs predominantly in the cytoplasm. As histone mRNA transport to the cytoplasm occurs within minutes, *in vivo* pulse-labeling experiments are unlikely to yield a definitive indication of transcription rates. Despite these limitations, we have measured the specific activity of histone mRNAs labeled *in vivo* with ^{3}H-uridine for one hour at various times in synchronized HeLa cells. Our choice of a one-hour labeling interval was influenced primarily by the specific activity of the total cellular RNA thus produced. The high specific activity ($2–3 \times 10^5$ cpm/μg of RNA) permitted the detection of radiolabel incorporation into distinct histone mRNA size classes, and therefore we might detect the differential regulation of member(s) of the heterologous histone mRNA population. It must be emphasized, however, that these results reflect both transcriptional and post-transcriptional events.

Radiolabeled total cellular RNA was hybridized to filter-immobilized plasmid DNAs containing cloned human histone gene sequences, and the eluted RNAs were resolved by denaturing polyacrylamide gel electrophoresis. To maximize the complexity of mRNAs hybridizing to the immobilized probe, the temperature of hybridization was kept at 45°c [in 50% (w/v) formamide, see Figure 16.10]. At this temperature we have observed some cross-hybridization with ribosomal RNAs, probably due to the high G:C content of both ribosomal RNA and histone mRNAs. This cross-hybridization can be eliminated by increasing the temperature of hybridization to above 48°C, but as shown in Figure 13.9, this reduces the complexity of detected histone mRNAs. For quantitation therefore, either RNA was hybridized at a higher temperature and the radioactivity bound to a pBR322 control was subtracted, or, radioactivity was monitored in specific areas of the polyacrylamide gel after electrophoretic resolution and fluorography of RNAs.

As shown at the top of Figure 16.14, there are no obvious qualitative differences in the histone mRNAs synthesized at different times during S phase, either for H4 (pF0108A), H2A + H2B (pFF435B) or H3 (pFF435C). Moreover, the specific activity of any one mRNA size

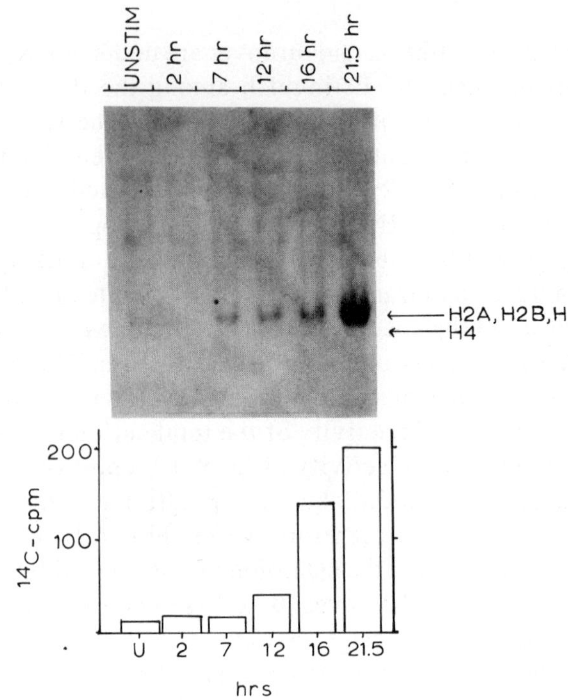

(a)

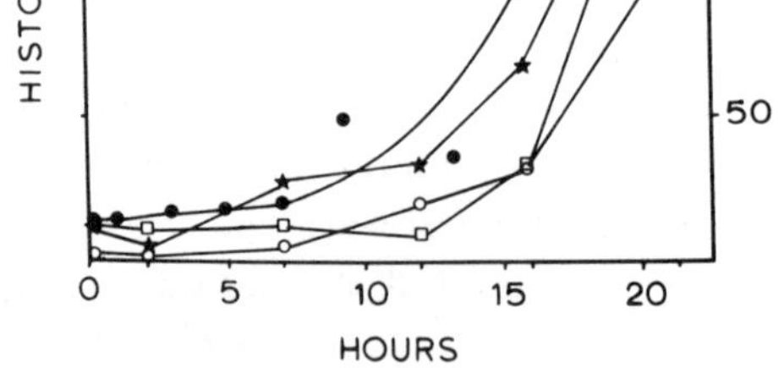

(b)

FIGURE 16.13. (*a*) Hybridization of total cellular RNAs from WI38 human diploid fibroblasts with a cloned human histone DNA sequence (pFF435) encoding histones H2A, H2B, and H3. The signals shown were obtained using 50 μg of electrophoretically fractionated, nitrocellulose-immobilized total cellular RNAs from quiescent WI38 human diploid fibroblasts isolated at various times after serum stimulation

class relative to another in the same sample does not change appreciably as cells traverse S phase. In total, five histone clones [pST512 (H3 + H4), pST519 (H3), pF0108A (H4), pFF435B (H2A + H2B), and pFF435C (H3), see Figure 16.9] were used to hybridize total cellular RNA, and the same pattern was observed at two hybridization temperatures (45 and 48°C) and over a threefold range of RNA con-

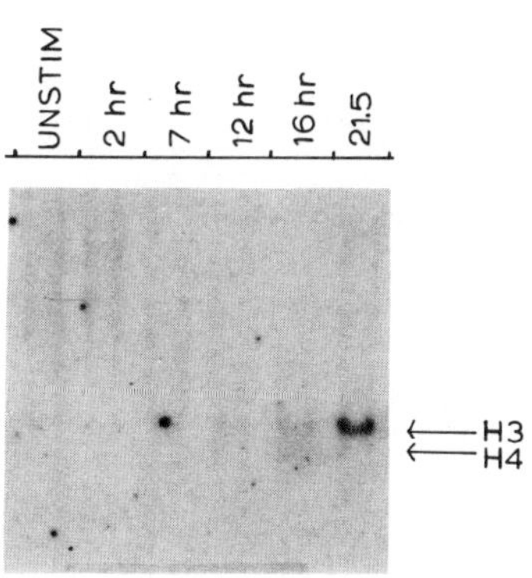

(c)

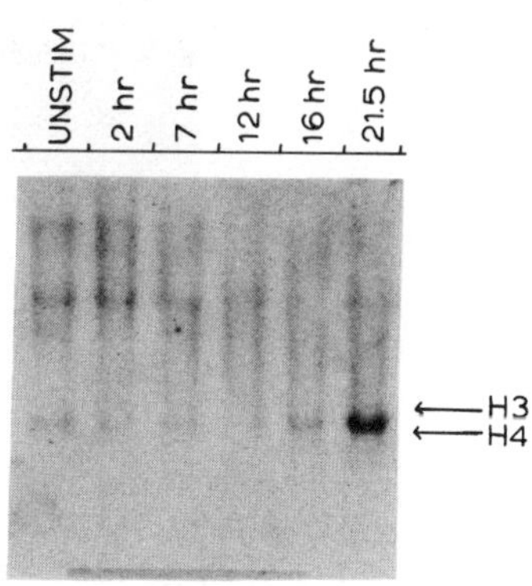

(d)

(2, 7, 12, and 21.5 hours) and from unstimulated cells. The lower bar graph depicts the levels of DNA synthesis at the various times after serum stimulation (2, 7, 12, 16, and 21.5 hours) and in unstimulated cells. The levels are expressed as ^{14}C-cpm incorporated into DNA, determined by pulse-labeling duplicate cultures with ^{14}C-thymidine for 30 min followed by assaying TCA-precipitable radioactivity. (*b*) Histone mRNA levels and DNA synthesis at various times after serum stimulation of quiescent WI38 human diploid fibroblasts. The histone mRNA levels were determined by densitometric analysis of hybridization signals obtained when 50 μg of electrophoretically fractionated, nitrocellulose-immobilized WI38 total cellular RNAs were hybridized with cloned human histone DNA sequences: *——*, H3 (pF0422); □——□, H4 (pF0108A); ●——●, H2A, H2B, H3 (pFF435). The DNA synthesis levels are expressed as ^{14}C-cpm incorporated into DNA by pulse-labeling with ^{14}C-thymidine for 30 min followed by assaying TCA precipitable radioactivity, [○——○]. (*c* and *d*) Hybridization of total cellular RNAs from WI38 human diploid fibroblasts to cloned human histone DNA sequences (pF0535 and pF0108A) encoding histones H3 and H4, respectively. The signals shown were obtained using 50 μg of electrophoretically fractionated, nitrocellulose-immobilized total cellular RNA from WI38 cells isolated at various times after stimulation (2, 7, 12, 16, and 21.5 hours) and from unstimulated cells (*c*) H3, (*d*) H4.

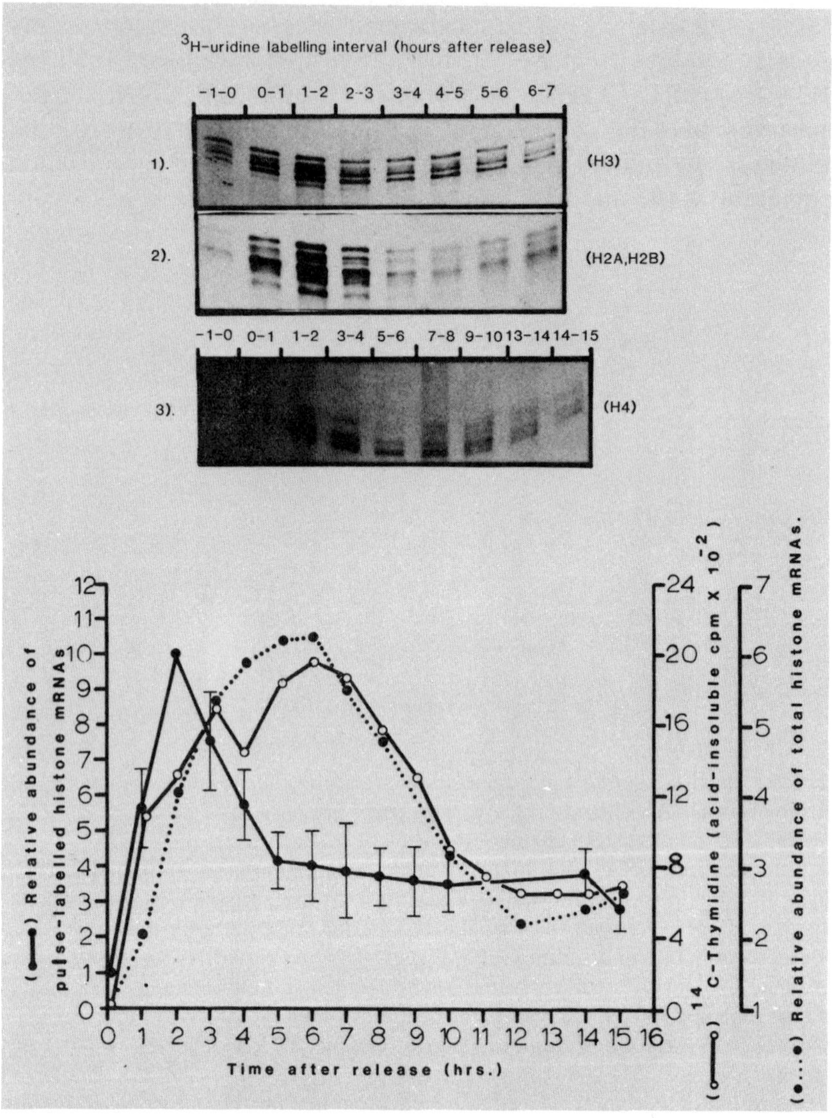

3H-uridine labelling interval (hours after release)
-1-0 0-1 1-2 2-3 3-4 4-5 5-6 6-7
1).
(H3)
2).
(H2A,H2B)
-1-0 0-1 1-2 3-4 5-6 7-8 9-10 13-14 14-15
3).
(H4)
( ) Relative abundance of pulse-labelled histone mRNAs
Time after release (hrs.)
( ) 14C-Thymidine (acid-insoluble cpm X 10-2)
(●...●) Relative abundance of total histone mRNAs.

centration (50–150 μg of total cellular RNA/50 μg of filter-immobilized plasmid DNA).

The results are quantitatively summarized at the bottom of Figure 16.14 along with the observed rates of DNA synthesis and the total histone mRNA levels detected by northern blot analysis. The maximal incorporation of ^{3}H-uridine into histone mRNAs of all size classes occurs between the first and second hour of S phase. As histone mRNA accumulation and DNA synthesis rates have only reached 50–60% of maximal levels after 2 hours, we conclude that transcription is the predominant regulatory mechanism modulating the concentration of cellular core histone mRNAs in the first half of S phase. We have observed a two- to threefold stimulation of total cellular transcription in the first 2–3 hours of S phase, but as this is significantly less than the apparent increase in ^{3}H-uridine incorporation into histone mRNAs (Figure 16.14), we conclude that there is a preferential stimulation of histone gene transcription as cells enter S phase. Five to six hours into S phase, a transition point is reached at which both DNA synthesis rates and total histone mRNA levels are maximal, but the relative abundance of pulse-labeled mRNAs is reduced (Figure 16.14). Thereafter total histone mRNA levels decrease with an apparent half-life of about 2.0 hours. As the apparent incorporation of ^{3}H-uridine into histone mRNAs remains low but relatively constant at this time, we conclude that both the transcription and the mRNA turnover rates are approximately constant in the second half of S phase, but the turnover rate exceeds the rate of synthesis. In the second half of S

FIGURE 16.14. (Top) Electrophoretic resolution of histone mRNAs pulse labeled *in vivo* for 1 hour (for the times indicated) after synchronization of HeLa cells by double thymidine blocks. One hundred fifty μg of one hour ^{3}H-uridine pulse labeled total cellular RNA was hybrid selected at 45°C with 50 μg of filter-immobilized plasmid DNA. Eluted RNAs were electrophoretically resolved on denaturing 6% (w/v) polyacrylamide, 50% urea gels, and visualized by fluorography. Histone mRNA selected by (1) pFF435C (H3), (2) pFF435B (H2A + H2B), and (3) pF0108A (H4). (Bottom) (●——●, ± S.E.M.) Quantitation for three hybrid selections using three probes [pFF435B (H2A + H2B), pFF435C (H3), and pF0108A (H4)]. Values were plotted at the end of the hour pulse-label and represent the mean relative abundances of pulse-labeled histone H4, H3, or H2A + H2B mRNAs. The DNA synthesis rates (○——○) and the relative abundance of total histone mRNAs (● - - - ●) are reproduced from Figure 16.11.

phase, therefore, the concentration of cellular histone mRNAs is largely regulated at the post-transcriptional level.

It is unclear whether the reduced concentration of pulse-labeled histone mRNAs in the second half of S phase (Figure 16.14) represents an S-phase-specific "maintenance level" of transcription, or whether it reflects synchrony decay. We therefore examined ^{3}H-uridine incorporation into histone mRNA over 1 hour labeling intervals in cells progressing from G1 to S phase after synchronization by selective mitotic detachment. HeLa cells were pulse-labeled at various times after mitotic detachment, and total cellular RNA was hybrid selected for histone mRNAs, which were then electrophoretically resolved as shown in Figure 16.15. We observed no obvious qualitative differences in the histone mRNA size classes as cells progressed from G1 to S phase, or was there a significant change in the specific activity of one mRNA size class relative to another in the same sample. As these histone mRNAs are qualitatively equivalent to those from cells synchronized by double thymidine treatments (Figure 16.14), we conclude that all the detected histone mRNA size classes are coordinately regulated in the HeLa cell cycle.

The *in vivo* incorporation of ^{3}H-uridine during 1-hour labeling intervals into histone mRNAs in cells progressing from G1 to S phase is quantitatively summarized at the bottom of Figure 16.15. ^{3}H-labeled histone mRNAs are detected in G1 (0–7 hours after mitosis), but these have a significantly lower radiolabel content than those found in S phase (8–12 hours after mitosis) (Figure 16.15). There is therefore a detectable but reduced apparent level of transcription

FIGURE 16.15. (Top) Electrophoretic resolution of histone mRNAs pulse labeled *in vivo* for 1 hour (at the times indicated) after synchronization of HeLa cells by selective mitotic detachment: 150 μg of ^{3}H-uridine pulse-labeled total cellular RNA was hybridized at 45°C with 50 μg of filter-immobilized plasmid DNA. Bound RNAs were eluted and electrophoretically resolved on denaturing 6% (w/v) polyacrylamide, 50% urea gels, and visualized by fluorography. RNA selected by (1) pFF435C (H3) and (2) pFF435B (H2A + H2B). (Bottom) (●——●, ± S.E.M.) Quantitation for three hybrid selections using three probes [pFF435B (H2A + H2B), pFF435C (H3) and pF0108A (H4)]. Values are plotted at the end of the hour pulse label, and represent the mean relative abundances of pulse-labeled histone H4, H3, or H2A + H2B mRNAs. The DNA synthesis rates (○——○) and the relative abundance of total histone mRNAs (● - - - ●) are reproduced from Figure 16.12.

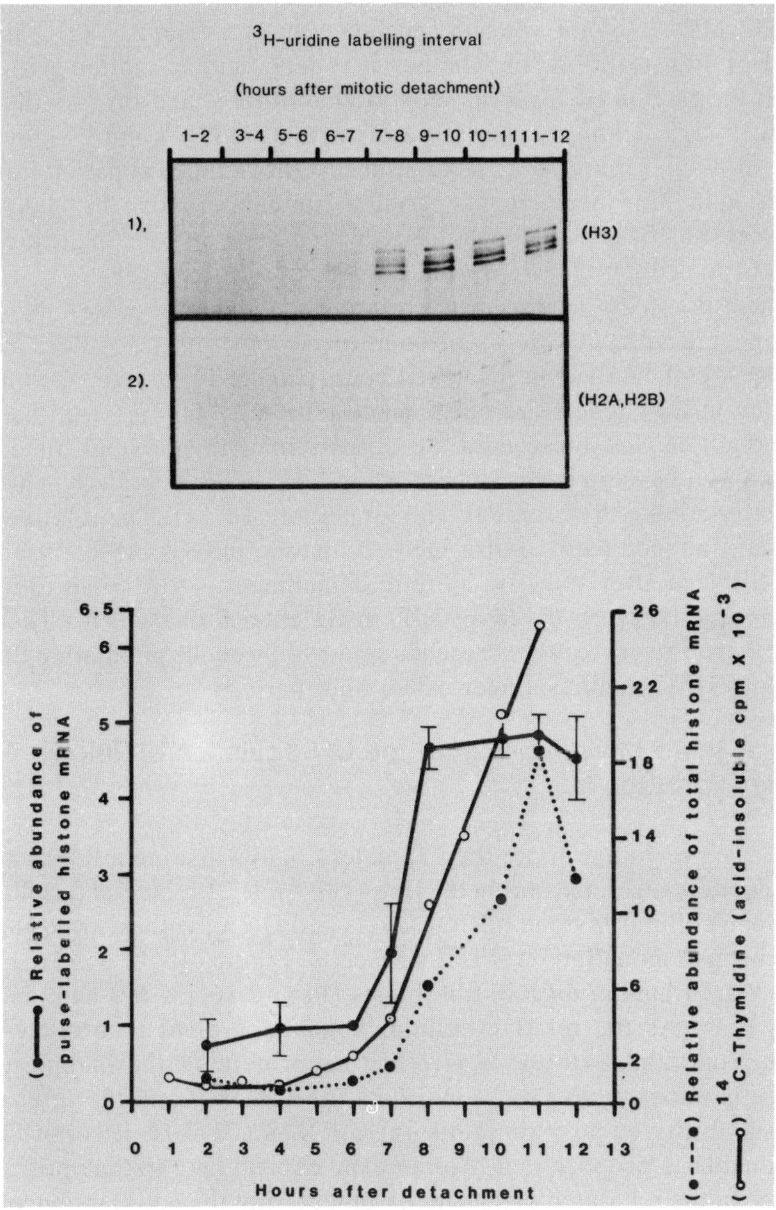

^{3}H-uridine labelling interval
(hours after mitotic detachment)
1-2 3-4 5-6 6-7 7-8 9-10 10-11 11-12
1),
2).
(H3)
(H2A,H2B)
(●—●) Relative abundance of pulse-labelled histone mRNA
(●---●) Relative abundance of total histone mRNA
(○—○) ^{14}C-Thymidine (acid-insoluble cpm X 10^{-3})
Hours after detachment

in G1 cells. It is not yet clear whether this represents a G1 basal level of transcription, or whether it reflects contamination with a small proportion of S-phase cells after mitotic shake-off. As there is not a significant accumulation of histone mRNAs until 8 hours after mitosis (Figure 16.15), we infer that the reduced apparent transcription in G1 is insufficient to permit accumulation above the putative G1-specific "basal" histone mRNAs which are translated in the presence of hydroxyurea as described by Wu and Bonner (58).

There is a sharp increase in the relative abundance of pulse-labeled histone mRNAs 7–8 hours after mitosis, and this level does not change significantly over the next 4 hours (Figure 16.15). As maximal histone mRNA levels are not reached until 11 hours after mitosis, we conclude that these data are consistent with maximal histone mRNA synthesis in early S phase, as was observed in cells synchronized by double thymidine treatment (Figure 16.14). The relatively constant abundance of pulse-labeled histone mRNAs in early and mid-S phase after selective mitotic detachment is attributed to the synchrony decay observed as cells in G1 enter S phase (76). These cells therefore probably represent a semisynchronous population that initiates DNA synthesis over a 3–4 hour period.

3.3. Is There More Than One Type of Regulation of Histone Gene Expression?

We have shown that more than 15 heterologous histone mRNAs are coordinately regulated during the HeLa cell cycle. A temporal coupling between cellular histone mRNA levels and DNA synthesis rates has now been demonstrated for HeLa S3 cells (Figures 16.11 and 16.12; 36), WI-38 human diploid fibroblasts (Figure 16.13; 67) and yeast cells (75). As the relative cellular abundance of histone mRNAs closely parallels both the *in vivo* histone protein synthesis reported for cells under equivalent conditions (55, 56, 62) and the *in vitro* translatability of polysomal histone mRNAs (77–79), it is not unreasonable to propose that histone gene expression (histone protein synthesis) is regulated primarily by modulating the cellular concentration of histone mRNAs available for translation. Histone mRNA synthesis apparently precedes maximal accumulation of mRNAs in both synchronized HeLa (Figure 16.16) and yeast (80) cells, indicating

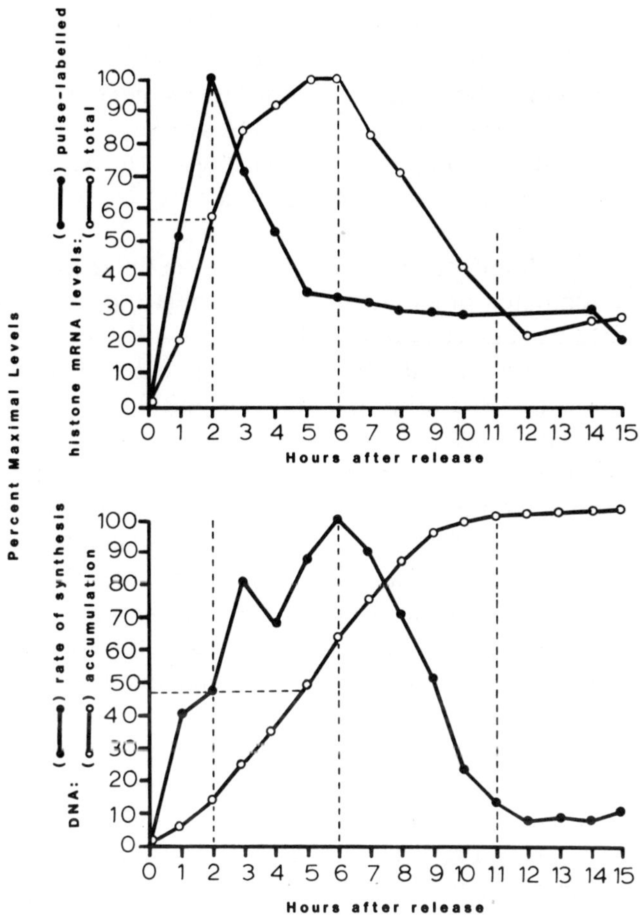

FIGURE 16.16. Summaries for the data obtained for cells synchronized by double thymidine blocks. Data are expressed as the percentage maximal levels detected above those observed in cells just before release from the second thymidine block. (Top) (○——○) The relative abundance of total histone mRNA. (●——●) The relative incorporation of ^{3}H-uridine into histone mRNA in 1-hour labeling intervals, plotted at the end of the pulse. (Bottom) (●——●) Relative rates of DNA synthesis monitored by pulse labeling with ^{14}C-thymidine. (○——○) Accumulation of DNA calculated from the rates of DNA synthesis.

that the cellular abundance of histone mRNAs in early S phase is predominantly under transcriptional control. For the moment, we cannot determine whether the onset of histone gene transcription in HeLa cells precedes the initiation of DNA synthesis, as is apparently the case in yeast (80), although our data (Figure 16.15) would suggest that the two occur within 30 min of each other (~ 6% of the HeLa S phase).

In the second half of the HeLa S phase, cellular histone mRNA levels are regulated predominantly at the post-transcriptional level as suggested by the decay of cellular histone mRNAs with apparent half-lives of 2.0 hours, and by the reduced and relatively constant incorporation of ^{3}H-uridine into histone mRNAs. These data are compatible with constant rates of both synthesis and turnover of histone mRNAs in the second half of S phase, but with the turnover rate exceeding the rate of synthesis. As studies in HeLa and yeast cells (36, 80) suggest that histone gene transcription rates vary during S phase, it is now of interest to determine whether the turnover of histone mRNAs varies during the cell cycle, in order to determine whether only one level of regulation (regulated transcription) or two (regulated synthesis and turnover) is responsible for the control of histone gene expression.

As a working model, we propose that the turnover of histone mRNAs remains constant throughout the cell cycle of HeLa cells. This model is consistent with Perry and Kelley's observation (81) that the mean lifetime of polysomal histone mRNAs in exponentially growing mouse L cells is equal to the length of S phase. As discussed later, however, histone mRNA destabilization can be induced, but this may reflect a secondary regulatory mechanism that is only applied under extreme biological circumstances. Our working model therefore relies predominantly on the regulation of histone mRNA synthesis. In the first half of S phase an elevated rate of synthesis permits histone mRNA accumulation, and in the second half of S phase a reduced transcription allows the decay of histone mRNA in the presence of a constant turnover rate. Furthermore, the kinetics of synthesis and accumulation of histone mRNAs in early S phase are suggestive of a feedback regulatory mechanism. One possibility currently being investigated is that histone gene replication is temporally and functionally related to transcription of the histone genes.

3.4. Can Histone Gene Expression Be Uncoupled from DNA Replication?

Other potential regulatory aspects of human histone gene expression are suggested by observations reported for other biological systems. For example, the differential expression of "early" and "late" histone gene clusters in developing sea urchin embryos (82–84) has stimulated the search for an analogous situation in higher eucaryotes. It is becoming evident that differential translation of variant histone proteins occurs during the cell cycle (58, 59), and differential expression of histones during development of higher eucaryotes has also been observed (Zweidler, Chapter 14; Cohen, Chapter 15). In lower eucaryotes, histone gene expression immediately prior to and shortly after fertilization is regulated almost exclusively at post-transcriptional levels. During the extremely rapid rate of cell division just after fertilization, it appears that developing *Xenopus* oocytes rely predominantly on a maternal store of histone proteins (85), and sea urchin embryos rely predominantly on a maternal store of histone mRNAs (84, 86–88), to provide the histone proteins required for packaging the rapidly replicating DNA. Although the rate of cell division of embryos in higher eucaryotes is significantly slower, it is of some interest to determine whether such post-transcriptional regulation occurs during early development and whether there are equivalents of the "early" and "late" histone genes of lower eucaryotes.

4. RELATIONSHIP BETWEEN HISTONE GENE EXPRESSION AND DNA REPLICATION

As has been described, there is a plethora of information that supports temporal coupling of histone gene expression and DNA replication. Although the molecular basis for the coupling remains to be elucidated, it is becoming clear that the inhibition of DNA synthesis by a variety of unrelated exogenous cellular perturbations results in a parallel decrease in cellular histone mRNA levels, and an equivalent decrease in histone protein synthesis. Early studies, for example, established that the inhibition of DNA replication by metabolic inhibitors [hydroxyurea (58, 59, 78, 79, 89–94), cytosine arabinoside (55, 56, 77, 79, 81, 94–96), and actinomycin D (90, 91, 93, 95, 97)] resulted in the rapid reduction of histone protein synthesis and the loss of polysomal histone mRNAs. More recently we have monitored total cellular levels of histone mRNAs following the inhibition of DNA replication by: (1) metabolic inhibitors (109, 110), (2) infection with adenovirus (Flint, Plumb, Stein and Stein, submitted for publication), and (3) after shifting temperature-sensitive cell cycle mutants to the nonpermissive temperature (Hirschhorn, Marashi, Baserga, Stein and Stein, submitted for publication). In all cases, the perturbation of distinct cellular processes resulted in both the inhibition of DNA replication and an equivalent reduction in total cellular histone mRNA levels.

When DNA replication is completely ($> 95\%$) inhibited by compounds that disrupt the metabolism of deoxynucleotide precursors (100 μM cytosine arabinoside or 1 mM hydroxyurea, see Figure 16.17) or that inhibit DNA polymerase α (2 μg/ml aphidicolin) (74, 98 and see Figure 16.17), the cellular representation of histone mRNAs is reduced to an inhibitor-resistant "basal level" corresponding to approximately 10% of the levels observed in exponentially growing cells, and reduced to approximately 6% of the peak levels observed in synchronized S-phase HeLa cells. These estimates for "basal levels" of histone mRNAs are quantitatively consistent with the basal levels of histone protein synthesis observed in G1 and S-phase cells in the presence of 1 mM hydroxyurea (58, 59). Both the rates of inhibition of DNA synthesis and the rates of histone mRNA decay occurred with apparent half-lives of less than 10 min. These kinetics differ significantly from the reported mean lifetime of 9 hours for polysomal histone mRNAs in exponentially growing L cells (81), and differ from the estimated apparent histone mRNA half-lives of 2.0 hours described for the latter part of S phase in synchronized HeLa cells (36; Figure 16.16). The inhibitors therefore apparently cause the rapid destabilization of histone mRNAs. Such a post-transcriptional coupling is supported by the observation that after transcriptional inhibition with 5 μg/ml actinomycin D, cellular histone mRNA levels decay in parallel with inhibition of DNA synthesis [95% inhibition within 2 hours (95, 109)] despite complete inhibition of transcription within approximately 6 min.

At present we are examining the coupling of DNA synthesis and histone mRNA levels under conditions that result in the incomplete inhibition of DNA synthesis—using reduced concentrations of the metabolic inhibitors hydroxyurea (10 μM–1 mM), cytosine arabinoside (7 nm–100 μM), and aphidicolin (10 ng/L–2 mg/L). As shown in Figure 16.17, partial inhibition of DNA synthesis results in the stoichiometric reduction of total cellular histone mRNAs irrespective of

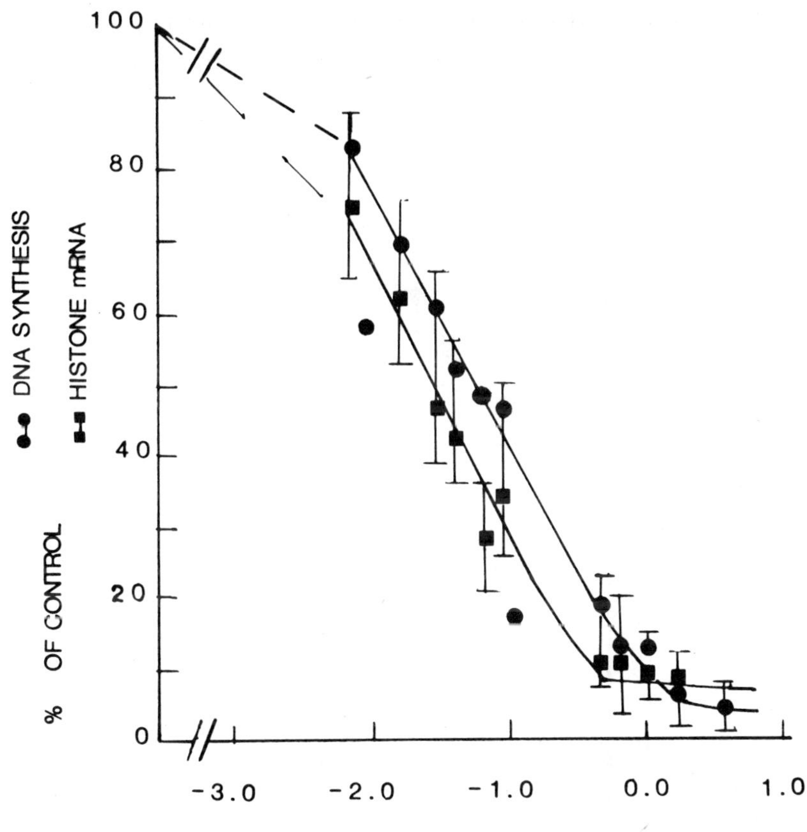

(a)

FIGURE 16.17. The effects of (*a*) cytosine arabinoside.

(*Figure continues on p. 446.*)

the inhibitor or its concentration. Time course experiments have indicated that both the rate of DNA synthesis and the level of histone mRNAs equilibrate within 30 min of the metabolic perturbation, implying that the primary coupling mechanism occurs at the post-transcriptional level.

As our studies on the effects of metabolic inhibitors discussed previously only yielded a generalized indication of the fate of the core histone mRNAs after complete or partial inhibition of DNA synthesis, it was of interest to determine whether coordinate regulation of the expression of the polymorphic histone gene family was operative. This question was also relevant to the observations that specific histone variant proteins are differentially synthesized in G1, $G1_0$,

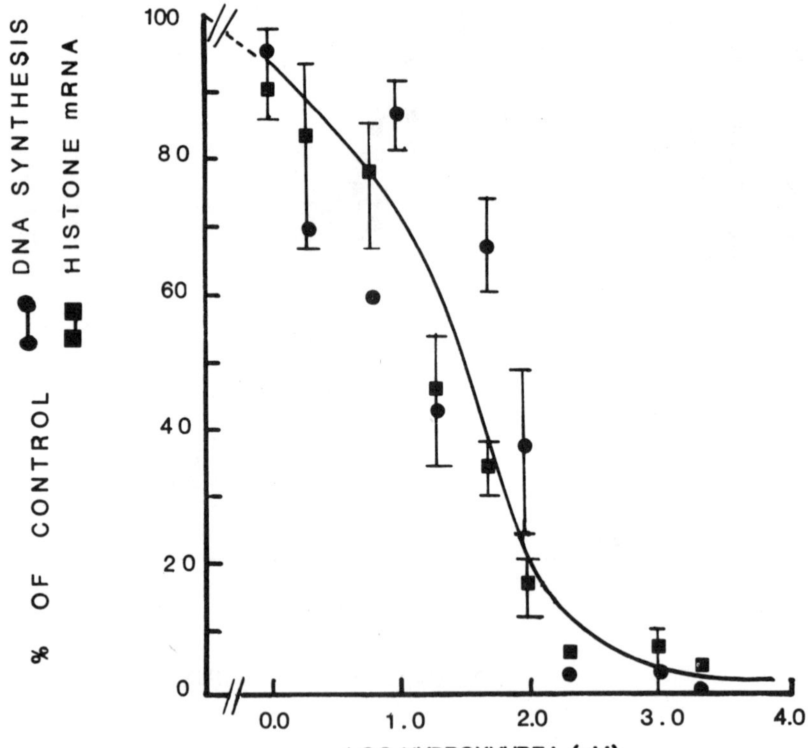

(b)

FIGURE 16.17. (*continued*). (*b*) hydroxyurea.

and S-phase cells exposed to 1 mM hydroxyurea (58, 59). We have initially approached this problem by posing three questions: (1) Are S-phase newly synthesized multiple histone mRNAs differentially turned over in the presence of 1 mM hydroxyurea? (2) Are specific subspecies differentially transcribed in S phase cells exposed to 1 mM hydroxyurea? (3) Are specific histone mRNAs differentially

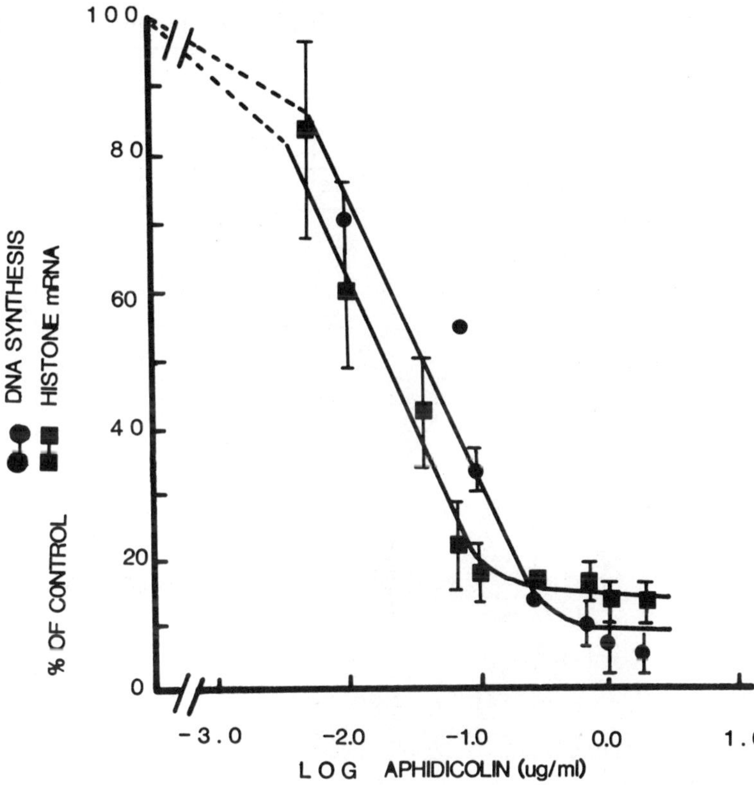

(c)

FIGURE 16.17. (*continued*). (*c*) aphidicolin on histone mRNA levels and DNA synthesis in exponentially growing HeLa cells. DNA synthesis was monitored by incorporation of ^{3}H-thymidine (2 μCi/ml and 4 × 10^5 cells/ml) into 10% TCA precipitable material. Cells were pretreated for 30 min with each of the various inhibitors and then pulse labeled for 30 min. Total cellular RNA was prepared and histone mRNA steady-state levels determined as described in Figure 16.11. Each value represents the mean of five northern blots using various histone gene probes: pF0108A (H4), pFF435B (H2A + H2B), pFF435C (H3), pST519D (H3), and pF0535 (H3).

sensitive to destabilization by 1 mM hydroxyurea at different times in the cell cycle? Using the hybrid selection of ^{3}H-uridine pulse-labeled histone mRNAs to visualize the incorporation of radiolabel into more than 15 distinct histone mRNA size classes, we have found: (1) The levels of these multiple histone mRNA species are apparently coordinately controlled at the post-transcriptional level when HeLa cells with prelabeled histone mRNAs are exposed to hydroxyurea either before, during, or after S phase (see Figure 16.18*a*). (2) The incorporation of ^{3}H-uridine into the multiple histone mRNA species

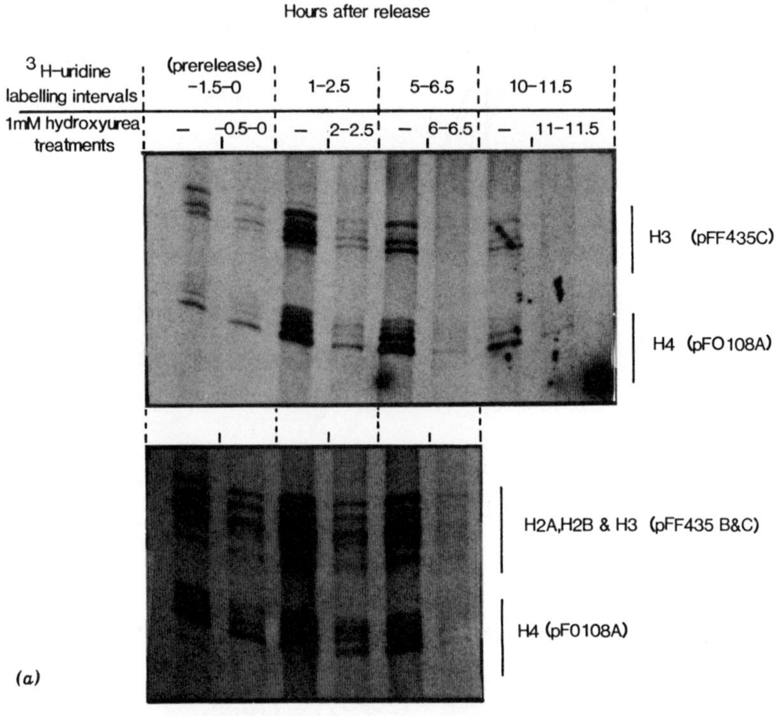

FIGURE 16.18. (*a*) The effect of hydroxyurea on S-phase HeLa histone mRNA species prelabeled with ^{3}H-uridine at various times after release from double thymidine block synchronization. At −1.5, 1, 4.5, and 9.5 hours after release from thymidine, cells were labeled with ^{3}H-uridine (0.1 mCi/ml and 2.5 × 10^6 cells/ml) for 1 hour and then exposed to 1 mM hydroxyurea for 30 min in the presence of the radiolabel. Control cells (not treated with hydroxyurea) were labeled for 90 min over the same time intervals. Total cellular RNA was prepared and 300 μg hybridized at 43°C to filter-immobilized plasmid DNA as described in Figure 16.7. Bound RNAs were eluted, resolved electrophoretically, and visualized by fluorography.

is reduced, but both qualitatively and quantitatively similar, after cells at various times in the cell cycle have been pretreated with 1 mM hydroxyurea for 30 min (see Figure 16.18*b*).

Several cautionary provisos must be maintained when interpreting these results. As alluded to earlier (Figure 16.10), the detection of ^{3}H-uridine pulse-labeled histone mRNAs is largely governed by the choice of probe and the hybridization conditions used. It is therefore possible that, with the probes we have used, we are not detecting all histone mRNA species and that those not detected may be regulated

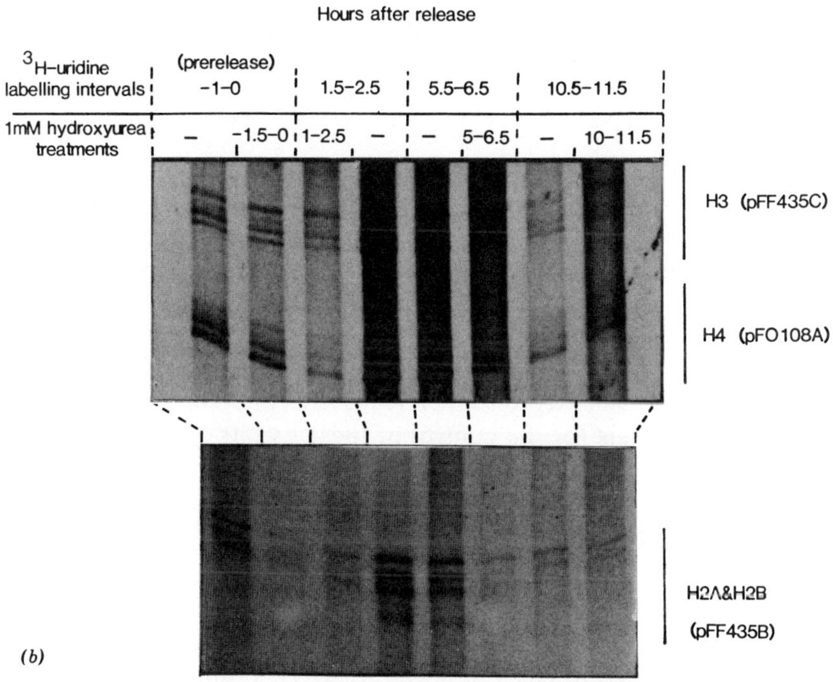

FIGURE 16.18. (*continued*). (*b*) The incorporation of ^{3}H-uridine into S-phase HeLa histone mRNA species after 30 min pretreatment with hydroxyurea at various times after release from double thymidine block synchronization. At −1.5, 1, 4.5, and 9.5 hours after release cells were treated for 30 min with 1 mM hydroxyurea, and then labeled for 1 hour with ^{3}H-uridine (0.1 mCi/ml and 2.5×10^{6} cells/ml) in the presence of hydroxyurea. Control cells were labeled for 1 hour of the 1.5-hour time interval. Total cellular RNA was prepared and 300 μg hybridized at 43°C to filter-immobilized plasmid DNA as described in Figure 16.7. Bound RNAs were eluted, resolved electrophoretically, and visualized by fluorography.

by a distinct mechanism. Similarly, we have relied to a greater or lesser degree on the assumption that the turnover of all histone mRNA species is sufficiently high to detect incorporation of ^{3}H-uridine in 1-hour pulse periods, but this assumption may not be valid. Finally, it must be emphasized that the interpretation of data obtained from perturbed cells may in fact reflect a secondary (fail-safe) regulatory mechanism which is not normally operative during the G1-S-G2 transitions. It is also possible that histone gene expression is functionally coupled to DNA replication by means of the histone gene products themselves, as it is not unreasonable to predict that the constitutive synthesis of histone proteins would severely affect cell viability.

5. CONCLUSION

In this chapter we have restricted our considerations to the regulation of histone genes, a specific set of repetitive genetic sequences which are differentially expressed during the cell cycle. Of broader biological relevance is the relationship between expression of histone genes and of other genetic sequences required for execution of the proliferative process. Although it is well documented that cell proliferation involves a complex and interdependent series of biochemical events, the molecular basis for the control of these events or for their interrelationship remains to be resolved. An intriguing possibility is that genetic sequences whose expression is related to specific cell cycle processes are coordinately controlled. Within this context it is tempting to postulate that expression of genes functionally related to DNA replication shares common structural and/or regulatory elements. Do histone genes have regulatory sequences similar to those, for example, of genes encoding enzymes required for DNA replication or deoxynucleotide metabolism? Do histone genes and other S-phase-specific sequences respond to similar cellular signals or regulatory molecules? These questions are being approached experimentally and will establish the credibility of such speculation.

ACKNOWLEDGMENT

Studies described in this chapter were supported by grants PCM80-18075, PCM81-18951, and PCM83-18177 from the National Science

Foundation, 1-813 from the March of Dimes Birth Defects Foundation and GM32010 from the National Institutes of Health.

REFERENCES

1. R. M. Lawn, E. F. Fritsch, R. C. Parker, G. Blake, and T. Maniatis, *Cell* **15,** 1157 (1978).
2. R. P. Lifton, M. L. Goldberg, R. W. Karp, and D. S. Hogness, *Cold Spring Harbor Symp. Quant. Biol.* **42,** 1047 (1977).
3. K. Gross, W. Schaffner, J. Telford, and M. Birnstiel, *Cell* **8,** 479 (1976).
4. M. Wu, D. S. Holmes, N. Davidson, R. H. Cohn, and L. H. Kedes, *Cell* **9,** 163 (1976).
5. M. C. Wilson and M. J. Melli, *J. Mol. Biol.* **110,** 551 (1977).
6. L. C. Yu, P. Szabo, T. W. Borun, and W. Prensky, *Cold Spring Harbor Symp. Quant. Biol.* **42,** 1101 (1977).
7. M. E. Chandler, L. H. Kedes, R. H. Cohn, and J. J. Yunis, *Science* **205,** 908 (1979).
8. F. Sierra, A. Lichtler, F. Marashi, R. Rickles, T. Van Dyke, S. Clark, J. Wells, G. Stein, and J. Stein, *Proc. Natl. Acad. Sci. USA* **79,** 1795 (1982).
9. S. Clark, Ph.D. Thesis, University of Adelaide, Australia.
10. N. Carozzi, F. Marashi, F. Sierra, J. Stein, and G. Stein, manuscript submitted.
11. N. Heintz, M. Zernik, and R. G. Roeder, *Cell* **24,** 661 (1981).
12. L. Hereford, K. Fahrner, J. Woolford, M. Rosbash, and D. B. Kaback, *Cell* **18,** 1261 (1979).
13. J. W. Wallis, L. Hereford, and M. Grunstein, *Cell* **22,** 799 (1980).
14. M. Zernik, N. Heintz, J. Boime, and R. G. Roeder, *Cell* **22,** 807 (1980).
15. A. F. M. Moorman, R. T. M. deLaaf, and O. H. J. Destree, *Gene* **10,** 185 (1980).
16. W. Van Dongen, L. deLaaf, R. Zaal, A. Moorman, and O. Destree, *Nucl. Acids Res.* **9,** 2297 (1981).
17. E. C. Stephenson, H. P. Erba, and J. G. Gall, *Cell* **24,** 639 (1981).
18. J. G. Gall, E. C. Stephenson, H. P. Erba, M. O. Diaz, and G. Barsacchi-Pilone, *Chromosoma* **84,** 159 (1981).
19. R. Britten and E. Davidson, *Science* **165,** 349 (1969).
20. R. Britten and D. E. Kohne, *Science* **161,** 3841 (1968).
21. E. H. Davidson, G. A. Galau, R. C. Angerer, and R. J. Britten, *Chromosoma* **51,** 253 (1975).
22. W. R. Jelinek and C. W. Schmid, *Ann. Rev. Biochem.* **51,** 813 (1982).
23. C. M. Houck, F. P. Rinehart, and C. W. Schmid, *J. Mol. Biol.* **132,** 289 (1979).

24. F. Sierra, A. Leza, F. Marashi, M. Plumb, R. Rickles, T. Van Dyke, S. Clark, J. Wells, G. S. Stein, and J. L. Stein, *Biochem. Biophys. Res. Comm.* **104,** 785 (1982).
25. I. Isenberg, *Ann. Rev. Biochem.* **48,** 159 (1979).
26. A. C. Lichtler, F. Sierra, S. Clark, J. R. E. Wells, J. L. Stein, and G. S. Stein, *Nature* **298,** 195 (1982).
27. A. C. Lichtler, S. Detke, I. R. Phillips, G. S. Stein, and J. L. Stein, *Proc. Natl. Acad. Sci. USA* **77,** 1942 (1980).
28. M. Grunstein, S. Levy, P. Schedl, and L. Kedes, *Cold Spring Harbor Symp. Quant. Biol.* **38,** 717 (1973).
29. W. Schaffner, G. Kunz, H. Daetwyler, J. Telford, H. O. Smith, and M. L. Birnstiel, *Cell* **14,** 655 (1978).
30. M. Busslinger, R. Portmann, J. C. Irminger, and M. L. Birnstiel, *Nucl. Acids Res.* **8,** 957 (1980).
31. Y. Ohe, H. Hayashi, and K. Iwai, *J. Biochem.* **85,** 615 (1979).
32. F. Marashi, K. Prokopp, G. Stein, and J. Stein, *Proc. Natl. Acad. Sci. USA* **81,** in press (1984).
33. C. C. Hentschel and M. L. Birnstiel, *Cell* **25,** 301 (1981).
34. E. F. Fritsch, R. M. Lawn, and T. Maniatis, *Cell* **19,** 959 (1980).
35. J. Lauer, C. K. J. Shen, and T. Maniatis, *Cell* **20,** 119 (1980).
36. M. Plumb, G. Stein, and J. Stein, *Nucl. Acids Res.* **11,** 2391–2410 (1983).
37. J. D. Engel, B. J. Sugarman, and J. B. Dodgson, *Nature* **297,** 434 (1982).
38. F. Sierra, G. Stein, and J. Stein, *Nucl. Acids Res.* **11,** 7069 (1983).
39. A. Efstratiadis, J. W. Pasakony, T. Maniatis, R. M. Lawn, C. O'Connell, R. A. Spritz, J. K. DeRiel, B. G. Forget, S. M. Weissman, J. L. Slightom, A. E. Blechl, O. Smithies, F. E. Baralle, C. C. Shoulders, and N. J. Proudfoot, *Cell* **21,** 653 (1980).
40. C. Benoist, K. O'Hare, R. Breathnach, and P. Chambon, *Nucl. Acids Res.* **8,** 127 (1980).
41. L. H. Kedes, *Ann. Rev. Biochem.* **48,** 837 (1979).
42. R. Tjian, *Cell* **26,** 1 (1981).
43. J. Banerji, R. Rusconi, and W. Schaffner, *Cell* **27,** 299 (1981).
44. L. J. Korn and D. D. Brown, *Cell* **15,** 1145 (1978).
45. C. Birchmeier, R. Grosschedl, and M. L. Birnstiel, *Cell* **28,** 739 (1982).
46. J. L. Manley, A. Fire, A. Cano, P. A. Sharp, and M. L. Gefter, *Proc. Natl. Acad. Sci. USA* **77,** 3855 (1980).
47. S. Detke, J. L. Stein, and G. S. Stein, *Nucl. Acids Res.* **5,** 1515 (1978).
48. R. Grosschedl and M. L. Birnstiel, *Proc. Natl. Acad. Sci. USA* **77,** 7102 (1980).
49. B. C. Grosveld, C. K. Shewmaker, P. Jat, and R. A. Flavell, *Cell* **25,** 215 (1981).

50. R. Grosschedl and M. L. Birnstiel, *Proc. Natl. Acad. Sci. USA* **79,** 297 (1982).
51. J. Pan, J. T. Elder, D. H. Duncan, and S. M. Weissman, *Nucl. Acids Res.* **9,** 1151 (1981).
52. J. L. Manley and M. T. Colozzo, *Nature* **300,** 376 (1982).
53. P. K. Ghosh, J. Reddy, J. Seinscoe, P. Lebowitz, and S. M. Weissman, *J. Mol. Biol.* **126,** 813 (1978).
54. N. J. Proudfoot, M. H. M. Shander, J. L. Mauley, M. L. Gefter, and T. Maniatis, *Science* **209,** 1329 (1980).
55. F. Marashi, L. Baumbach, R. Rickles, F. Sierra, J. L. Stein, and G. S. Stein, *Science* **215,** 653 (1982).
56. J. Spalding, K. Kajiwara, and G. C. Mueller, *Proc. Natl. Acad. Sci. USA* **56,** 1535 (1966).
57. G. Stein and T. M. Borun, *J. Cell Biol.* **52,** 292 (1972).
58. R. S. Wu and W. M. Bonner, *Cell* **27,** 321 (1982).
59. R. S. Wu, S. Tsai, and W. M. Bonner, *Cell* **31,** 367 (1982).
60. M. A. Tarnowka, C. Baglioni, and C. Basilico, *Cell* **15,** 163 (1978).
61. V. E. Groppi and P. Coffino, *Cell* **21,** 195 (1980).
62. E. Robbins and T. W. Borun, *Proc. Natl. Acad. Sci. USA* **57,** 409 (1967).
63. A. M. Delegeane and A. L. Lee, *Science* **215,** 79 (1982).
64. J. L. Stein, G. S. Stein, and P. McGuire, *Biochemistry* **16,** 2207 (1977).
65. B. Moss, A. Gershowitz, L. A. Weber, and C. Baglioni, *Cell* **10,** 113 (1977).
66. R. Rickles, F. Marashi, F. Sierra, S. Clark, J. Wells, J. Stein, and G. Stein, *Proc. Natl. Acad. Sci. USA* **79,** 749 (1982).
67. L. Green, G. Stein, and J. Stein, *Mol. Cell. Biochem.*, in press.
68. G. S. Stein, J. L. Stein, W. D. Park, S. Detke, A. L. Lichtler, E. A. Shepard, R. L. Jansing, and I. R. Phillips, *Cold Spring Harbor Symp. Quant. Biol.* **42,** 1107 (1977).
69. S. Detke, A. Lichtler, I. Phillips, J. L. Stein, and G. S. Stein, *Proc. Natl. Acad. Sci. USA* **76,** 1995 (1979).
70. I. Parker and W. Fitschen, *Cell Diff.* **9,** 23 (1980).
71. I. M. Chiu, D. Cooper, and W. F. Marzluff, *Abstracts of the Second Annual Meeting of the American Cancer Society (Florida Division)* **36** (1979).
72. G. S. Stein, W. D. Park, C. L. Thrall, R. J. Mans, and J. L. Stein, *Nature* **257,** 764 (1975).
73. J. L. Stein, C. L. Thrall, W. D. Park, R. J. Mans, and G. S. Stein, *Science* **189,** (1975).
74. N. Heintz, H. L. Sive, and R. G. Roeder, *Mol. Cell Biol.* **3,** 539 (1983).
75. L. M. Hereford, M. A. Osley, J. R. Ludwig, and C. S. McLaughlin, *Cell* **24,** 367 (1981).

76. D. M. Prescott, R. M. Liskay, and G. M. Stancel, in C. Nicolini, Ed., *Cell Growth*, Vol. 38, 305 (1980).
77. T. W. Borun, F. Gabrielli, D. Ajiro, A. Zweidler, and C. Baglioni, *Cell* **4,** 59 (1975).
78. M. Breindl and D. Gallwitz, *Mol. Biol. Rep.* **1,** 263 (1974).
79. M. Jacobs-Lorena, F. Gabrielli, T. W. Borun, and C. Baglioni, *Biochim. Biophys. Acta* **324,** 275 (1973).
80. L. Hereford, S. Bromley, and M. A. Osley, *Cell* **30,** 305 (1982).
81. R. P. Perry and D. E. Kelley, *J. Mol. Biol.* **79,** 681 (1973).
82. N. S. Kunkel, K. Hemminki, and E. S. Weinberg, *Biochemistry* **17,** 2591 (1975).
83. P. A. Hieter, M. B. Hendricks, K. Hemminki, and E. S. Weinberg, *Biochemistry* **18,** 2707 (1979).
84. R. Maxson, T. Mohan, G. Gormezano, G. Childs, and L. Kedes, *Nature* **301,** 120 (1983).
85. H. R. Woodland and E. D. Adamson, *Develop. Biol.* **57,** 118 (1977).
86. A. Skoultchi and P. R. Gross, *Proc. Natl. Acad. Sci. USA* **70,** 2840 (1973).
87. D. E. Woods and W. Fitschen, *Cell Diff.* **7,** 103 (1978).
88. R. Maxson and F. Wilt, *Develop. Biol.* **94,** 435 (1982).
89. H. Stahl and D. Gallwitz, *Eur. J. Biochem.* **72,** 385 (1977).
90. M. Breindl and D. Gallwitz, *Eur. J. Biochem.* **45,** 91 (1974).
91. D. Gallwitz and G. C. Mueller, *J. Biol. Chem.* **214,** 5947 (1969).
92. E. A. Shephard, I. R. Phillips, J. Davis, J. L. Stein, and G. S. Stein, *FEBS Letters* **140,** 189 (1982).
93. W. Butler and G. Mueller, *Biochim. Biophys. Acta* **294,** 481 (1973).
94. G. Stein, J. Stein, E. Shephard, W. Park, and I. Phillips, *Biochem. Biophys. Res. Comm.* **77,** 245 (1977).
95. T. W. Borun, M. Scharff, and E. Robbins, *Proc. Natl. Acad. Sci. USA* **58,** 1977 (1967).
96. L. Kedes and P. Gross, *Nature* **223,** 1335 (1969).
97. N. Craig, D. E. Kelley, and R. R. Perry, *Biochim. Biophys. Acta* **246,** 493 (1971).
98. J. Huberman, *Cell* **23,** 647 (1981).
99. R. P. Harvey, A. J. Robins, and J. R. E. Wells, *Nucl. Acid Res.* **10,** 7851 (1982).
100. C. C. Hentschel, J. C. Irminger, P. Bucher, and M. L. Birnstiel, *Nature* **295,** 147 (1980).
101. G. S. Stein, J. L. Stein, L. Baumbach, A. Leza, A. Lichtler, F. Marashi, M. Plumb, R. Rickles, F. Sierra, and T. Van Dyke, in R. Baserga, Ed., *Proceedings of the New York Academy of Sciences Conference on Cell Proliferation, Cancer and Cancer Therapy*, Vol. 397, p. 148 (1982).

102. J. D. Engel, B. J. Sugarman, and J. B. Dodgson, *Nature* **297,** 434 (1982).

103. G. Childs, C. Noecente-McGrath, T. Lieber, C. Holt, and J. A. Knowles, *Cell* **31,** 383 (1982).

104. A. Seiler-Tuyns and M. L. Birnstiel, *J. Mol. Biol.* **151,** 607 (1981).

105. P. C. Turner and H. R. Woodland, *Nucl. Acid. Res.* **10,** 3769 (1982).

106. A. F. M. Moorman, P. A. J. de Boer, R. T. M. de Laaf, W. M. A. M. Van Donen, and O. H. J. Destree, *FEBS Lett.* **136,** 45 (1981).

107. E. C. Stephenson, H. P. Erba, and J. G. Gall, *Nucl. Acids Res.* **9,** 2281 (1981).

108. M. Grunstein, K. E. Diamond, E. Knoppel, and J. E. Grunstein, *Biochemistry* **20,** 1216 (1981).

109. L. Baumbach, F. Marashi, M. Plumb, G. Stein, and J. Stein, *Biochemistry*, in press.

110. M. Plumb, J. Stein, and G. Stein, *Nucl. Acids Res.* **11,** 7927 (1983).

CHAPTER 17

PATTERNS OF HISTONE GENE EXPRESSION IN HUMAN AND OTHER MAMMALIAN CELLS

ROY S. WU
MICHAEL H. P. WEST
WILLIAM M. BONNER
Laboratory of Molecular Pharmacology
Division of Cancer Treatment
National Cancer Institute
National Institutes of Health
Bethesda, Maryland

CONTENTS

1. INTRODUCTION

2. HISTONE VARIANTS
   - 2.1. Analysis
     - 2.1.1. One-Dimensional Polyacrylamide Gels
     - 2.1.2. Two-Dimensional Polyacrylamide Gels
   - 2.2. Characterization
   - 2.3. Evolution
   - 2.4. Functional Significance

3. HISTONE SYNTHESIS DURING THE CELL CYCLE
   - 3.1. Basal Histone Synthesis
   - 3.2. Cell Cycle Coordination
   - 3.3. Linkage of Histone Synthesis to DNA Synthesis

4. QUIESCENT CELL STUDIES
   - 4.1. Quiescent Histone Synthesis
   - 4.2. Relation of Basal and Quiescent Histone Synthesis

5. MODELS OF CELL BEHAVIOR
   - 5.1. The G_1 Phase
   - 5.2. The Quiescent State G_0

6. IMPLICATIONS FOR TRANSCRIPTIONAL CONTROL

REFERENCES

1. INTRODUCTION

The best understood function of the histones is to package the chromosomal DNA into nucleosomes (1). The nucleosome contains two molecules of four histone species, H4, H3, H2B, and H2A, known as core histones. A fifth histone, H1, is found associated with the DNA of the internucleosomal or linker region. Each of the histones except H4 is composed of isoprotein species also known as nonallelic variants. However, the histones also exhibit a great variety of posttranslational modifications the functional significance of which is still obscure (2). These mysteries suggest that a great deal remains to be learned about the functions of the histone proteins.

In mammalian cells, the genes coding for the histone proteins seem to be dispersed throughout the genome. The best characterized histone genes are unusual compared to most other eucaryotic genes in that they lack introns (3, 4). The mRNAs for the well characterized histones are also unusual in that they lack poly A (5). However, a few exceptions to both these properties have been reported recently (6).

It was also generally accepted until recently that in cycling cells, histone synthesis is coordinated with DNA replication and therefore takes place during S phase (7). New methods of analysis (8) have shown that the relationship of histone to DNA synthesis is somewhat more complex. It now seems that significant histone synthesis may be found all during the life of a cell, not just when its DNA is being replicated (9, 10). Evidence indicates that these histones are incorporated into chromatin without DNA replication. However, the majority of histone synthesis in cycling cells is still linked to DNA replication. The mechanism involved in this coordination is unknown. The four core histones always seem to be synthesized in approximately equimolar amounts; the mechanism for this is also unknown. During replication the histones are distributed to the daughter chromatids, although there is some disagreement as to how this is done (11–13). Once histones are incorporated into the chromatin, there are also many as yet unexplained phenomena, particularly concerning modifications. Certain modifications such as ubiquitination of H2A and H2B (14, 15) and phosphorylation of H1 (16, 17) have cycles that correlate with the cell cycle, but the significance, much less the functionality of most modifications, is still obscure. In this chapter we consider primarily questions concerning histone synthesis, the ultimate result of gene expression, rather than post-translational modifications.

However, because the analysis of histone variants has been instrumental in the recent results concerning histones synthesis, we first describe in some detail the analysis of variants and then review the evidence for histone synthesis in cells that are not replicating their DNA.

2. HISTONE VARIANTS

The existence of histone isoproteins that differ by a small number of amino acid substitutions has been known for sometime, but the evidence at first was usually indirect; sequence analysis would yield two amino acids at a particular position (18–20). Most fractionation methods could barely resolve the main histone species, not to mention isoproteins or variants of these species.

Electrophoresis in polyacrylamide gels containing acetic acid and urea (21) is capable of separating the main histone species. Addition of Triton X-100 to this gel system (22, 23) allows the separation of several histones into isoprotein species. Triton X-100, a nonpolar detergent, forms uncharged micelles within the polyacrylamide gel which have different affinities for each of the histone isoproteins. The resolution can be quite striking considering that some of these isoproteins differ by a single amino acid residue.

2.1. Analysis

The technique for analyzing histone isoproteins or variants described is one that we have developed over the last few years. It substantially modifies the acetic acid–urea gels containing Triton X-100 by the addition of a stacking gel and discontinuous buffer system. After neutralization, HCl extracts of nuclei, whole cells, and homogenized tissues can be loaded directly onto this gel, thereby greatly decreasing sample preparation time. The first-dimension gel can be stained, destained, and then loaded onto a second-dimension gel without Triton

X-100. This procedure completely separates the core histones from nonhistone proteins, as well as resolving all the known variants.

2.1.1. One-Dimensional Polyacrylamide Gels. The recipe for the preparation of acetic acid–urea–Triton X-100 (AUT) gels can be found in Bonner et al. (8). Figure 17.1 shows the pattern of histones from progressively more complex samples on AUT gels. With the HCl extract of salt-washed chromatin, the pattern consists predominantly of histones (Figure 17.1*a*). The bands behind each front band are modified forms of the histone, and in crowded regions of the gel

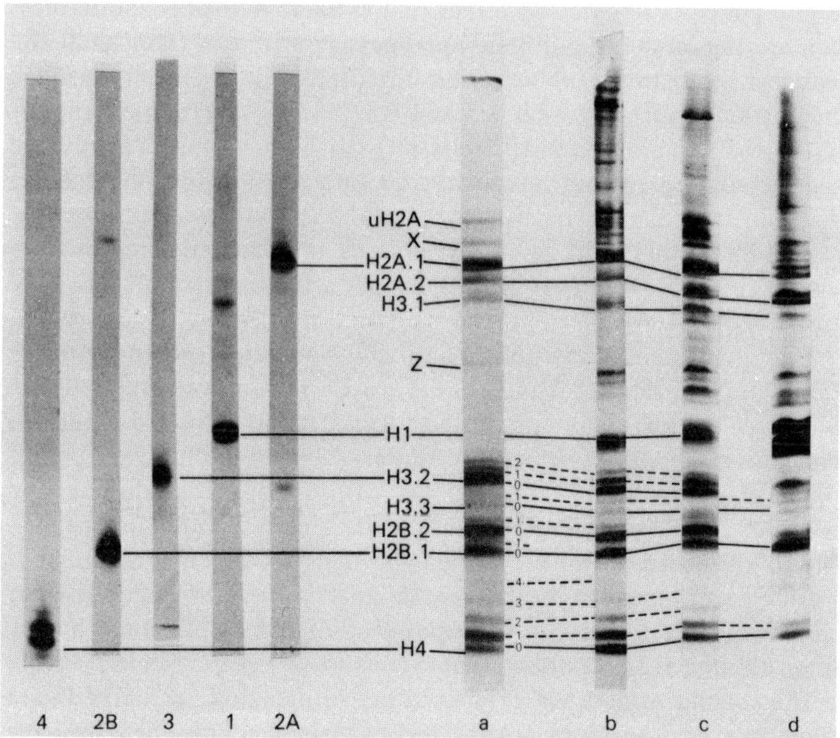

FIGURE 17.1. Acetic acid–urea–Triton X-100 gel of HCl extracts of (*a*) L1210 salt-washed chromatin, (*b*) L1210 nuclei, (*c*) L1210 cells, (*d*) and mouse liver homogenate. The left panels show the migration of purified mouse histone species in this gel system. From Bonner et al. (8).

these modified forms of one histone may overlap other variants or histone species. There are also some faint bands of nonhistone proteins. With whole nuclei (*b*), most of the histone species are easily discernable but nonhistone proteins interfere with others. With whole cells (*c*) or tissues (*d*), the histone pattern becomes decreasingly discernable. One-dimensional gels can be quite useful with highly purified samples or where the identity of the bands is already known.

2.1.2. Two-Dimensional Polyacrylamide Gels. Many of the problems associated with the purification of histone samples, such as selective loss, proteolysis, oxidation, and loss of modification can be minimized or avoided by direct HCl extraction of cells and tissues, then relying on the two-dimensional gel to purify and fractionate the histones. For example, pellets of tissue culture cells can be resuspended and extracted directly with 0.5 M HCl, 0.1% 2-mercaptoethanol, 1 mM PMSF (no more than 10^8 cells/ml) at 4°C for 1 hour. The residue is pelleted; the extract is transferred to a 1.5-ml microfuge tube, frozen in dry ice, and freeze dried. The mercaptoethanol helps prevent oxidation. In this method (8) the histone is extracted directly from a living cell into 0.5 *N* HCl in the cold, allowing virtually no time for proteolysis or other enzyme-mediated reactions.

A two-dimensional pattern from L1210 nuclei is shown in Figure 17.2. The overlapping bands in Figure 17.1 are now separated into diagonals behind each unmodified form. The histones are separated from the nonhistones which are mainly but not exclusively on the diagonal. The diagonal is composed of proteins which do not bind to Triton X-100 and therefore migrate identically in both dimensions. The relationship of bands, unclear on the one-dimensional gels, is much clearer on the two-dimensional gels. In fact, West and Bonner (24) were able to pick out the pattern (Figure 17.2) made by the ubiquitinated H2As as being analogous to .X, .1, .2, and .Z.

The second-dimension gel uses a first-dimension gel stained with Coomassie brilliant blue R-250. Using a stained gel has the advantage that only certain regions need be cut out for two-dimension analysis. In fact, single bands can be excised and pooled on the second dimension, as was done for peptide analysis (15, 24, 25). Because the first dimension is stained, all buffers and other small molecules are

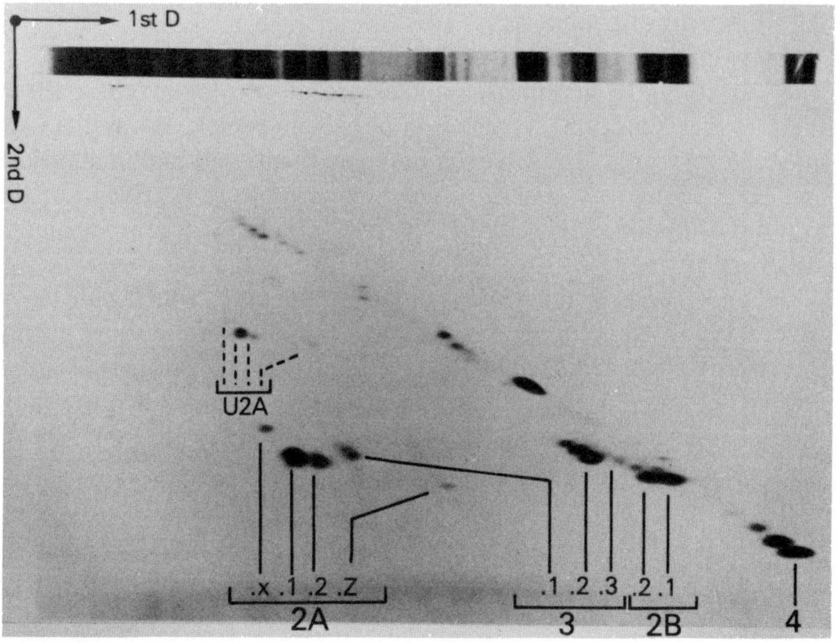

FIGURE 17.2. Two-dimension gel of HCl extract of L1210 nuclei. The identity of each histone isoprotein or variant species is noted. Adapted from Bonner et al. (8).

removed, so that not only Triton X-100, but also SDS gels can be used in the first dimension (8).

The second-dimension gel uses the cationic detergent, cetyltrimethylammonium bromide (CTAB). The CTAB dissociates the dye from the proteins, allowing them to migrate into and concentrate in the stacking gel, then to fractionate in the resolving gel. Although CTAB is an ionic detergent, it evidently does not bind to the histones under these conditions. If it did bind to proteins as SDS does, then modified forms of the histones would not separate in the second dimension (see Figure 3 of Reference (24)). However, Figure 17.2 shows that the separation of the modified forms of histones is the same in both dimensions, hence the diagonals behind each unmodified form. These diagonals also show that these modifications, mainly acetylation of lysine and phosphorylation of serines, do not alter the Triton X-100 binding of the histones.

2.2 Characterization

The variations in histone sequences are of two main types. First, there is the evolutionary variation between species. H4 and H3 sequences vary only slightly even between plants and mammals; H2A and H2B sequences generally vary much more (26). Even so, the binding of histones H2A and H2B from widely divergent species is still very strong (27–30).

Second, there is the variation within species, which can be of several types. Developmental variants occur at different times during the life of an organism (31–33). Tissue-specific variants are also known (26, 34, 35). Variants also occur in the same cell at the same time. The question whether or not these isoproteins are functionally equivalent is discussed later.

Figure 17.3 shows the histone pattern from a human cell line, IMR-90. A comparison with Figure 17.2 shows that the variant pattern of human and mouse histones is very similar. Mouse has an H2B.2

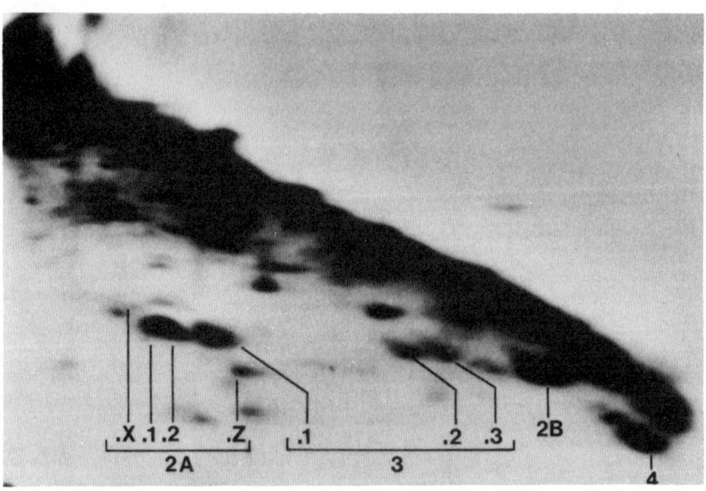

FIGURE 17.3. Two-dimension gel of exponentially growing IMR-90 fibroblasts. The cells were labeled with ^{14}C-arginine and the gel fluorographed. Note that histones 4 and 2B are better resolved from the diagonal of nonhistone proteins in this recipe which uses 6 M urea in the first dimension. However the variants H2A.1 and H2A.2 are less well resolved.

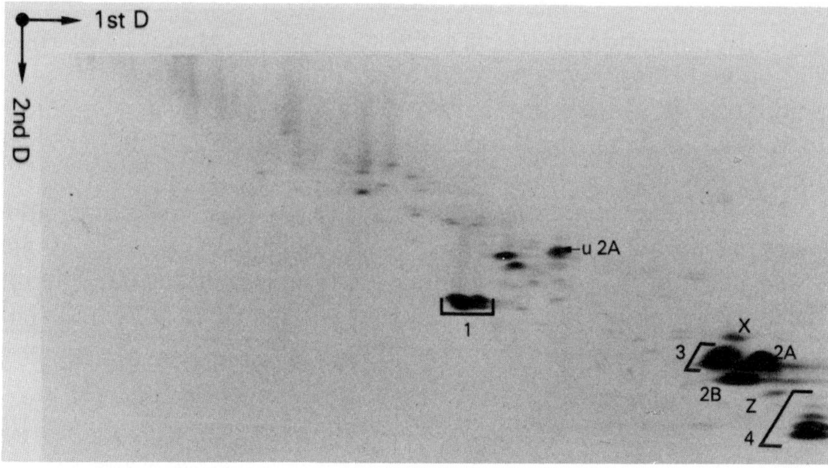

FIGURE 17.4. Sodium dodecyl sulfate (1st D)/acetic acid–urea–cetyltrimethylammonium bromide (2nd D) gel of mouse L1210 nuclei. Z and X denote the histone 2A variants H2A.Z and H2A.X. From Bonner et al. (8).

variant which seems to be absent in human cells, although perhaps not totally (P. Pantazis, personal communication). The same three H3 variants are present in both species, although in human cells H3.1 is the major variant, while H3.2 is the major variant in most mouse cells. The H2A pattern is similar in both.

The identification of H3 and H2B variants was straightforward in that in both these cases, the isoproteins comigrate on SDS or acetic acid–urea gels. Peptide analysis showed that they differ by a small number of amino acid substitutions (36). The variants H2A.1 and H2A.2 also comigrate in these gel systems and differ by just a few amino acids (36).

The identification of the other H2A variants was more involved since these variants separate from H2A.1 and H2A.2 on SDS or acetic acid–urea gels (Figure 17.4). One hint as to the relationship of proteins H2A.X and H2A.Z to H2A.1 and H2A.2 was the finding that these two proteins also seemed to have ubiquitin adducts, thus making a similar pattern of uH2As with H2A.X, H2A.1/2, H2A.Z as seen in Figure 17.2. Ubiquitin adducts of H2A had been discovered and characterized by Goldknopf and Busch (37). Ubiquitin itself is a protein of about 8000 daltons (38) which becomes attached through a gly-gly dipeptide on its carboxyl terminal to lysine 119 of H2A. The adduct is rapidly reversible to free ubiquitin and free H2A (15). At the time it was thought that ubiquitination of histones was unique to H2As, but later small amounts of ubiquitinated H2B were also found (39).

The final identification of the H2A variants was made by peptide mapping on 50% polyacrylamide gels (Figure 17.5). H2A.X and H2A.1 have many arginine-containing tryptic peptides in common, the only apparent difference being in H2A.1 peptide R5 (leu-leu-gly-arg). The pattern of lysine-containing tryptic patterns has only

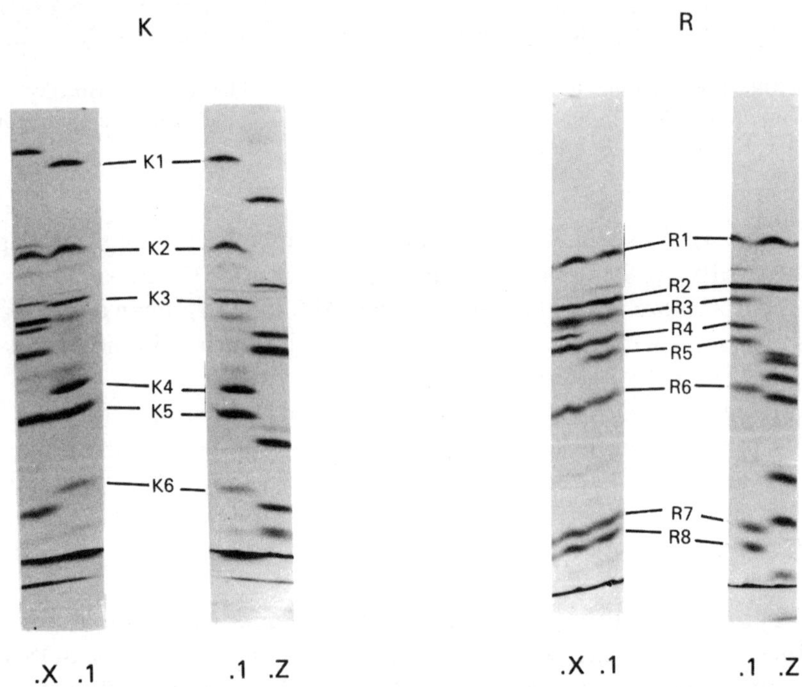

FIGURE 17.5. Peptide analysis of histone 2A species on 50% polyacrylamide gels. L1210 cells were labeled with ^{14}C-lysine(K) or ^{14}C-arginine (R). The histones were extracted and purified by two-dimension gel electrophoresis. The histone 2A species were recovered from the gel, digested with trypsin, and the digest electrophoresed on 50% polyacrylamide gels. The gel was autoradiographed. Left panel, lysine-containing peptides. Right panel, arginine-containing peptides. Peptides range in size from 19 amino acids (K1) down to three amino acids (R7, R8).

two identities between H2A.X and H2A.1. These are peptides K2 (asn-asp-glu-glu-leu-asn-lys) and K5 (asp-asn-lys-lys). Even though there are significant differences between the peptides of H2A.X and H2A.1, they are clearly closely related.

The situation for H2A.Z is different. Only two peptides are common to H2A.1 and H2A.Z. These are R1 (ala-gly-leu-gln-phe-pro-val-gly-arg) and R2 (his-leu-gln-leu-ala-ile-arg). There are no lysine-containing tryptic peptides on these gels common to both H2A.Z and H2A.1. Arginine peptide R1 is found in every sequenced H2A from animals as well as plants (26). Wheat germ H2A has this peptide, although all the other sequenced tryptic peptides differ from those in mammalian H2A (40). Therefore, although H2A.Z is quite different in sequence from H2A.1, it can still be classified as an H2A.

H2A.Z and H2A.X were found in mononucleosomes isolated by micrococcal nuclease digestion (8). In addition, both variants were also found in histone octamers isolated in 2 M NaCl and in the H2A:H2B dimer (40a). Thus, H2A.X and H2A.Z must substitute for a fraction of the H2A.1 and H2A.2 in the nucleosome.

2.3. Evolution

The histone variant patterns of the mammals and birds are very similar on the two-dimensional AUT-AUC gel system. As animals further down the evolutionary scale are analyzed, the patterns also vary more. Figure 17.6 shows the sea urchin pattern with the mouse H2As superimposed (33). Notice that the main sea urchin H2A variants α and β, as well as the one labeled .X, migrate somewhat differently from the mouse H2As. However, sea urchin contains a protein that seems to comigrate with mouse H2A.Z. When all these proteins were trypsin-cleaved and their peptides analyzed on 50% polyacrylamide gels, it became clear that the H2A.Z sequences from the two species are much more conserved than those of the other H2As. In addition, the H2A.Z peptide maps were very different from the maps of both the other urchin H2As and the mouse H2As. Proteins possibly related to H2A.Z have been isolated and partially characterized from *Drosophila* (41) and *Tetrahymena* (42). Therefore it seems that evolutionary selective pressure on H2A.Z sequences has been significantly more stringent than on the sequences of other H2As.

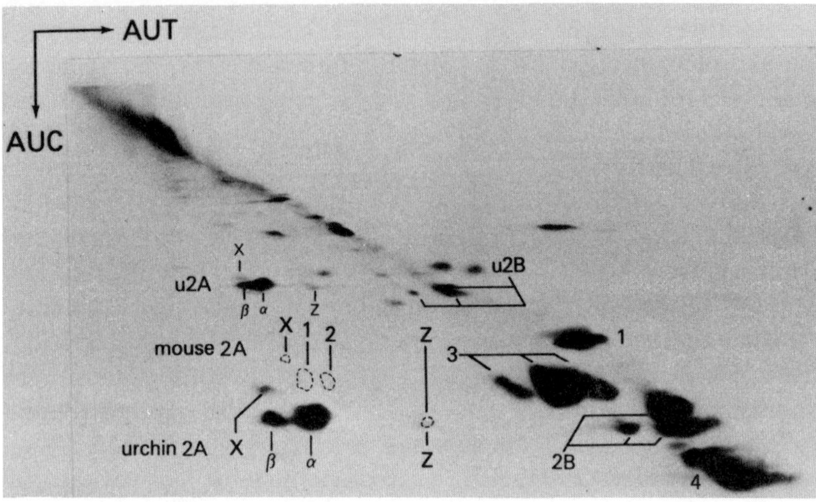

FIGURE 17.6. Comparison of histone 2A species from mouse cells and sea urchin embryos. The latter were labeled between 14 and 18 hours post-fertilization with ^{14}C-arginine, extracted with HCl, and the extract analyzed by fluorography of two-dimension gels. The dotted circles denote the position of added mouse histones (mass). From Wu et al. (33).

2.4. Functional Significance

The question of functional significance of histone variants is relevant but difficult. One extreme position is that variants are merely the reflection of allowable sequence diversity for the identical set of functions. The opposite position is that variants have significantly different functions, although these have yet to be characterized.

It is quite likely that both positions are correct depending upon the variant being considered. As has already been pointed out, the range of sequence divergence between variants can be large. This is particularly true of the H2As. H2A.1 and H2A.2 differ by only four positions, H2A.X differs by somewhat more, and H2A.Z by many more. With H2A.1 and H2A.2 one must search for the differences; with H2A.Z and H2A.1, one must search for the identities.

In some cases there is good evidence that a pair of variants is not functionally different. One example is the yeast H2Bs, composed of two variants. Mutants can be made that have only one H2B variant and these survive (43). Mouse H2B.2 (36) seems to be absent in most other mammals, although the validity of this last piece of data rests on the sensitivity of the assay. One could argue that other mammals may have a tiny amount of H2B.2, enough for some critical function. However, because H2B.1 and H2B.2 differ by a gly-ser substitution at position 75, and because of the variability of the relative amounts of the H2Bs in various tissues, it is difficult to imagine functional differentiation for these two isoproteins.

On the other hand, there is also evidence favoring functional differentiation. Simpson (44) showed that sea urchin chromatin containing early and late histone variants had different nuclease digestion and thermal parameters. Blankstein and Levy (45) showed that the H2A.2/H2A.1 ratio in different lines of Friends erythroleukemia cells correlated with their DMSO inducibility.

Perhaps the best argument can be made for H2A.Z and is based on the different selective pressure on the evolution of H2A.Z and other H2As as discussed previously (33). Certainly if a variant did have a critical and fundamental function, differentiated in some way from that of other variants, one would expect this variant to be recognizable in all cells and all species. H2A.Z does seem to fulfill these criteria in that it has been found in every cell type or species that we have analyzed. Another curious feature about H2A.Z is that it always seems to be present in rather small amounts. This differs from all other histones except H2A.X, where in a cell type or species one or another variant might be found as a predominant type. This small but rather constant fraction of total H2A also supports the other evidence in assigning a differentiated role for H2A.Z. Also, as discussed in the next section, the regulation of H2A.Z (and H2A.X) synthesis differs from that of H2A.1 and H2A.2. In conclusion then, it seems likely that with questions of functionality, variants should be considered individually rather than collectively.

3. HISTONE SYNTHESIS DURING THE CELL CYCLE

Ever since the realization that DNA replication occurs in a discrete time segment between mitoses (46), the relationship between histone and DNA synthesis has been investigated. It has generally been accepted that most histone synthesis is linked to DNA synthesis (7), although there have been reports to the contrary (47). Investigators have reported some histone in the non-S-phases of the cell cycle (48, 49), but these results could not be conclusive because of possible contributions from S-phase cells that got through the selection or induction procedures.

3.1. Basal Histone Synthesis

Wu and Bonner (9) reinvestigated the question of cell cycle regulation of histone synthesis using the two-dimensional gel system which separates variants, and found a significant amount of histone synthesis in the G_1 and G_2 phases of the cell cycle. Because its variant pattern was qualitatively different from the pattern in S phase (Figure 17.7), the synthesis in G_1 could not be attributed to contamination by S-phase cells. This synthesis was called basal synthesis because it occurred throughout the cell cycle. S-phase synthesis was superimposed on basal synthesis when DNA replication commenced. In basal synthesis the major H2A variants H2A.1 and H2A.2 as well as the major H3 variants H3.1 and H3.2 are not synthesized, while the synthesis of H2A.X and H2A.Z and H3.3 continues around the cell cycle. Figure 17.8 shows typical patterns of histone synthesis in S phase and G_1 obtained with synchronized CHO cells; similar results were obtained with HeLa cells.

3.2. Cell Cycle Coordination

How cells cycle is a central question of cell biology. One common view (50) is that the cell cycle is composed of two cycles, a chromosome replication (S, G_2, M) cycle and a growth cycle. G_1 is that part of growth cycle left over after the chromosome cycle is completed.

FIGURE 17.7. Histone synthesis during G_1 and S phases in Chinese hamster ovary (CHO) cells. G_1 cells were obtained by mitotic shakeoff. One hour after replating, they were labeled for 1 hour with ^{14}C-arginine. Histones were analyzed after HCl extraction. S-phase cells were labeled for 1 hour starting 5 hours after replating. (*A*) mass. (*B*) S-phase label. (*C*) G_1 phase label. From Wu and Bonner (9).

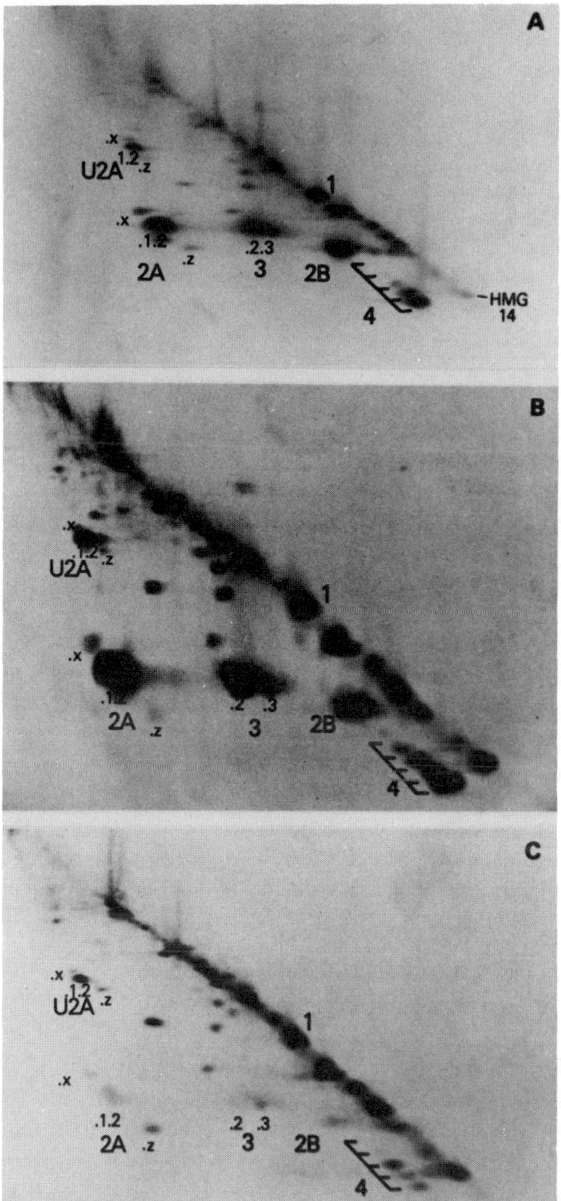
A
.x
U2A
1.2
.z
1
.x
.1.2
.z
.2.3
2A
3
2B
4
HMG
14
B
.x
.1.2
.z
U2A
1
.x
.1.2
2A
.z
.2
.3
3
2B
4
C
.x
1.2
U2A
.z
1
.x
.1.2
2A
.z
.2
.3
3
2B
4

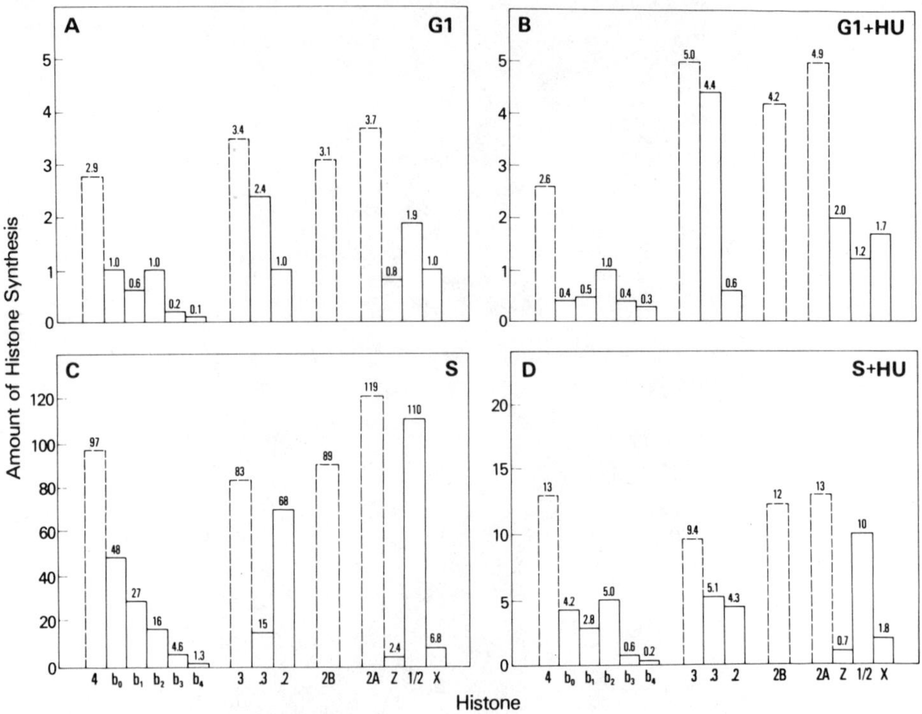

FIGURE 17.8. Quantitation of histones synthesized in the presence and absence of hydroxyurea. Spots of the appropriate histone, labeled with ^{14}C-arginine were excised from the dried gel, digested with H_2O_2/NH_4OH overnight, and analyzed by scintillation counting. The radioactivity in each spot was divided by the arginine content (13 for H2As, 8 for H2Bs, 18 for H3s, and 14 for H4s) of that histone to obtain relative moles of histone synthesis. Hydroxyurea (+HU) treated samples were incubated with 1 mM hydroxyurea. From Wu and Bonner (9).

However, the chromosome cycle and the growth cycle cannot work totally independently; they must be coordinated. In one model (51), a hypothetical initiator protein is synthesized at a rate proportional to cell growth rate and then incorporated into the chromosome. When the sites are full, the chromosome is triggered to initiate another replication cycle. It is striking that basal histone synthesis has the necessary properties for such an initiator. Its synthesis is independent of DNA synthesis, is dependent on the rate of cell growth, and is not limited to G_1. Basal histones seem to be stably incorporated into chromosomes. It should be mentioned that basal histone synthesis is unusual in that it includes only the minor H2A variants H2A.X and H2A.Z while excluding the major variants H2A.1 and H2A.2. Also, H2A.Z is more highly conserved than are the other H2As. Thus basal histone synthesis does seem to link the growth cycle with the chromosome cycle and may therefore be an attractive candidate to coordinate these two cycles.

3.3. Linkage of Histone and DNA Synthesis

The presence of basal histone synthesis in G_2/G_1 suggests that it is not linked to DNA synthesis. A substantiating experiment is shown in Figure 17.8, in which the amount of histone synthesis in G_1 and S-phase cells with and without hydroxyurea treatment is quantitated. Notice that in each case approximately equal amounts of each histone species is synthesized, although this synthesis may be composed of different variants in each case. In the G_1 cells the presence of hydroxyurea does not inhibit basal histone synthesis, even though the residual ^{3}H-thymidine incorporation is reduced another 80%. From the data presented in Figure 17.8, one can calculate that in hydroxyurea-treated S-phase cells, the synthesis of S-phase histone variants is inhibited to 6–9% of their level in untreated cells, while the comparable values for basal variants are 26–34%. Clearly, the synthesis of basal variants even in S-phase cells is much more resistant to the inhibition of DNA synthesis. In G_1 cells, the synthesis of basal variants even seems to be stimulated 1.7–2.5-fold. We have no ready explanation for this phenomenon. Note also that in comparing G_1 and S-phase cells, the synthesis of S-phase variants is stimulated 58–68-fold, while the synthesis of basal variants is stimulated only 3.0–6.8-fold. The two histones with only one variant in CHO cells, H2B and H4, are stimulated 29–35-fold. Total (averaging the different variants together) H2A and H3 synthesis is also stimulated 23–32-fold. These results indicate that H2B and H4 have subsets the synthesis of which is controlled like basal variants, even though the protein sequences may be identical.

A related question concerns the linkage between S-phase histone synthesis and DNA synthesis. Generally histone synthesis was reported to be confined to S phase (52, 53). But if basal histone synthesis is not linked to DNA synthesis, could S-phase histone synthesis become unlinked to DNA synthesis? There have been reports in certain cells that histone synthesis may be partially or even totally unlinked to DNA synthesis (47–49).

Cell growth in continuous exponential phase, the way many cell lines are maintained, inherently selects for faster-growing cell subpopulations. One adaptive change might be to accelerate the growth cycle so that the cell could go directly into S phase after mitosis. A lack of G_1 has been reported for some nonmammalian (54) as well as mammalian cells (55, 56) with short cell cycle times and has been used to argue for the nonessential nature of G_1.

Is it possible that cells that are always in exponential growth could also unlink histone and DNA synthesis for faster growth (47)? Perhaps one possible disorder of certain tumor lines in culture is a partial or total unlinking of S-phase histone synthesis and DNA replication. Such a possibility cannot be excluded. However, it is curious that such cells are reported to still maintain a G_1, even though cell lines are known with no G_1. It is not clear that unlinking histone and DNA synthesis would in fact be beneficial to the cell, since it is the growth cycle and not the chromosome cycle that is longer. Also, histones are one of the, if not the major protein products during S phase, so that toxic effects could be expected rather rapidly unless there is a specific storage mechanism of rather large capacity, as seems to exist in *Xenopus* oocytes (57, 58). Clearly there are many questions yet to be answered in the areas of chromosome replication and cell cycle control.

4. QUIESCENT CELL STUDIES

Many cells stop dividing when they become confluent on a surface or when nutrients are depleted (59). These cells can remain viable for a long time and resume cycling under the appropriate conditions. This nonproliferative state has been commonly called quiescence or G_0. Its reversability distinguishes it from some differentiated states; however, the ability to cycle upon a given stimulus can also be a property of differentiated cells, such as lymphocytes (60).

The relationship of quiescent (G_0) cells to cycling cells and to particular phases of the cell cycle such as G_1 is an area of research with views as divergent as those concerning the relationship of G_1 with the rest of the cell cycle. Some view quiescence as an extended G_1, whereas others view quiescence as a separate and discrete state.

One reason for these divergent views is the scarcity of well-defined biological and biochemical criteria for distinguishing G_0 cells from G_1 cells. The criteria that do exist are usually quantitative changes that can be viewed as consistent with any of several very different models. In contrast, histone variant synthesis patterns show qualitative differences that may be useful in distinguishing cells in different states.

4.1. Quiescent Histone Synthesis

Figure 17.3 shows the pattern of histone synthesis in the normal human fibroblast line IMR-90 during exponential growth. The histone synthesis pattern including the H3 pattern is qualitatively the same as the mass. Figure 17.9 shows the pattern of histone synthesis in quiescent cells. Note that H3.1 and H3.2 are not synthesized in quiescent cells, even though H3.1 and H3.2 are the major H3 variants by mass; only H3.3, a minor variant by mass, is synthesized. In

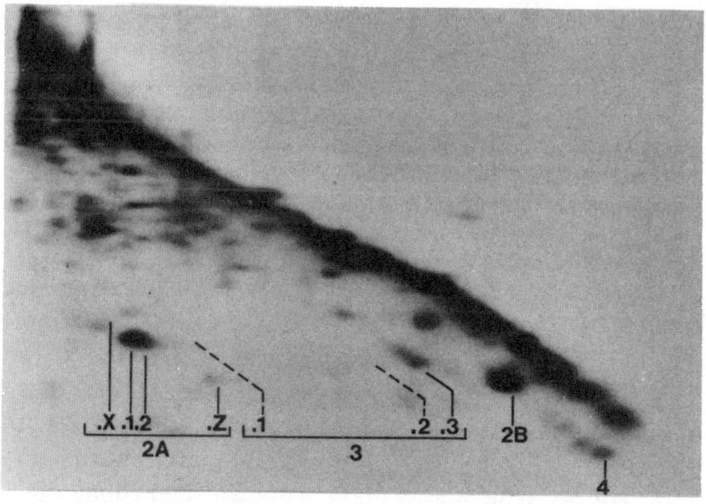

FIGURE 17.9. Two-dimension gel of quiescent IMR-90 fibroblasts. Compare with Figure 17.3; note that H3.1 and H3.2 are not synthesized in quiescent cells.

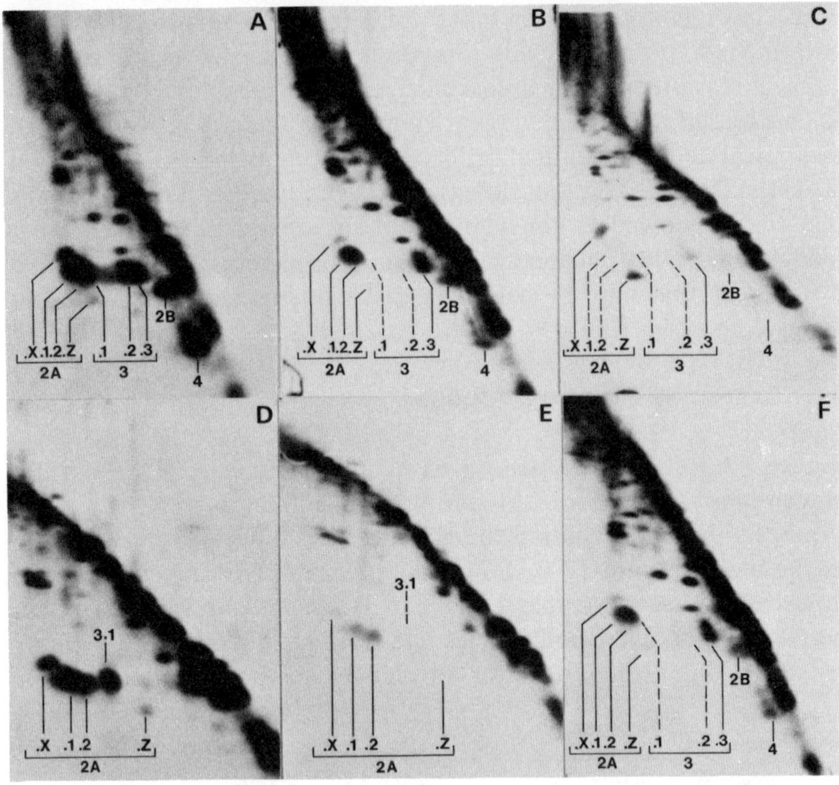

FIGURE 17.10. Histone synthesis patterns of CHO cells in various growth states. G_1 and S-phase CHO cells were obtained as in Figure 17.7. Quiescent cells were grown to confluency and maintained in medium containing 2% fetal bovine serum. The cells were labeled for 1 hour with ^{14}C-arginine and the histones analyzed as previously described. Fluorograms of gels are shown. Panels A, B, C, F, are 15% polyacrylamide in the first dimension; panels D and E are 10% polyacrylamide in the first dimension to increase the resolution of the H2A region. (A, D) exponential, (B, E) quiescent, (C) G_1 phase, (F) quiescent cells treated with 1 mM hydroxyurea 1 hour before and during the 1-hour labeling.

quiescent cells, the H2A ratio (H2A.1 + .2)/(H2A.X + .Z) is not significantly different from that in exponential cells, but the H3 ratio (H3.1 + .2)/H3.3 changes strikingly from 2.8 in exponential cells (Figure 17.3) to 0.2 in quiescent cells (Figure 17.9).

In quiescent IMR-90, ^3H-thymidine incorporation has decreased to 0.6% of the exponential control; ^{14}C-arginine incorporation into histones has decreased to 7.9% of the exponential control. This difference that in quiescent cells the rate of incorporation is higher for the histones than for the DNA, is also seen in other cell types tested, including CHO (10), and human lymphocytes (60a).

4.2. Relation of Basal and Quiescent Synthesis

Figure 17.10 shows the patterns when CHO cells in different states are analyzed. Like IMR-90, quiescent CHO do not synthesize H3.1 and H3.2, even though exponentially growing CHO synthesize all three H3 variants. However, when the quiescent CHO pattern is compared to the G_1 pattern, it is clear that there is a major difference in the H2A pattern. The H2A pattern in quiescent cells is similar to the H2A pattern in exponentially growing cells, but in G_1 cells the major H2A variants are not synthesized. These qualitative differences in histone synthesis pattern suggest that quiescent or G_0 cells are not just in an extended or arrested G_1 but are in a discrete state.

5. MODELS OF CELL BEHAVIOR

There are many unanswered questions concerning cell behavior. Two that we discuss here involve the importance of G_1 in the cell cycle and the relationship of quiescence or G_0 to the cell cycle phases, particularly G_1.

5.1. The G_1 phase

G_1 is the most variable portion of the cell cycle. In one view it contains unique steps essential for the initiation of DNA replication. Cell-cycle-specific temperature-sensitive mutants (61) should block at some of these steps. There may be a decision point during G_1 to become quiescent and enter G_0 (62). One model (63) postulates the synthesis of a rapidly turning over protein that is synthesized at a rate proportional to the growth rate. If the cell is growing fast enough in a favorable environment, then this protein can build up to a level that would trigger a series of events leading to the initiation of DNA replication. Inherent in this model is a mechanism for coordinating the chromosome cycle and the growth cycle.

In a different view, G_1 is merely the time difference between the end of the chromosome cycle at mitosis and the end of the growth cycle. In one model, there are no G_1 specific events (51). The two cycles are coordinated again by the synthesis of a protein, but in this case a stable protein that is synthesized throughout the cell cycle. This protein gradually fills sites in the chromatin during G_2, M, and G_1. When a critical number of sites are filled, a new round of DNA replication is initiated which generates new sites. As mentioned previously, basal histone synthesis has the proposed properties for such a timer. However, such a mechanism is still purely hypothetical, and the cell may have a totally different mechanism for coordinating the two cycles.

5.2. The Quiescent State G_0

Another question concerns the relationship of quiescence of G_0 to G_1. The models that view G_1 as nonessential can easily view G_0 as an extended (64) or arrested G_1 (51). One prediction of this model is that there should not be qualitative differences in protein synthesis between G_0 and G_1, but only quantitative ones. However, the results reviewed earlier do not support this relationship since they indicate that there are qualitative differences in H2A synthesis patterns between these two states. These data support other models in which quiescence or G_0 is a separate state from G_1.

However, in these models G_0 usually enters and exits from G_1 (65, 66). The results from histone synthesis patterns suggest that quiescence may even be a partially discrete cycle, sharing only S phase with the cell cycle (Figure 17.11). The reason for this proposal concerns what happens when quiescent cells are stimulated to divide. Resting lymphocytes when stimulated with phytohaemagglutin start DNA replication after 12 hours. During this time the lymphocytes maintain the quiescent pattern and do not switch to the basal (G_1) pattern (60a). This result suggests that the quiescent state does not traverse G_1.

It may be then that a cell decides ("decides" here includes both the deterministic and probabilistic senses) during S phase whether or not to become quiescent after mitosis. In that case, the cell may exhibit the quiescent pattern directly after S phase. However, it has not been excluded that a cell could adopt the basal pattern in G_2 and

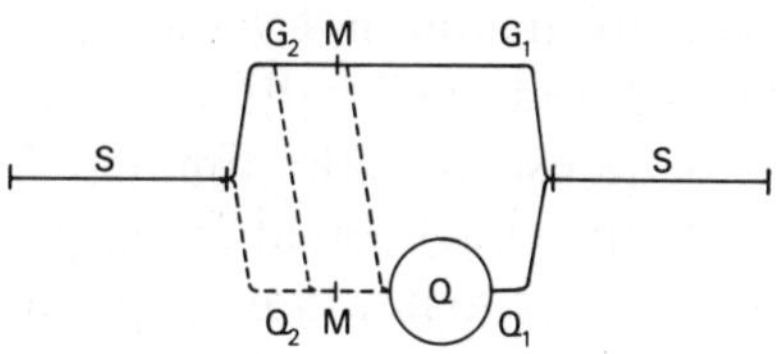

FIGURE 17.11. Possible relationships of quiescent and basal (G_1/G_2) cells.

then switch that pattern to a quiescent one in G_2 or G_1 depending on the environment (as shown in Figure 17.11 by dotted arrows). The probabilistic model of Brooks et al. (67) includes a similar possibility. The exponentially growing cell may prepare for the next cycle during the present one. If it does not prepare, then it must become quiescent until it can undergo the necessary preparation.

One possible interpretation of these patterns of histone synthesis is that the basal pattern is present in cells already committed to another cycle. In this case basal histone synthesis could function as a "timer" for the initiation of the following S phase. Since basal histone synthesis includes the two heteromorphous H2A variants and excludes the major ones, it may be that basal synthesis cannot be maintained indefinitely. To date, H2A.Z or H2A.X have not been found as major variants in any cell type.

If the environment is not conducive to exponential growth owing to nutrient deprivation, absence of growth factors, or high cell density, a cell may enter quiescence, adopting that pattern of histone synthesis. Quiescent histone synthesis has all four of the H2A variants, but only one H3 variant, H3.3. However, the differences between H3.1, H3.2, and H3.3 are small compared to the differences between the major and minor H2A variants, and in contrast to the H2As, H3.3 can be found as the major variant in certain cells and tissues, particularly tissues that have remained quiescent for a long time (8, 68, 69). This could be explained by the H3.3 slowly displacing the H3.1 and H3.2 variants. Commerford et al. (70) measured a histone half-life in mouse liver of 117 days while the DNA had a half-life of 318 days, suggesting that histone turnover on preexisting DNA does exist.

6. IMPLICATIONS FOR TRANSCRIPTIONAL CONTROL

The qualitative differences in protein synthesis are likely to be reflected by qualitative changes in the activity of the genes. In cell cycle studies, it has been shown that the levels of histone mRNA are increased dramatically in S-phase cells compared to G_1 cells, both in continuously cycling cells (71–77) and in quiescent cells stimulated to proliferate (75–77). These results certainly suggest that the large increase in histone mRNA levels in S phase is due to new transcription (71–77). It is probable that the histone RNAs measured in these studies are the replication-variant RNAs. The histone mRNAs responsible for basal histone synthesis have yet to be identified and could either be rapidly turning over or fairly stable; there is little information on this point.

Qualitative differences in the control of histone variant genes suggest that the genes may differ structurally. The well-characterized histone genes and mRNAs have generally been considered unusual in two ways. First, until recently, the working hypothesis has been that histone mRNA lacks poly A. However, increasing evidence has suggested that at least a portion of some histone mRNAs may contain poly A. Such subsets have been reported in yeasts (78), HeLa cells (79), amphibians (80), sea urchins (81), and clams (82). Since specific mRNAs for variants were not identified, it may be possible that basal or quiescent histone mRNA contains poly A; such a difference could be expected to lead to differences in control. However, reports of poly A on histone mRNA must be treated with some caution. Engel et al. (6) recovered an H3 mRNA in the poly A containing fraction, but this affinity was due to a transcribed A-rich sequence ($A_{11}GA_9$) rather than to post-transcriptionally added poly A.

Second, no histone gene with introns had been reported until recently when Engel et al. (6) reported a chicken H3 mRNA with introns. These reports indicate that histone genes as a group may be heterogeneous with regard to poly A attachment sites and introns. These differences may provide for many possibilities in differential control. Conversely, histone genes may be useful in helping to understand the purposes of introns and poly A.

Because histone synthesis patterns change with changes in cell proliferation state, studying the processes controlling the transcription of histone genes, particularly the variant genes used for basal and quiescent synthesis may lead to insights for the control of cell proliferation at the molecular level.

REFERENCES

1. J. D. McGhee and G. Felsenfeld, *Ann. Rev. Biochem.* **49,** 115 (1980).
2. I. Isenberg, *Ann. Rev. Biochem.* **48,** 159 (1979).
3. I. Sues, A. Maxam, R. H. Cohn, and L. H. Kedes, *Cell* **9,** 495 (1976).
4. M. L. Birnstiel, W. Schaffner, and H. O. Smith, *Nature* **266,** 603 (1977).
5. M. Adesnik and J. E. Darnell, *J. Mol. Biol.* **67,** 397 (1972).
6. J. D. Engel, B. J. Sugarman, and J. B. Dodgson, *Nature* **297,** 434 (1982).
7. S. C. R. Elgin and H. Weintraub, *Ann. Rev. Biochem.* **44,** 725 (1975).
8. W. M. Bonner, M. H. P. West, and J. D. Stedman, *Eur. J. Biochem.* **109,** 17 (1980).
9. R. S. Wu and W. M. Bonner, *Cell* **27,** 321 (1981).
10. R. S. Wu, S. Tsai, and W. M. Bonner, *Cell* **10,** 367 (1982).
11. V. Jackson and R. Chalkley, *J. Biol. Chem.* **256,** 5095 (1981).
12. I. M. Leffack, R. Grainger, and H. Weintraub, *Cell* **12,** 837 (1977).
13. A. T. Annunziato, R. K., Schindler, M. G. Riggs, and R. L. Seale, *J. Biol. Chem.* **257,** 8507 (1982).
14. S. I. Matsui, B. K. Seon, and A. A. Sandberg, *Proc. Natl. Acad. Sci. USA* **76,** 6386 (1979).
15. R. S. Wu, K. W. Kohn, and W. M. Bonner, *J. Biol. Chem.* **256,** 5916 (1981).
16. E. M. Bradbury, R. J. Inglis, H. R. Matthews, and N. Sarner, *Eur. J. Biochem.* **33,** 131 (1973).
17. L. R. Gurley, J. A. D'Anna, S. S. Barham, L. L. Deaven, and R. A. Tobey, *Eur. J. Biochem.* **84,** 1 (1978).
18. L. Patthy, E. L. Smith, and J. Johnson, *J. Biol. Chem.* **248,** 6834 (1973).
19. W. F. Marzluff, Jr., L. A. Sanders, D. M. Miller, and K. S. McCarty, *J. Biol. Chem.* **247,** 2026 (1972).
20. P. Sautiere, *Ciba Found. Symp.* **28,** 77 (1975).
21. S. Panyim and R. Chalkley, *Arch. Biochem. Biophys.* **130,** 337 (1969).
22. C. Alfageme, A. Zweidler, A. Mahowald, and L. Cohen, *J. Biol. Chem.* **249,** 3729 (1974).
23. A. Zweidler, *Methods Cell Biol.* **17,** 223 (1978).
24. M. H. P. West and W. M. Bonner, *Biochemistry* **19,** 3238 (1980).
25. P. Pantazis and W. M. Bonner, *J. Biol. Chem.* **256,** 4669 (1981).
26. C. von Holt, W. N. Strickland, W. F., Brandt, and M. S. Strickland, *FEBS Lett.* **100,** 201 (1979).
27. J. K. W. Mardian and I. Isenberg, *Biochemistry* **17,** 3825 (1978).
28. J. A. D'Anna and I. Isenberg, *Biochemistry* **13,** 2098 (1974).

29. S. Spiker and I. Isenberg, *Cold Spring Harbor Symp. Quant. Biol.*, **42,** 157 (1977).
30. S. Spiker and I. Isenberg, *Biochemistry* 16, 1819 (1977).
31. K. M. Newrock, L. H. Cohen, M. B. Henricks, R. F. Donnelly, and E. S. Weinberg, *Cell* **14,** 327 (1978).
32. G. Childs, R. Maxson, and L. H. Kedes, *Dev. Biol.* **73,** 153 (1979).
33. R. S. Wu, D. Nishioka, and W. M. Bonner, *J. Cell Biol.* **93,** 426 (1982).
34. L. S. Hnilica, *Experientia* **20,** 13 (1964).
35. J. M. Neelin, P. X. Callahan, D. C. Lamb, and K. Murray, *Can. J. Biochem.* **42,** 1743 (1964).
36. S. G. Franklin and A. Zweidler, *Nature* **266,** 273 (1977).
37. I. L. Goldknopf and H. Busch, *Cell Nucleus*, Vol. 6, Academic Press, N.Y., 1978, pp. 149–180.
38. D. H. Schlesinger, G. Goldstein, and H. D. Niall, *Biochemistry* **14,** 2214 (1975).
39. M. H. P. West and W. M. Bonner, *Nucl. Acids Res.* **8,** 4671 (1980).
40. J. deA. Rodrigues, W. F. Brandt, and C. von Holt, *Biochim. Biophys. Acta* **578** 196 (1979).
40a. C. Hatch, W. M. Bonner, and E. N . Mourianakis, *Science* **221,** 468 (1983).
41. D. Palmer, L. A. Snyder, and M. Blumenfeld, *Proc. Natl. Acad. Sci. USA* **77,** 2671 (1980).
42. C. D. Allis, C. V. C. Glover, J. K. Bowen, and M. A. Gorovsky, *Cell* **20,** 609 (1980).
43. M. Rykowski, J. Wallis, J. Choe, and M. Grunstein, *Cell* **25,** 477 (1981).
44. R. T. Simpson, *Proc. Natl. Acad. Sci. USA* **78,** 6803 (1981).
45. L. A. Blankstein and S. B. Levy, *Nature (London)* **260,** 638 (1976).
46. A. Howard and S. R. Pelc, *Heredity, Suppl.* **6** 261 (1953).
47. V. E. Groppi, Jr. and P. Coffino, *Cell* **21,** 195 (1980).
48. A. Sadgopal and J. Bonner, *Biochim. Biophys. Acta* **186,** 349 (1969).
49. D. Gallwitz and M. Brendl, *Biochem. Biophys. Res. Commun.* **47,** 1106 (1972).
50. J. M. Mitchison, *The Biology of the Cell Cycle*, Cambrige University Press, London (1971).
51. S. Cooper, *Nature* **280,** 17 (1979).
52. J. Spalding, K. Kaijwara, and G. C. Mueller, *Proc. Natl. Acad. Sci. USA* **56,** 1535 (1966).
53. F. Marashi, L. Baumbach, R. Rickles, F. Sierra, J. L. Stein, and G. S. Stein, *Science* **215,** 683 (1982).
54. R. T. Hinegardner, B. Rao, and D. E. Feldman, *Exp. Cell. Res.* **36,** 53 (1964).
55. R. M. Liskay and D. M. Prescott, *Proc. Natl. Acad. Sci. USA* **75,** 2873 (1978).
56. E. Robbins and M. D. Scharff, *J. Cell Biol.* **34,** 684 (1967).

57. R. A. Laskey, B. M. Honda, A. D. Mills, and J. T. Finch, *Nature* **275,** 416 (1978).
58. A. D. Mills, R. H. Laskey, P. Black, and E. M. DeRobertis, *J. Mol. Biol.* **139,** 561 (1980).
59. R. W. Holley, *Nature* **258,** 487 (1975).
60. M. Hartog, M. J. Cline, and G. M., Grodsky, *Clin. Exp. Immunol.* **2,** 217 (1967).
60a. R. S. Wu, S. Tsai, and W. M. Bonner, *Biochemistry* **22,** 3868 (1983).
61. C. Basilico, *Adv. Cancer Res.* **24,** 223 (1977).
62. A. B. Pardee, *Proc. Natl. Acad. Sci. USA* **71,** 1286 (1974).
63. P. Rossow, V. G. H. Riddle, and A. B. Pardee, *Proc. Natl. Acad. Sci. USA* **76,** 4446 (1979).
64. J. A. Smith and L. Martin, *Proc. Natl. Acad. Sci. USA* **70,** 1263 (1973).
65. R. Baserga, *Multiplication and Division in Mammalian Cells*, Dekker, New York (1976).
66. C. Nicolini, *Biophysics* **2,** 271 (1980).
67. R. F. Brooks, D. C. Bennett, and J. A. Smith, *Cell* **19,** 493 (1980).
68. A. Zweidler, *Life Sci. Res. Rep.* **4,** 187 (1976).
69. A. Zweidler, *Histone Genes and Histone Gene Expression* (This volume) Wiley, New York (1984).
70. S. L. Commerford, A. L. Carsten, and E. P. Cronkite, *Proc. Natl. Acad. Sci. USA* **79,** 1163 (1982).
71. R. Rickles, F. Marashi, F. Sierra, J. Wells, J. Stein, and G. Stein *Proc. Natl. Acad. Sci. USA* **79,** 749 (1982).
72. M. Plumb, G. Stein, and J. Stein, *Nucl. Acids Res.* **11,** 2391 (1983).
73. N. Heintz, H. L. Sive, and R. G. Roeder, *Mol. Cell. Biol.* **3,** 539–550 (1983).
74. D. B. Sittman, R. A. Graves and W. F. Marzluff, *Proc. Natl. Acad. Sci. USA* **80,** 1849 (1983).
75. G. S. Stein, J. L. Stein, W. D. Park, S. Detke, A. C. Lichtler, E. A. Sheppard, R. L. Jansing, and I. R. Phillips, *Cold Spring Harbor Symp. Quant. Biol.* **42,** 1107 (1978).
76. L. Green, J. Stein, and G. Stein, *Mol. Cell. Biochem.*, manuscript submitted.
77. M. L. Parker and W. Fitschen, *Cell Differentiation* **9,** 23 (1980).
78. K. Fahrner, J. Yarger, and L. Hereford, *Nucl. Acids Res.* **8,** 5725 (1980).
79. T. W. Borun, R. Ajiro, A. Zweidler, T. W. Dolby, and R. E. Stephens, *J. Biol. Chem.* **252,** 173 (1977).
80. R. Levenson and K. Marcu, *Cell* **9,** 311 (1976).
81. J. V. Ruderman and M. L. Pardue, *J. Biol. Chem.* **253,** 2018 (1978).
82. F. Gabrielli and C. Gablioni, *Dev. Biol.* **43,** 254 (1975).

INDEX

ACCA motif, 413
Acetic acid-urea-Triton X (AUT), 461–462
Acetylation, 171, 463
Actin mRNA, 148
Actinomycin D, 109, 319, 450
Active transcription, 67
Allelic variation, 344
Ambystoma mexicanum, 232
Amido Black, staining with, 347
Amphibians, 480
Aphidicolon, 311–313
"Approach to steady-state," 305
Arbacia punctulata, 121
Ars functions, 16
Ars plasmids, 15–16
Artemia:
  core histones, 194
  dormant embryos, 183
  early larvae, histone gene expression, 188–190
  5S-related sequences, 190–193
  histone genes:
    cloning, 184–185
    isolation, 184–185
    organization, 185–188
  "library," 184
  mRNA, 188–190
Arthropods, 194
A + T regions, 22–23

AUT-AUC gel system, 467
Autoradiography, 322, 323, 333
Autoregulation, 312–313

Bacteriophage vector lambda Charon 4A, 184
Bal-31 nuclease digestion, 414
Basal synthesis, 470
Base composition, 21–22
Bigene expression:
  embryogenesis, 211
  oogenesis, 211
Brine shrimp, *see Artemia*
Buffer system, discontinuous, 460

CAAT boxes, 413, 414
CAP initiation site, 251
CAT box, 23, 253
CAT sequence, 250–254
Cdc mutants (cell-division-cycle mutants), 319
Cell, exponentially growing, 479
Cell behavior:
  G_1, 477–478
  G_0 phase, 478–479
  models, 477
Cell cycle, 58, 459
  control, 474
  exponential phase, 474
  histone mRNA, 423–428
    synthesis, 432–444

Cell cycle (*Continued*)
histone protein synthesis, 422–423
mammalian, 319–336
mRNA, 308–311
Cell-division-cycle mutants (cdc), 319
Centrifugal elutriation, 322
Cetyltrimethylammonium bromide (CTAB), 463
CG doublets, 296–299
λCH-01, 253–257
Chicken:
H5 genes, 270
histone genes:
clusters, 267–270
embryonic, 270–271
evolution, 248–249
organization, 248–254, 266–270
sequences, 249–254
stage-specific, 271
histone variants, differential expression, 355
intervening DNA sequence, 273–275
somatic histone H_3, 273–275
CHO cells, 376, 470, 473, 477
Chromatin, 67, 148–149, 155–156, 341, 459
histone content, 98
Chromatin structure, *Tetrahymena*:
macronuclei, 165–166
micronuclei, 165–166
Chromatin substructure, isohistones, 96–101
compaction, 96
differentiation, 96
Chromatography:
ion exchange, 69–72
molecular sieve, 69–72
Chromosomal walks, 269
Chromosome cycle, 470, 477
Chromosome replication, 470
Ciliates, *see Tetrahymena*
Clams, 480
Cloned histone genes, 202
Coding sequences, human specific, 405–409
Codons:
ATG initiation, 409
mouse histone genes, 296–299
Compaction, 67
Condensation, 98
Consensus sequences, 5′ and/or 3′, 203
Copy number, 266, 271
Core histones, 266, 400, 459
Artemia, 194
chicken, variants, 256–258
components, characterization of, 347–355
high resolution analysis, 345–347
isolation methods, 346–347
non-allelic:
characterization, 347–355
resolution, 344–347
Tetrahymena, 166–168
Core histone variants, mouse, 341–367
Coulter Channelyzer size analysis, 333
CTAB, *see* Cetyltrimethylammonium bromide
C-value paradox, 234
Cycloheximide, 319
Cytosine arabinoside, 109–305, 311–313, 450

Deletion mutants, 414
Developmental regulation, *in situ* hybridization, 111–115, 127–130
Diploid GY109D-2, 47
Disperse repeated genes, 12
Distribution of bases, 22
DNA replication:
Drosophila embryo, 138–140
phases, 139
histone mRNA, 444–450
rate, 149
DNA sequences:
Alu family, 403
human, 405
intervening, chicken, 273–275
DNA supercoil, 88
DNA synthesis:
inhibition, 473
inhibitors, 303–308
rates:
Drosophila embryo, 138–140
histone mRNA, 430–432
Dot matrix analysis, 20–21
Doublet frequency, 298
Drosophila, 6, 8, 10, 12
Artemia and, 184–195
embryo:
DNA replication, 138–140
DNA synthesis rates, 139–140
embryogenesis:
histone mRNA changes, 141–148
levels of control, 141–148

histone genes:
expression, 137–159
organization, 266–267
histone variants, evolution, 467
oocyte, 143
O-phosphorylation, 386
Drosophila melanogaster, 155, 156, 218
Drosophila virilis, 155, 156

Echinolampas crassa, 75, 95
Echinus esculentus, 8, 12
Electrophoresis, preparative, 69
Erythroleukemia cells, Friends, 469
Escherichia coli, 341
Evolutionary clock, 29–30

False positive clones, 10–12
F9 cells, 376
Fibroblasts, mouse embryo, 376
"Fingerprints," 403
Flanking regions, 297–299
Flow cytometry (FCM), 322
Fluorodeoxyuridine, 311, 312
Fractionation methods, 460
Frameshift mutations, in yeast subtypes, 45–49
Friends erythroleukemia cells, 469

Gallus domesticus, 82
G_1 cells:
histone synthesis, 326–329
larger, 323
small, 322–323
Gene cluster, 202, 207
Gene conversion, 28–29
Gene duplication, 26–29
Gene number, 203
Genome, genes in, 16–17
G_0 phase, 474–479
G_1 phase, 470, 477–479
arrested, 478
decision point, 477
G_2 phase, 470
Growth curves, embryos, 216–218
Growth cycle, *see* G_1 phase

H1, 459
and chromatin, 98
coding region, 400
embryo, PAGE analysis, 93
lack of, 6
phosphorylated, 379–382
Tetrahymena, 169
H3, 38, 91, 202, 287–288, 459, 473
mRNA, 480
somatic, 273–275
sequences, 463
Tetrahymena, 167–168, 173–176
yeast *vs.* calf thymus, 18–20
H3.1, 291–293, 465, 470, 477, 479
H3.2, 291–293, 465, 470, 477, 479
H3.3, 475, 479
H4, 38, 91, 202, 294–295, 459, 473
histone genes, polymorphism, 403
sequences, 464
Tetrahymena, 168, 173–176, 177
H5:
chicken, 270
chromosomal organization, 256
relationship to H1°, 254–255
sequences, 254–256
H1A protein, in oogenesis, 211
H2A, 40, 91, 202, 349–350, 459, 467–470, 477, 479
frameshift mutations, 45–49
in vivo interaction with H2B, 54–60
pattern, 465
protein sequences, 40–45
sequences, 463
Tetrahymena, 167, 173–176
ubiquination, 459, 462
H2a, 290, 294
H2A.2, 465, 467–470, 473. *See also* M1
$H2A^F$, 167, 173–176
H2AF, 257–258
$H2A^S$, 167, 173–176
H2A1 protein sequences, comparison to other eukaryotes, 43
H2A.1, 465, 467–470, 473
H2A2 protein sequences, comparison to other eukaryotes, 43
H2A.2, 465, 467–470, 473
H2A:H2B dimer, 467
H2A subtypes, 18
H2A.X, 465, 467–470, 473
H2B, 38–40, 202, 353–354, 459, 468–469, 473
frameshift mutations, 45–49

H2B (*Continued*)
- *in vivo* interaction with H2A, 54–60
- lack of, 51–52
- protein sequences, 40–45
- repeating pentapeptide structure, 44–45
- sequences, 403
- structure, 82–84
- subtypes, 18
- *Tetrahymena*, 167, 173–176

H2b, 287–289, 291–294
H2B.2, 464, 469
H2B 5′-specific sequence, chicken, 251–254
H5 cDNA clones, 255
H1 complements:
- dividing cell, 376
- non-dividing cell, 377–383

HeLa cells, 470, 480
- histone mRNA synthesis, 426
- histone synthesis, 332–334

Hepatocytes, 361–362
Heteroduplex analysis, 13–14
Heterologous probes, 269
λHh2, 400
λHh7, 400
λHHG5, 400
λHHG6, 400
λHHG17, 400
λHHG22, 400
λHHG55, 400
λHHG415, 400
λHHG555, 407
High mobility group, 61
H1 Isotype 1, 101
H1 Isotype 2, 101
Histone:DNA ratio, 148–149
Histone gene expression:
- *Artemia*, early larvae, 188–190
- DNA replication, 443–450
  - inhibition, 444–450
- *Drosophila*, mechanisms of control, 150–158
- human regulation, 421–423
- newt, 236–240
- regulation, 149
- *Tetrahymena*, 175–176
- types, 440–443
- *Xenopus*, oogenesis-embryogenesis, 210–213

Histone genes:
- *Artemia*:
  - organization, 185–188
  - 5S-related sequences, 190–193
- batteries, 68–69
- chicken:
  - clusters, 267–270
  - evolution, 248–249
  - H2B 5′-specific, 251–254
  - organization, 248–254, 266–270
  - reiteration, 248–249
  - sequence, 249–251
- chromatin, 155–156
- clustered, 402
- comparisons among organisms, 266–267
- deficiencies, 156
- *Drosophila*, 266–267
  - organization, 154–156
- evolution, *Artemia*, 184
- human:
  - arrangement types, 400
  - cloned, restriction maps, 401
  - coding sequences order, 400
  - non-transcribed sequence, 399
  - organization, 399–404
  - restriction sites, 400
  - sequence variations, 403
- late, 266–267
- mouse, 283–314
  - expression control, 301–308
  - isolation, 283–284
  - organization, 284–287
  - sequence, 287–290
  - transcription signals, 295–296
- newt:
  - expression, 236–240
  - heterogeneity, 228–230
  - organization, 227–236
  - reiteration, 228
- polymorphic organization, 402
- region, genetic analysis, 158
- reiteration, chicken, 248–249
- sea urchin, 266–267
- sequences:
  - chicken, 249–254
  - conservation, 301–308
  - species variations, 463
- switching, 150

trans-regulatory elements, 159
urodeles, reiteration, 231–233
Xenopus, reiteration, 233
yeast, 266
Histone-histone interaction, 83
Histone homogeneity, newt, 234
Histone messages, *Tetrahymena*, 173–176
Histone modification:
acetylation, 171
phosphorylation, 171–172
proteolytic processing, 172–173
secondary, 170–173
Histone mRNA synthesis:
relative rate of DNA synthesis, 430–432
S-phase, 427–428
Histone mRNA variants, absence in *Drosophila*, 151
Histone protein, 148–149
evolution, 29–30
Histone protein stores, oocytes, 219
Histone protein synthesis, cell-cycle specific, 422–423
Histones:
cleavage stage, 80
evolution, 341–344
minor, non-competing, 361
primary function, 341–344
properties, 341–344
regulatory function, 67
replacement, 366
synthesis, *see* Histone synthesis
Tetrahymena, 165–177
turnover, 360
Xenopus, 201–220
expression, 201–209
genomic organization, 203–209
organization, 201–209
yeast, 5–31
Histone subtyping, 18
Histone synthesis:
basal, 470–473, 479
cell cycle, 469–474
coordination, 470–473
regulation, 470
DNA synthesis, 473–474
G_1 cells, 326–329
HeLa cells, 332–334
quiescent, 474–477
basal, 477
regulation, 109
S-phase coupling, 321–322
S phase, DNA synthesis, 473–474
Histone transcripts, maternal:
half-life, 124
pronuclear, 115–127
time of accumulation, 122–127
translation control, 125–127
Histone variants:
analysis, 460–463
one-dimensional polyacrylamide gels, 461–462
two-dimensional polyacrylamide gels, 462–463
characterization, 464–467
chicken:
core, 256–258
linker, 254–256
developmental, 464
differential expression, 355–358
evolution, 467–468
functional significance, 468–469
late, 469
metabolism, erythroleukemia cells, 358–360
partially replication dependent, 361
replication independent, 361
spermatocyte-specific, 363–366
S-phase, 473
strictly replication dependent, 361
synthesis rates, 358
Tetrahymena, macronuclear specific, 169–170
tissue specific, 464
HMG (High mobility group), 61
H1mRNA variants:
erythroblasts, 215–216
oocytes, 215–216
H4mRNA, noncoding region, 409
Homologous probes, 269
Homology blocks, 5′ and/or 3′, 203
HR9, 376
H1 subtypes:
developmental characteristics, 381
evolutionary properties, 377
metabolic properties, 377
mouse, other mammals, 386–389

H1 subtypes (*Continued*)
- *O*-phosphorylation, 384–386
- spermatogenesis, 383–384
- synthesis, 377–378

HTA, organization, 38
HTA1:
- frameshift mutation, 55
- inferred protein sequences, 40–41

HTA2, inferred protein sequences, 40–41
hta1-strains, growth rates, 56
HTA1-HTB1, 38
HTA2-HTB2, 38
hta2-strains, growth rates, 56
HTAT, frameshift mutation, 55
HTB organization, 38
HTB1:
- frameshift mutations, 51
- inferred protein sequences, 38–39

HTB2, inferred protein sequences, 38–39
htb1⁻:
- isolation, 52–54
- meiosis, 49–51
- mutation, 47–48

htb2⁻:
- isolation, 52–54
- meiosis, 49–51

hv1, 169–170, 173–176
hv2, 169–170, 173–176
H1 variants:
- embryogenesis, 215
- erythroblasts, 215

H3 variants, 354–355
Hybridization, 109–110
Hybridization, *in situ*:
- early-late histone gene transcriptional switch, 127–130
- pronuclear signals, 117–119
- sea urchin embryo, 110–131
  - efficiency, 111–114
  - noise, 112–114
  - probe, 111–112
  - sensitivity, 111
  - specificity, 114–115

Hydroxyurea, 109, 305, 307, 311–313, 330–331, 450, 473

IMR-90, 464, 475, 477
Introns, 459
Isohistone H1, 91–96
Isohistone H2B, sperm cell, 75–77
Isohistones:
- acetylation, 100
- identification, 68–69
- purification, 69–75

Isoprotein species, 459–460

"Label chase," 305
Lampbrush chromosomes, 227, 236
Liver regeneration, 361–362
Lymphocytes, 474, 478
Lymphoma, mouse, *see* S49 cells

M1, 350–351
Macronuclei, 165–166
2-Mercaptoethanol, 462
Micronuclei, 165–166
Mitoses, 469
Message stability, 145
- changes in, 150–153

MM221, 284–299, 303
MM614, 299–301
Mouse:
- histone genes:
  - isolation, 283–284
  - organization, 284–287
  - sequence, 287–290
- histone variants, differential expression, 355–358

Mus musculus domesticus, 363
Mutants, cell-cycle-specific temperature sensitive, 477
Mycophenolic acid, 311–313

Necturus masculosus, 213–232
NEPHGE gels, *see* Nonequilibrium pH gradient electrophoresis gels
Newt, histone genes:
- expression, 236–240
- heterogeneity, 228–230
- organization, 227–236
- reiteration, 228

Nonallelic variation, 344, 459
Nonequilibrium pH gradient electrophoresis (NEPHGE) gels, 326, 329, 423
Nonrandom sequences, 413
Northern blot analysis, 175, 271–272, 275–277
Notophythalmus viridescens, *see* Newt

Nucleases, 98
Nucleic acid packing, 87–88
Nucleoplasmin, 313
Nucleosomes, 67
  core particles, 98–99
  di-, 99
  evolution, 97
  heterogeneity, *Tetrahymena*, 173
  mono-, 99
  nuclease susceptibility, 98
  number, 98
  tri-, 99
Nucleotide sequence comparisons, 408–411

Oocyte(s):
  Drosophila, 143
  histone pattern stores in, 219
  Xenopus, 474

Pachytene stage, 383–384
PAGE analysis, 69–72
PAGEND, *see* Polyacrylamide gel electrophoresis in presence of nonionic detergents
Parenchinus angulosus, 40, 75, 78–79, 80, 91
Pentapeptide, repeating, 75
  calf thymus cells, 86
  DNA supercoil, 88
  erythrocyte, 86
  nucleic acid packing, 87–88
  phosphorylation, 84
  proline, 87
  significance, 84–87
  yeast H2B, 44–45
Phosphorylation, 171–172, 463
  interphase, 385
  sites, 384–385
  S-phase, 385
c-Phosphorylation, 386
O-Phosphorylation:
  Drosophila and, 386
  of H1 subtypes, 384–386
Plasmid pMG101, 47
Plasmid pYH18, 46
Poly A, 459, 480
Polyacrylamide gel(s):
  one-dimensional, 461–462
  two-dimensional, 462–463, 467
    50%, 466–467
    mapping, 466
Polyacrylamide gel electrophoresis in presence of nonionic detergents, 344–345, 350, 354
Polymorphism:
  histone variants, genetics, 363
  Xenopus, 203
Prepachytene stage, 385
Primer extension experiments, 414–421
Pronuclei, 115–127
  mRNA content, 119–120
Proteolytic processing, 172–173
Psammechius miliaris, 8, 12, 75, 91, 201, 404
"Pseudogene," 407
Pulse labelling, 98
PYS, 376

Quiescence, 474–477. *See also* G_0 phase
Quiescent pattern, 478

Rana catesbeiana, 255
Recombination, 28–29
Regulation variants, 301–308
Regulatory function, histones, 67
Reiteration, 201
  chicken, 248–249
  newt, 228
  urodeles, 231–233
  Xenopus, 233
Repeat unit, *Xenopus*, 205–206
Replacement variants, 356. *See also* Replication variants
Replication, autonomous, 15–16
Replication-variant RNA's, 486
Replication variants, 307, 356
Repression:
  permanent, 67
  temporary, 67
R loop hybridization, 7
  analysis, 13
cRNA, 111–112
mRNA, 308, 459, 480
  Artemia, 188–190
  cell cycle, periodic synthesis, 432–440
  changes during embryogenesis, 141–148
  changing mass, 141–142
  coding sequences, putative regulatory, 409–421

mRNA (*Continued*)
- concentration posttranscriptional level, 438
- de novo transcription, 190
- DNA replication, 428–432
- 3′-ends, 413
- export, 154
- fingerprints, 403
- genomic organization, 154–156
- H3, 214, 480
- half-life, apparent, 437
- H2B, 214
- levels:
  - in arrested cells, 308–311
  - evaluation, 329–332
  - signal regulating, 311–313
- masked, 188
- maternal, 149
  - *Drosophila*, 148–150
- message stability, 145
- mouse:
  - half-life, 301–308
  - number, 299–300
  - structure, 290–295
- oogenesis-embryogenesis pools, 212
- polysomal association, 141
- processing, 154
- pulse-labelled, 440
- rate of accumulation, 141
- rate of synthesis, 150
- sea urchin clones, 151
- S-phase, early, 442
- stability, 150
  - changes during embryogenesis, 153
- stores, oocytes, 219
- synthesis, 143–148
- synthesis rates, 109
  - changes, 144, 149
- temporal coupling, 441
- 3′-termini, 23–26
- 5′-termini, 23–26
- *Tetrahymena*, 175–176
- transcribed histone, 109–131
- transcription, 153–154
- translational control, 143
- turnover, 144, 149
  - rate, 109
- *Xenopus*, 213

RNA gel blots, 151

Saccharomyces cerevisiae, 5, 12, 16, 17, 26, 37

Satellite DNA histone genes, 230

Sc 191, 13

Sc 202, 13

Sea urchin:
- histone amounts, cellular, 69
- histone variants:
  - differential expression, 355
  - evolution, 467

Sea urchin eggs:
- histone transcripts, 115–127
- mRNP particles, 110
- pronuclear, 115–127

Sea urchin embryo, 68–101
- developmental regulation, 110–115, 127–130
- histones:
  - amino acid composition, 73
  - chromatography, 69–72
- histones, PAGE analysis, 69–72
- isohistones, synchronized, developing, 77–84

Sea urchin sperm histones, amino acid composition, 74

Selective extraction, 69–70

S49 cells, 322–326

Signal noise ratios, 112–114

SMT, 13–15

Somatic histone H3, 273–275

S1 nuclease mapping, 287, 290–295, 299–300

Southern blot analysis, 176, 187
- hta 2^- htb1^- haploids, 56–60

Spacer DNA, *Xenopus*, 205–206

Spacer sequence, 8, 13
- non-transcribed, 399

Spaerechinus angulosus, 75, 95

Sperm cells:
- cleavage stage histones, 80
- developmental program:
  - hatching blastula, 90
  - late blastula, 79
- isohistone H2B, 75–77
- isohistones:
  - distributions, 78
  - PAGE analysis, 77

Sperm histone H1, 93–94
- tetrapeptide, 93

Sperm histone H2B, 88–91
echinoderms, 90
pentapeptides, 90
protamines, 88
Sphaerechinus granulosus, 95
S-phase, 323–326, 470, 478, 480
coupling, 321–322
5S-related sequences, *Artemia*, 190–193
5S ribosomal RNA, 191–193
Stacking gel, 460
Stagespecific mRNA, 68
Strong coupling, 321–322
Strongylocentrotus purpuratus, 78–79, 218, 284
Subtypes, protein, yeast histones, genetic analysis, 37–61
Symmetry:
dyad hyphenated, 413
T-hyphenated, 409
Synthesis rates:
actin mRNA, 145–148
total poly(A)$^+$RNA, 145–148

TATA box, 23, 250–255, 413, 414
3T3 cells, 376
Tetrad mapping, 16–17
Tetrahymena, 37
core histones, 166–168
histone genes, 176–177
histone messages, 173–176
histones, 165–171
histone variants, evolution, 467
nucleosome heterogeneity, 173
Tetrahymena pyriformis, 177
Thymocytes, large, 376
Tm's, *in situ*, 114
T1 ribonuclease oligonucleotides, 403
Transcription:
control, 479–480
polarity, 402
rates, 303–311
signals, mouse, 295–296
unit, 237
Translational control, 150
Translation assays, *in vitro*, 109
Triturus cristatus, 231, 239
TRT1, 8, 13–15
TRT2, 8, 13–15
TRT3, 8, 13–15

Ubiquination, H2A, 459
Ubiquitin, 465
uH2A, 465

Variant genes, *Xenopus*, 208–209

Weak coupling, 321–322

Xenopus:
gene clusters, maps, 205–207
gene expression, 201–209
gene number, 203
gene organization, 201–209
genus, genomic organization, 209–210
H1, 204
histones:
amino acid sequences, 202
evolution, 202
nucleotide sequences, 202
orientation, 202
nucleosomal histones, 204
oocytes, 474
oogenesis-embryogenesis, histone gene expression, 210–213
polymorphism, 203
reiteration, 201, 233
repeat unit, 206
mRNA accumulation, 213
spacer DNA, 205–206
variant genes, 208–209
Xenopus borealis, 209
Xenopus fraseri, 209
Xenopus laevis, 26, 201, 210–216, 320, 402
Xenopus l. victorianus, 209

Yeast, 480
histone gene expression, organization, 266
Yeast histones, 5–31
base composition, 21–22
DNA sequences, 17–26
coding, 17–20
features, 20–21
H3 and H4, 21–23
specific motifs, 23–26

Yeast histones (*Continued*)
  duplications, 26–29
  evolution, 26–30
    protein, 29–30
  false positive clones, 10–12
  isolation, 6–10
  lack of H1, 6
  loci, 7
  organization, 12–17
    autonomously replicating sequences, 15–16
    genome, 16–17
    local gene arrangement, 13–15
  subtyping, 18